T0136753

THE DOCTOR WHO WASN'T THERE

THE DOCTOR WHO WASN'T THERE

Technology, History, and the Limits of Telehealth

JEREMY A. GREENE

THE UNIVERSITY OF CHICAGO PRESS
CHICAGO AND LONDON

The University of Chicago Press, Chicago 60637
The University of Chicago Press, Ltd., London

Published 2022
Printed in the United States of America

31 30 29 28 27 26 25 24 23 22   1 2 3 4 5

ISBN-13: 978-0-226-80089-9 (cloth)
ISBN-13: 978-0-226-82152-8 (e-book)
DOI: https://doi.org/10.7208/chicago/9780226821528.001.0001

Library of Congress Cataloging-in-Publication Data

Names: Greene, Jeremy A., 1974– author.
Title: The doctor who wasn't there : technology, history, and the limits of telehealth / Jeremy A. Greene.
Other titles: Technology, history, and the limits of telehealth
Description: Chicago ; London : The University of Chicago Press, 2022. | Includes bibliographical references and index.
Identifiers: LCCN 2022011197 | ISBN 9780226800899 (cloth) | ISBN 9780226821528 (ebook)
Subjects: LCSH: Telecommunication in medicine. | Telecommunication in medicine—United States—History. | Medical telematics—United States—History.
Classification: LCC R119.9 .G74 2022 | DDC 610.285—dc23/eng/20220404
LC record available at https://lccn.loc.gov/2022011197

♾ This paper meets the requirements of ANSI/NISO Z39.48-1992 (Permanence of Paper).

To my grandparents
Marilyn Freedman and Gerard Buter

CONTENTS

INTRODUCTION
Disrupting Care, Continuing Care 1

1. On Call 13

2. The Wireless Body 49

3. The Electronic Leash 79

4. The Amplified Doctor 105

5. The Wired Clinic 141

6. The Push-Button Physician 179

7. The Automated Checkup 211

CONCLUSION
The Medium of Care 241

Acknowledgments 257

List of Manuscript Collections 261

Notes 263

Index 309

Introduction

DISRUPTING CARE, CONTINUING CARE

The early months of the pandemic hit my urgent care clinic in East Baltimore with confusion and crisis. As the map of the outbreak spread from China to Italy to everywhere, our rules and protocols for COVID-19 screening changed weekly, yet somehow still seemed to lag behind common knowledge. Testing was hard to come by, personal protective equipment even more so. There was no known treatment, no prospect of a vaccine. Soon anyone with any fever, cough, or shortness of breath was being screened outside the clinic doors and sent to the hospital—until we realized that people *without* symptoms could spread the virus as well. As clinic staff began getting ill, and we feared our community health center might become a source of community infection, we turned to a technological solution. Within a matter of weeks, all urgent care services shunted to telemedical visits, and I became a teledoctor.

Like other professionals making the shift from in-person to remote work, I appreciated both the advantages (no need to wear dress pants, more time to help my children adapt to their own remote schoolwork) and the challenges (much harder to establish rapport with new patients, new difficulties with wonky Wi-Fi). Some of my telepatients had crisp, well-framed video connections in professional-grade home offices, and could provide me with readouts from at-home blood pressure cuffs, pulse oximeters, and other remote sensors that effortlessly transmitted clinical data across a distance. They experienced this new ability to see a doctor through their own phone or laptop, in their own domestic space, as a form of liberation. It freed them of the hassle of a drive to the clinic and an uncomfortable period in a waiting room with other sick people. But for people with more complex urgent care issues, like acute asthma exacerbations, who did not have access to these home health technologies, telemedicine posed severe constraints. Had these patients been able to walk into my clinic, I might have saved them the longer wait in the emergency department—but now all I could do was to send them right back there. Telehealth had its limits as well as its advantages, and they were not felt equally by all people.

For me, for my patients, and for the millions of others suddenly engaging in clinical practice through electronic devices, telemedicine was a new medium of care. Yet it was not new to the medical electronics and device industries, which had been lobbying for this transition for decades. A decade earlier, the US Congress had passed the Health Information Technology for Economic and Clinical Health (HITECH) Act as part of the American Recovery and Reinvestment Act of 2009, providing federal incentives to encourage the use of telemedical systems and other forms of electronic health records and wireless "smart" medical devices. Over the next five years, the global telehealth market would more than double, from $11 billion to more than $27 billion. By the end of 2016, more than 600 companies entered the private telemedicine market, with more than $4 billion in new investments in the first nine months of the year. "Telemedicine is so white hot right now it makes *Shark Tank* look like an aquarium in a dentist's office," Robert Calandra wrote in *Managed Care Magazine* in 2017, as nine out of ten healthcare executives were rolling out telehealth plans, with an anticipated $36.2 billion in value by 2020 in the United States alone.[1] Telemedicine (the direct provision of clinical care through telecommunications technologies) and telehealth (the broader use of electronic and digital media for health and healthcare) were understood by tech firms and equity investors to be lucrative, revolutionary platforms that promised to transform the face of clinical care as we knew it. And that was before the pandemic hit.

Digging a bit deeper into the history of telehealth, one finds another set of promises connecting information technology to health equity instead of equity markets. In the mid- to late twentieth century, a series of new platforms for practicing medicine at a distance were developed with the intention of flattening disparities in access to healthcare. Early forms of remote medicine by closed-circuit television were tested in the early 1960s to link mental healthcare services over hundreds of miles of Nebraska farmland. The term "telemedicine" itself was coined in 1971 by a physician in Boston who built a microwave link connecting a remote urgent care clinic to the emergency room of Massachusetts General Hospital. Within a few years, the concept of telemedicine was picked up by the Rockefeller Foundation and the US Department of Health, Education, and Welfare as a means of reducing barriers to accessing primary healthcare. Federally funded demonstration projects for this technology of community care were set up in Harlem in New York City, on the West Side of Chicago, in rural Vermont and New Hampshire, and on American Indian reservations in Arizona. Telemedicine made a lot of promises, to a lot of people, over a lot of years. While the recent pandemic growth market for telehealth technologies yielded high returns for investors, it is far less clear whether it also increased equity in access to care.

Telemedicine has clearly helped many Americans, especially in rural counties, access care that otherwise would have been unavailable.[2] In the context of the current pandemic, telemedicine made possible care that otherwise had become too risky. But healthcare via video did not provide access equally, at least not for the patients in my care. While established, insured patients found it relatively easy to transition their care from in-person care to video visits through the portal of the electronic medical record, this was trickier for uninsured patients and those who were new to the system. Before the pandemic, anyone could walk into the community health center and be seen on a sliding-scale fee basis, without needing to show documentation of citizenship or insurance status. During the pandemic, new electronic forms of access presented new barriers to care. Some people could not reliably connect to the video interface, and others could not connect at all. Like so many other aspects of the COVID-19 pandemic, these disparities in access to care were far from color-blind.

A bitter historical irony was at work here. The community health center where I practice in East Baltimore was established by a group of neighborhood activists in the late 1960s to set up a preferential option for primary care for the largely African American neighborhoods surrounding it. In the 1990s the clinic expanded its mission to providing a safety net of medical care for Baltimore's expanding Latino community, many of whom had no access to formal health insurance.[3] Yet in my first full month as a telepractitioner, not a single African American or Latino patient was able to successfully access the full telemedical suite in my clinic sessions. I was not the only clinician to notice this paradox. Similar challenges of equity in access to telemedicine were reported in community health centers and other primary care practices in Philadelphia, New York, and Boston. Video visits were repeatedly found to be less common in telemedical encounters among Black and Latino patients, and in households earning less than $50,000 per year.[4] Telemedicine, a technology that initially promised greater access to care to patients of color in poor urban areas, had in the crisis of the early pandemic come to serve more well-to-do, white patients who needed assistance least.

How could a technology with the potential to provide greater equity in healthcare serve instead to widen gaps between haves and have-nots? The fate of telemedicine in the COVID-19 crisis poses a fundamental problem for those who would see new information technology as a revolutionary means of providing better healthcare for all. But this story, like the story of telemedicine, long predates the pandemic. It does not start in the twenty-first century, or even in the twentieth. Indeed, a repeating cycle of promises and limitations of electronic medicine can be found well before the television was even invented. The history of healthcare information technology is full of revolutionary promises that did not come to pass, and more mundane ones that did.

* * *

Consider the telephone. Only a few years after Alexander Graham Bell's first demonstration in 1876, medical journals carried exuberant reports of the possible clinical applications of this experimental technology. Soon, many assumed, telephone medicine would become a new medical specialty. Doctors would listen to the hearts, lungs, or abdomens of their patients over the wires using new, telephonic stethoscopes. The problems of access to medical care in sparsely settled areas would be resolved by long-distance lines that instantly linked even the most remote rural residents to urban specialists. The democratic ideal was admirable. But not everyone had access to a telephone in 1880—or in 1900, or in 1920, or even 1940. The telephone user in all of these early narratives was invariably both middle-class and white.

Over and over again, across the twentieth century, new communications technologies promised to democratize access to healthcare. Two-way radio and other wireless devices, interactive cable and community access television (CATV), the "electronic brains" of networked mainframe computers: each of these new platforms promised a radical reformation of the healthcare landscape. Telephone medicine, radio medicine, television medicine, mainframe medicine: each suggested new pathways to improve access to care. If we have forgotten that none of them quite produced the more inclusive, more accessible system of healthcare they initially promised, we have also largely forgotten the transformations they *did* bring about.

The medium of care is never neutral. New communications technologies continuously transform the practice of healthcare, but they rarely deliver on promises of increased health equity. Nor do they tend to produce the singular acts of disruption celebrated in popular accounts that praise innovators and innovations as the driving force of American medicine, or in the initial public offerings of tech start-ups that monetize their worth. New platforms arrive wreathed in the language of revolution: every year a parade of new devices promise a paradigm shift that will creatively disrupt or radically transform healthcare through sudden and total change.[5] Yet when electronic communications devices *do* drive change in medical practice, the changes they bring about often just as readily entrench existing power relations as overturn them.

This is a book about the history of electronic communications in American medicine, old and new.[6] It argues that the medium in which healthcare takes place—by which I mean the social as well as technical context in which sick people seek help and receive medical advice—matters a great deal. The history of media teaches us that any new means of producing, recording, transmitting, or circulating information quickly becomes an object of cultural as well as financial speculation: a new vehicle for generating possible futures.[7] The history of technology teaches us that when stories are played forward from the past rather than backward from the present, the fate of any given device can be understood as a much more open-ended affair: a speculative repository for broader hopes and fears of designers and users.[8] In the American medical system, where health policy is so deeply entwined with market speculation, the adoption of health communications technologies can carry very different stakes for manufacturers and marketers than they do for practitioners and patients.

History teaches us as much about forgetting as about remembering.[9] This paradox was already apparent within the field of medical electronics as early as 1956. In that year, standing before a group of technological enthusiasts gathered in New York to speculate about the future of this young field, Vladimir Zworykin paused to consider its recent past. The celebrated innovator of modern television asked his audience to consider how the X-ray tube, a new and experimental electronic technology at the turn of the twentieth century, had since "become so familiar that few people think of it as an electronic device."[10] In just a few decades, the new technology had become invisible: not because it had become obsolete, but because it had become so useful that people had come to accept it as part of everyday life.

Zworykin, who had recently retired as vice president for research and development at the RCA Corporation to invest all of his effort into the Center for Medical Electronics at the storied Rockefeller Institute for Medical Research, was in a position to speculate further. If X-rays were electronic, and electrocardiograms (ECGs) were electronic, and electroencephalograms (EEGs) were electronic, why not patch their data, along with patient histories and physical examinations, directly into an electronic medical record that also contained all relevant data from the world's medical literature? Why couldn't the digital computer someday become as familiar a feature of medicine as the X-ray?

Just as the X-ray machine had already become a pedestrian technology by 1956, many of the technologies considered speculative in Zworykin's present—the wireless summoning of a physician by radio-pager, the long-distance evaluation of patients' bodies by radiotelemetry, remote medical encounters by closed-circuit television, or the automated evaluation of an electrocardiogram by computer algorithm—have by the early twenty-first century become everyday aspects of clinical medicine. We no longer include them when we project new visions of digital medicine into the future. Nonetheless they were, in their own time, every bit the objects of financial, cultural, and medical speculation that our smartphones, neural nets, and wearable devices are now.

Then, as now, the role of electronic communications devices in medicine was also vocally contested by physicians who thought their risks would outweigh any benefits. One doctor in the room stood up after Zworykin's speech to challenge his depiction of medicine's electronic future. An "artificial computer," they warned, could never develop bedside manner, or make meaningful connections with patients in intimate matters where life or death might hang in the balance of a single conversation. After this exchange was covered in the *New York Times* and other prominent newspapers, Zworykin argued that electronic medicine would humanize rather than dehumanize

American healthcare. "Freed of much of the routine effort of physical examinations as well as the necessity of keeping abreast of new developments in the diagnosis and therapy of physical disease," he elaborated, the computer-enhanced doctor would be "increasingly concerned with [the] patient's emotional well-being and social adjustment . . . assuming to a greater degree the role of the family physician, a role which had almost vanished before the advent of the central diagnostic computer."[11]

The same argument is taken up by new adversaries today.[12] A continuous debate over how electronics will disrupt medicine can be traced back to the mid-twentieth century, if not earlier. These arguments are not abstract. The medium of care is always contested by different parties with very real professional, political, and financial stakes at play. The source of contention has always been an exchange about technology and power. In the name of empowering the consumer of healthcare, technologists present their new platforms as essential passage points for the future of medicine. In the name of defending the humanity of the patient, physicians assert that no technology should displace the doctor from the bedside. This is as visible in the exchange between Zworykin and his physician critic in 1956 as it is in exchanges between boosters and detractors of digital care platforms today. Disrupting care, continuing care.

In these contests the best interests of patients are repeatedly invoked by those who claim to speak for them, without necessarily providing a space for patients themselves to have their say. When technologists promote the health benefits of a new, disruptive technology, they are placing their own proprietary devices and algorithms at the center of a new system in which they become more relevant, lucrative, and powerful. When physicians resist a new information technology, they are restating fundamental moral concerns of the medical profession and resisting the perceived loss of their own control over the nature of medical work.[13] Early twenty-first-century concerns linking the use of electronic medical records with physician burnout can be traced back to early twentieth-century concerns that the use of telephones was doing the same thing. But neither party should be credited with representing the true interests of the patient.

This book reframes our current understanding of new forms of digital healthcare in the twenty-first century by examining the continuity—and change—in disputes surrounding earlier forms of electronic telecommunications that promised to transform health over the course of the twentieth century. Today's telehealth devices are far more sophisticated than the hook-and-ringer telephones that became widespread by the 1920s, the FM radio technologies used to broadcast health information in the 1940s, the televisions used to pioneer telemedical evaluation in the 1950s, or the first full-scale attempts to establish electronic medical records in the mid-1960s. But the ethical, economic, and logistical concerns they raise are prefigured in these earlier episodes, as are the gaps between what was promised and what was delivered. Each of these platforms in turn produced more subtle transformations in health and healthcare that we have learned to forget, as promises of newer communications platforms take their place. This forgetting, too, is a consequence of the power dynamics at play when supposedly revolutionary technologies become part of everyday life.

* * *

History is about what we forget as much as it is about what we remember. The stories we tell about the history of medical technology tend to be progressive and triumphant: the advances in surgery enabled by anesthesia and aseptic techniques, the conquest of infectious disease by antibiotics. These stories often focus on the impact of diagnostic machines, like the X-ray, or major therapeutic shifts, like the development of new anticancer drugs or implantable devices like the cardiac pacemaker. The role of information technologies in health and healthcare receives far less attention.[14]

Yet the practice of medicine has been shaped by information technologies for a long time, and physicians have long fretted over what is lost in these exchanges. In early modern Europe, a great deal of medical practice was conducted at a distance, as patients corresponded with their physicians, described their symptoms, received diagnoses and prescriptions, and reported the outcomes through long exchanges of letters and the occasional shipment of a flask of urine. Doctors often worried about the risk of being deceived in these exchanges, but the practice of "epistolary medicine" was still a widely accepted norm.[15] In turn, in the early nineteenth century, when the stethoscope was introduced into practice as a tool for practicing "mediate auscultation"—that is, listening to the body of a patient through an amplifying device, rather than with the naked ear alone—many physicians expressed concern that the distance between the listening ear and the body of the patient would produce opportunities for dangerous artifacts and misdiagnosis.[16] Although the stethoscope may seem more readily identifiable as a medical technology than a written letter, both should be understood as media of medical care. Both were powerful technologies that shaped and reshaped the experience of health and disease.

Most doctors still carry stethoscopes today and use them to listen to the hearts, lungs, and abdomens of their patients, even when that information may be irrelevant to the case at hand. While patients tend to experience the touch of the stethoscope as a cold lump of metal, physicians have warmer feeling toward the device. Older physicians frequently complain that in an era of portable ultrasound machines, younger physicians aren't learning how to use their stethoscopes properly anymore. The stethoscope here is a humanizing technology, a metonym for an older, humanistic physician who *listens* to their patients. How easy it is to forget that in its own time of novelty the stethoscope was also feared as a technology that brought distance between doctor and patient.[17] It was a tool that (like the pocket ultrasound today) increased diagnostic power but did so at the expense of direct contact with patients. The fear in both cases—that the clinician might lose control of the diagnostic or therapeutic process through the intervention of newer, disruptive technology—is the same.[18]

Both the stethoscope and the pocket ultrasound—like the telephone, the radio, the television, and the computer—offer powerful new opportunities to the practicing physician. Both threaten to introduce new forms of distance and artifact between doctor and patient. And yet in our present concerns with the Internet of Things, the allure of artificial intelligence (AI) as applied to healthcare, and other forms of electronic care, we tend to selectively forget that similar concerns were also expressed over earlier technologies of care. Instead, the stethoscope, an object of concern in the early nineteenth century, becomes a technology of nostalgia for twenty-first-century physicians who lament those arts of bedside diagnosis that might soon become lost.

Nostalgia is just one of many forms of history, and it requires selective erasures. Historians work to investigate other narratives, to find traces of that which has been omitted or overlooked. We read other people's mail. We track ideas back through published and unpublished literatures. We look for the shape of the dominant discourse in mainstream media—and try to search out other voices not included in this narrative, especially those which might counter it. We look at promises made and ask: whom did they serve? Were they kept or were they broken? Did anyone notice? We study how formerly extraordinary things become commonsensical objects, how front-page news items become forgotten structures of the world we live in, the accepted reality of the world-as-it-is.

The stethoscope of the early nineteenth century was a new technology with many boosters and detractors. Now the stethoscope is a symbol of close presence and concerned listening, threatened by newer technological forms of care. The early telephone, too, had its boosters and its detractors. Like the stethoscope, it promised access to better care, while threatening to introduce diagnostic artifacts and to mechanize the life of physicians. Soon, however, being "on call" by telephone (and, a few decades later, by radio-pager) was how physicians defined themselves as an accountable human presence in the lives of their patients. The earlier promise and threat of the telephone had been forgotten. How quickly we erase that which was transformational a generation ago. How quickly we revise prior controversies into seamless narratives of progress.

Time, then, to take a closer look, or better yet, to *listen*. Our story starts with the ring of the telephone. Ring ring, little bell: it's the early twentieth century, and the doctor is on call.

1
ON CALL

CliniCloud, a smartphone-enabled health platform, was the darling "disruptive technology" of Silicon Valley's 2015 TechCrunch festival. Its creators, Australian physician-inventors Hong Wen Chin and Andrew Lim, promised an entirely new mode of medicine, "empowering every parent and carer to play an active role in healthcare."[1] By plugging an electronic stethoscope and an electronic thermometer into their proprietary CliniCloud app, parents could have their feverish, coughing children seen by a doctor without being coughed on in an overcrowded clinic waiting room. The initial pitch generated $5 million in venture capital funding and an exclusive co-marketing deal with the Best Buy big-box chain store to build a new Doctor on Demand service. The future of healthcare in the cloud seemed imminent.[2] Soon all patients could be seen in their own homes, on their own schedules, on their own terms.

Every new technology seems to promise a break from the past. Yet while CliniCloud may have seemed unprecedented, the application's promise was not altogether new. More than 130 years earlier, the *Cincinnati Lancet and Clinic* reported the same revolutionary potential in the landline telephone. Just three years after Alexander Graham Bell's first demonstration in 1876, the telephone was already being promoted as a game-changer in healthcare. Late one night a Cincinnati physician was summoned by a caller who feared his coughing child had an emergent case of the croup. Instead of making a midnight ride to the house of the caller, the technologically savvy physician requested the father to simply "hold his child for a few moments before his telephone."[3]

Listening carefully to the child's breathing by using the telephone as a long-distance stethoscope, the "practiced ear of the physician" determined the cough was not croup; there was no emergency. Father, child, and physician were all able to go back to sleep, and by the time the physician saw the patient in the morning "all symptoms of laryngismus stridulus had disappeared, and the child was apparently quite well."[4] As this story circulated widely through American and British medical journals, so did attempts to couple other newly domesticated medical technologies.[5] A few years later, the *Journal of the American Medical Association* argued that physicians could be spared needless disruption to their sleep if they encouraged their patients to pair another new home health device with their telephone: the thermometer. If more parents used thermometers before calling the doctor, "the record of the clinical thermometer transmitted by telephone would usually enable the physician to properly estimate the immediate gravity of the case."[6]

Though separated by more than a century, the advocates of both the Bell telephone and the CliniCloud shared an optimism that these innovations would overcome important clinical obstacles. Both prioritized the urgency of remote diagnosis. Both aimed to shorten the distance between domestic and clinical spaces. Both leveraged the convenience of telecommunications to convey medical data. Both promised a seamless, effortless, and more accessible vision of healthcare that could be achieved through a revolutionary new communications device.

At the same time, these innovations were designed to empower different users at very different historical moments. If the CliniCloud positioned the smartphone as a technology through which patients could demand more control over their own care, the *Cincinnati Lancet* episode positioned the early telephone as a technology through which doctors could amplify the reach of their own expertise. The earlier case is clearly centered on the figure of the physician. It resonated with an audience of readers concerned with the rising status of the American medical profession. The physician was the person to call in the middle of the night in a crisis, and if you happened to be wealthy enough to have a telephone in 1879, you were wealthy enough to access the expertise of a physician. And that expertise was unquestionably white, male, and middle-class.

CliniCloud, in contrast, turned the smartphone into a tool of patient empowerment: a means to subvert physicians' control over the when and where of clinical care. While access to the telephone in 1879 was coded by race, ethnicity, and class, access to smartphones by 2015 was increasingly universal across all sectors of American society. Yet patients were not the only figures empowered by the algorithmic care of CliniCloud. The new app also represented an opportunity for Silicon Valley investors and big-box stores like Best Buy to have their own say over when and where healthcare was accessed—and what new forms of economic control they themselves might come to exert over access to expertise.

To celebrate the CliniCloud as a medical innovation in 2015 is to forget the telephone as a medical innovation in 1879. This erasure is also a form of power. In her pathbreaking cultural history of the telegraph, Carolyn Marvin demonstrated the value of reconsidering older communications devices as the shining novelties they once were, "when old technologies were new."[7] But it is just as important to ask how these new technologies became old. How can we understand a recent novelty like the CliniCloud not merely as a new thing, but as something also destined to become old and forgotten in its own right, even if it succeeds? As historian Lisa Gitelman describes, successful new media have a paradoxical tendency to become invisible at the very moment they achieve widespread adoption. If telephones were front-page news when they were first being developed, their success could later be measured in how *little* attention telephones received once they became a part of everyday life. Innovations are most powerful when taken for granted. "Science and media become transparent," she concludes, "when scientists and society at large forget many of the norms and standards they are heeding, and then forget that they are heeding norms or standards at all."[8]

But before the widespread amnesia of the telephone as an innovation, telephone technology was enthusiastically taken up in medical research as well as practice. In the late nineteenth century, physiologists hooked up telephones to drums of paper, leaned their microphones upon their patients' chests and used them to record heart murmurs. Surgeons pressed telephones up against the skin of the abdomen to catch sonic evidence of kidney stones or other causes of abdominal distress. As the telephone network took shape in the early twentieth century as a standard communications platform, it became a symbol of the twenty-four-hour accessibility of the general physician. Indeed, the spread of telephones was so pervasive that a new form of malpractice emerged for the on-call physician: "culpable neglect of telephone." But by the close of the 1960s, mention of this older technology largely disappeared from clinical and popular accounts of electronic media in medicine. In the late twentieth century telephones became platforms for newer forms of information technologies—physiological sensors, magnetic data storage, and networked computers—to make new promises as medical innovations.

Along with the early flush of hopes for the telephone's diagnostic and therapeutic uses came a fear and anxiety for how this technology would change medical practice. As telephones spread throughout professional and middle-class America as symbols of modern medical practice, predictions regarding their ability to revolutionize the practice of healthcare became even more visible. So, too, did the rising tenor of concern that these devices would lead to automated, detached, and dehumanized forms of medical practice. Detractors feared how the telephone would change the nature of the relationship between doctors and patients. If the doctor was not actually *seeing* the patient, what was the art of medicine, after all? Would a physician be able to accurately diagnose a patient remotely? Would expectations of accessibility distort the quality of care? Would the telephone transform the American patient into an impatient consumer? These questions would set in motion strains of cultural resistance to information technology that would reverberate throughout the century.

Although these hopes and fears dominated public discourse about telephones in health and medicine, they too would become muted and forgotten. Once the telephone network settled into its seemingly natural role in the infrastructure of the American healthcare system, it became integral to care. The telephone changed the way we communicate and behave, in medicine and daily life. The magnitude of its contribution—the contribution of any communications technology—can only be seen in retrospect.

* * *

The first aspirations for the medical applications of the telephone emerged almost immediately. In the *Boston Medical and Surgical Journal* in 1878, Harvard physiologist Henry Pickering Bowditch expounded on the telephone's possibilities, which he was following in research reports from Europe. "Though so short a time has elapsed since the invention of the telephone," he wrote, "it has already found numerous applications as an instrument of physical and physiological research."[9] Like other elite American physicians of his generation, Bowditch pursued training in scientific medicine overseas in the laboratories of France and Germany.[10] Bowditch circulated among those on the cutting edge of laboratory medicine, studying alongside Étienne-Jules Marey under experimental physiologist Claude Bernard in France and in the laboratories of Carl Ludwig in Leipzig. These were towering figures in the emerging cohort of physician-investigators who sought to rebuild the status of medicine as a positive science. Bernard is perhaps most famous today for his iconic *Introduction to Experimental Medicine*, while Ludwig's new pressure-tracing 'kymograph' machine became the precursor to generations of medical devices that traced physiological parameters, such as blood pressure, temperature, heart rhythms, and brain waves, onto pieces of paper. Their work, and the work of students like Bowditch, looked to the laboratory for new forms of medical practice, new ways to make the hidden functions of the body legible on paper.

After Bowditch returned to Boston to establish the first physiological laboratory in the United States, he continued to read medical correspondence from France and Germany, noting with interest a report from Jacques-Arsène d'Arsonval at the Société de Biologie of Paris. The 1878 report claimed the telephone was "an infinitely more sensitive reagent" in determining electrical current than the previous metric, frog nerve tissue. d'Arsonval demonstrated that the telephone was a new precision tool, "by first placing an induced current in contact with the sciatic nerve of a frog, and showing that this current, when it can no longer excite the nerve, still makes the telephone vibrate."[11] Bowditch wanted to encourage more work like this among his American colleagues. The telephone, he argued, should be a standard item in physiology laboratories, just like flasks, test tubes, and microscopes.

Like the carbon microphone, the early telephone promised to amplify otherwise inaudible sounds from the body and thereby augment the senses of the average physician.[12] "Here is the secret of the matter," University College Hospital of London surgeon Henry Thompson told a crowded lecture hall, "that however tiny an acoustic wave is, it may be magnified the instant it becomes an electrical wave." For Thompson, the telephone was an experimental technology that could enable doctors to detect lesions they could not see or hear with unaided eye or ear. In one series of public experiments, Thompson coupled a microphone to a telephone to form a sonic probe, perhaps best understood as a sort of early ancestor of later forms of diagnostic ultrasound. Putting a stone in a washbasin (meant to simulate the human bladder) and covering it with leather (meant to simulate the flesh of the abdominal wall), he would invite audience members to see if they could feel the stone with their hands. They invariably could not, but the telephone probe could sense the stone and would emit an audible squawk when passed over the basin. "You may push it perhaps, but you cannot feel anything; but, with this instrument, if you touch with your sound a fragment, however small, *your telephone will at once speak.*"[13]

The telephone could also be used as a sort of electronic stethoscope to amplify the subtle abnormalities of lungs, heart, or abdominal organs into audible signs of disease. John McKendrick, Regius Professor of Physiology at the University of Glasgow, predicted that soon "even such feeble sounds as those produced by the heart and lungs might be transmitted for many miles and be correctly interpreted by an educated ear."[14] Using the telephone, physicians could detect objects like bullets and bladder stones, or they could amplify and transmit faint signals from the human body and make them legible to medical science and clinical practice. Once sound was made into an electric wave, in theory it could be sent anywhere you could connect with a wire.

The feebler the sound, however, the greater the risk of distortion, even for signals sent over the shortest of wires. And these wires were full of noise. Already by 1880, the Boston aural surgeon Clarence John Blake recalled with some humor how quickly his colleagues had conjured new, far-reaching applications for the telephone in medical practice. They had imagined a new specialty of telephonic consultants who "would each settle themselves down in the center of a web of wires" and listen to "the heart beats of a nation," diagnosing and treating patients at a distance. But more than four years after the introduction of the telephone, this fantastical picture of networked patients and faceless diagnosticians had not yet come to pass. "Several forms of microphone have been constructed especially for purposes of auscultation," Blake sighed, "but none have as yet, even in a slight degree, answered this purpose."[15]

Blake, a colleague of Alexander Graham Bell, had been one of the first physicians to use the telephone to amplify heart sounds. Yet he found little success in translating the theoretical abilities of the telephone-as-stethoscope into the practical demands of his clinical practice. After a series of experiments his work yielded no results, except "in one instance only, of the suspicion of a barely perceptible 'thud,' no sound which could be referred to the heart as its source was heard." Instead, he heard all sorts of artifacts. These ambient and distorted sounds came from the electrical grounding of the device: "the snapping and crackling noises indicative of earth currents, the clicking of the Morse instruments, and the sound of a 'fast speed transmitter' on the Western Union lines running along the Providence Railroad, and the ticking of the clock connected with the Observatory in Cambridge."[16] Even after the introduction of the microphone, Blake was disheartened that little signal could be discerned above the noise of the telephone wires. The problem of noise persisted. As late as 1910, when *Scientific American* carried an account of a functional telephone-stethoscope involving a special relay coupled directly to a transformer (fig 1.1), widespread concerns over electronic artifacts prevented it from being adopted in standard practice.[17]

Blake feared that transmission of signals over telephone wires degraded signal quality, rendering the instrument impractical outside the laboratory. Nonetheless, uses for the telephone in physiological laboratories continued to proliferate. As Dr. William Brown of Carlisle wrote in *The Lancet* in March 1878, a combination of telephone and phonograph could make "sound vibrations visible to the eye, registered on paper like a pulse-tracing, and kept for future study and reference."[18] By 1888 several similar laboratory devices incorporated the telephone as a tool to inscribe physiological processes onto paper. At the University of Edinburgh, Dr. Byrom Bramwell and Dr. R. Milne Murray devised an apparatus that connected a telephone

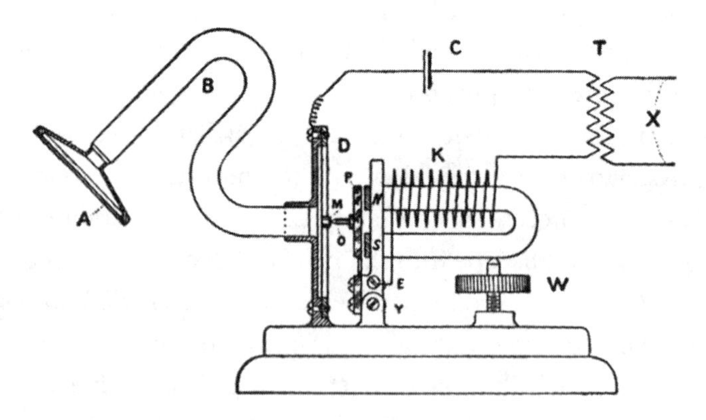

FIGURE 1.1. Diagram of a "telephonic stethoscope" designed to be connected to telephone wires in place of an earpiece.

receiver to a needle and rotating drum of paper to graphically record heart sounds. Encouraged by this successful visualization of a vital function of the living human heart, Bramwell sought to apply this technique to recording heart murmurs, and he asked Thomas Edison to direct the efforts.[19] When Edison declined, Bramwell and Murray resolved to build a do-it-yourself version on their own, which incorporated a telephone receiver among other standard objects found in the lab.

Telephones offered new opportunities to visualize the functioning of the human heart and make it legible on paper.[20] As historian Mara Mills has demonstrated, the early telephone also informed the study of the physiology and pathology of hearing.[21] The desire to build a tool to more precisely study hearing impairment had been a prime motivation behind Bell's collaborations with Blake.[22] By 1878, ear specialists at the University of Edinburgh speculated that the conversion of sound into electrical energy could be used to restore hearing. With adequate amplification, "the telephone may also be made so as to improve or entirely supersede the present artificial means used for conducting and intensifying the sound-waves to the auditory nerve" far more efficiently than conventional ear trumpets.[23] In the same year, Dr. A. J. Balmanno Squire described a "miniature telephone" to the readership of the *British Medical Journal*, one that was "so small as to fit, unperceived, into the external auditory meatus":

> With this stuck in my ear, I have been amusing myself this evening by listening in my kitchen to some songs sung by a friend shut up in my attics, my friend singing to a telephone in the attics connected by a wire down the staircase with the miniature telephone concealed in my ear in the kitchen. Here I was enabled to appreciate with relish his feeling rendering of "God Save the Queen" with *ad libitum* variations, by means of an instrument no bigger than the plug of wax which often occupies the auditory meatus.[24]

As the use of Squire's "miniature telephone" extended beyond the acoustic lab and into the research and development (R&D) process for new consumer technologies for hearing impairment, it somehow ceased to be a telephone—it became a hearing aid.

The early telephone presented many possible futures as a medical technology, although most of them did not come to pass. But some few made an indelible impact. The descendants of Squire's "miniature telephone" became the electronic hearing aid, and the more distant cousins of Thompson's "telephone-probe" became the technology we now call diagnostic ultrasound. These innovations, now integrated into medical practice, are no longer remembered to be connected to telephones at all. To the extent that these devices succeeded in becoming more specialized medical tools, their origins in early communication technologies have been forgotten. In turn, as the telephone became more successful as a consumer product for everyday communication, it gradually ceased to be thought of as a potential medical technology. As a general communication technology, however, connecting people in new ways, the telephone would play its most powerful role as an agent of change in American healthcare.

* * *

By the turn of the twentieth century, the many possible meanings of telephonic technology in medicine began to narrow. Diagnostic and laboratory applications of the telephone became less prominent compared to its general use as a communications platform. The telephone, in popular consciousness, was now understood not as an experimental electromagnetic technology but as a consumer good that connected people one-on-one through switchboards. In many towns, the physician's office (and home) was among the first to have a telephone. Physicians were often early investors in local telephone exchanges. As the increasingly familiar technology of the telephone became standard to office and home communication, the telephone took on new meaning. It became a symbol of the ethic of constant accessibility, technological currency, connectivity, and efficiency that characterized the modern physician.

When the prominent surgeon Dr. James Fairchild Baldwin purchased his first telephone in 1879, he was one of the earliest adopters in Columbus, Ohio. Baldwin balked at purchasing multiple phones for his home and office, and instead installed "an unusually long cord attached to his telephone" to link his medical office on the ground floor to his living quarters above. "In this way," the *Ohio State Medical Journal* later reported, "when his office hours were over in the basement of his home, he would place the instrument in his handy dumbwaiter and send it up to the first or second floor of his home as the need might be."[25] Dr. A. L. Brobeck also saw early on the advantage of being reachable by his patients, many of whom lived at some distance from his home and office in Wellington, Illinois. When Brobeck erected the first telephone line in Wellington, eight miles in length, it connected all the local farmhouses to his office in the center of town. The physician who owned a telephone exchange controlled the connection. As Brobeck noted, "he has the advantage of his city *confrères* in being able to converse directly with his clientage without a wordy struggle with the 'hello girl,'" without needing to pay out "his hard-earned dollars to a soulless and grinding monopoly."[26]

While early urban and rural adopters like Baldwin and Brobeck saw prosperity and progress in wiring their clinics, they were outliers. As the telephone encroached on clinical practice, many practitioners feared unintended consequences. The spread of telephones into everyday medical practice provoked skepticism and outright hostility. Concerned physicians saw the telephone as a symbol of the accelerating pace of modern life, as well as a threat to their labor. What good was the convenience of telephone communication if it opened up every bit of one's private life to the possibility of being called back into professional duty? "As if the Telegraph and the Post Office did not sufficiently invade and molest our leisure," the *British Medical Journal* complained in 1883, "it is now proposed to medical men that they should become subscribers to the Telephone company, and so lay themselves open to communications from all quarters and at all times." The telephone destroyed the possibility of leisure time, and it also fundamentally changed the nature of medical work itself. What value was the doctor's presence "when people can open up a conversation with you for a penny"?[27]

Dr. Theodor Schaefer, a Kansas general practitioner, claimed the telephone was the key link in a sequence of communications technologies that reduced the work of the physician to an automated commercial function. "The busy ring of the telephone, the familiar click-clack of the typewriter, and even of the telegraph instrument," Schaefer complained, "in short, the modern doctor's office is transformed into a mercantile establishment."[28] Schaefer saw a connection between the mechanization of human communications, on the one hand, and the demands of the marketplace, on the other. The human practice of medicine, he feared, would be trapped in between. The key features that distinguished a doctor's office from a stockbroker's office were being lost in a convergence of technologies that tethered all of these workplaces to an accelerating economy that demanded instantaneous communications.

The "annihilation of space" by instantaneous communications—a term used by the *Philadelphia Ledger* to report the first long-distance telegram in the United States—was linked in the late nineteenth-century mind to the acceleration of time. Karl Marx, in his critical efforts to describe capitalism as a world-shaping force, saw these communication devices as part of a system that demanded increasing efficiency, acceleration of workload, and automation of labor. "While capital must on one side strive to tear down every spatial barrier to intercourse," Marx wrote in the *Grundrisse*, ". . . it strives on the other side to annihilate this space with time, i.e., to reduce to a minimum the time spent in motion from one place to another."[29] Communications devices like the telegraph, typewriter, and telephone all played a role in this acceleration in the stock exchange, factory floor, and also in the hospital and clinic.

This analysis resonated with the Massachusetts physician John L. Hildreth, who warned his colleagues that the telephone risks "making available for a patient, who for any reason has become impatient, the services of a physician from a distance." Writing a generation after Marx, Hildreth raised concerns that "the work of the general practitioner is harder to-day than it was a generation ago" because of this speed-up. While the physician of the 1900s was not quite as brutally affected by the accelerating pace of industry as was the factory worker, the tethering of clinical practice to the telephone was its own form of proletarianization. The formerly independent professional was now "much less independent, and has less time to himself."

> The multiplication of telephones makes him more subject to interruptions by his patients, who find it so easy to call him by day or by night, to ask questions—some necessary, some unnecessary—that they leave him no time which he can call his own. He is on call at all times; and it is not surprising that, while he appreciates the uses of the telephone, he sometimes regards it as a device of the Evil One. He is a slave to the peremptory call of the instrument, and must constantly be ready to answer it, as in many cases no one can answer it for him.[30]

Hildreth saw the convenience of telephonic communication as a Faustian bargain. While the telephone made him more available to care for his patients, it risked reducing his professional work into a few words transmitted over the phone lines. Worse, it distorted patients' expectations. By making medical services available anytime, anywhere, the telephone served to transform patients into "impatient" consumers.[31]

During the quarter-century that spanned Baldwin's initial Columbus telephone line in 1879, Brobeck's rural telephone exchange in 1894, and Hildreth's critique of the "device of the Evil One" in 1906, the reach of telephonic networks expanded substantially. In 1900, only 5 percent of American households had telephones, many of which were on private exchanges. By 1915 that number had jumped to 30 percent and was rapidly consolidating through the AT&T/Bell monopolies. This last fact was not lost on Hildreth and other concerned practitioners.[32]

By the early 1920s, medical journals urged physicians not only to have a telephone in both the home and the office, but also to have a separate phone in the bedroom. "The man doing family practice, or emergency or industrial surgery," as Dr. A. E. Rockey of Portland, Oregon, noted in 1922, "must literally sleep with his telephone." Rockey recommended that readers of the *Journal of the American Medical Association* choose their bedside telephone carefully, suggesting specifically a model that "can comfortably be used with one hand without raising the head from the pillow."[33] Just as doctors like Rockey advertised home telephones to physicians, telephone companies used access to physicians to advertise home telephones to the general public. As one early twentieth-century advertisement for British Telecom advertised to British consumers: "She's suffering—you dash to the telephone—the Doctor will be round immediately. Later, when the crisis is happily over, you realize it was a matter of minutes . . . *What would have happened without the telephone?*"[34] These advertisements worked on at least two levels at once. On the one hand, they mobilized a sense of urgency around the accessibility of the modern physician, and fostered expectations of constant access through telephone medicine. On the other hand, they served to spread another sense of urgency around the telephone as a lifesaving device for those who could afford to set up a line in their own household.

The increased acceptance of telephones in clinical practice would have many unintended consequences. "It is a self-evident fact that the telephone has come to be a necessity and particularly to physicians," the editors of the *Boston Medical and Surgical Journal* stated in 1910, but "it is also true that not infrequently its very usefulness becomes a source of extreme annoyance."[35] Soon, it was also a source of comedy. In 1912, Ellen ("Mary") Firebaugh wrote *The Story of a Doctor's Telephone—Told by His Wife*, in which Firebaugh describes a series of unanticipated complications that follow as the telephones in the doctor's household multiply. For Mary, the doctor's wife and Firebaugh's protagonist, who is tasked with answering the telephone, its so-called convenience is the stuff of farce:

> When the first 'phone went up Mary soon accustomed herself to its call—three rings. When her husband connected it with the office the rings were multiplied by three. One ring meant someone at the office calling central. Two rings meant someone calling the office. Three rings meant someone calling the residence, as before. Mary found the three calls confusing. When the Farmers' 'phone was installed and the same order of rings set up, she found the original ring multiplied by six. This was confusion worse confounded. To be sure the bell on the Farmers' had a somewhat hoarser sound than that on the Citizens' 'phone, but Mary's ear was the only one in the household that could tell the difference with certainty. The clock in the same room struck the half hours which did not tend to simplify matters. When a new door-bell was put on the front door Mary found she had eight different rings to contend with.[36]

As manager of both a medical office and a household, Mary becomes a human-technological octopus contorting to manage eight different buzzers and bells. Firebaugh wasn't the only writer using humor to cope with the real frustration with the ways in which the telephone was changing how physicians lived and worked. "Appreciation of the Telephone," a poem of light verse published the following year in the *California State Journal of Medicine*, begins and ends with a curse:

> Tinkle, tinkle, little bell—
> How I wish you safe in h—l!
> Central on the job all night,
> Doctor sleeping sound and tight.
> "Baby's got the stomach ache,"
> Mama shaking like a quake;
> Papa running here and there,
> Barks his shins upon a chair!
> Doctor scooting though the air;
> Lights go out, gas all gone;
> Motor dead a mile from home;

Doctor cussing like a fiend
Baby, motor, gasoline!!
(He arrives.)
Baby's sleeping in his bed;
Papa's arm round mama's head.
Nothing happened after all!
Doctor on a useless call!
Tinkle, tinkle little bell!
I don't hear you. Go to h—l![37]

Was the telephone a labor-saving device, or a device that produced new forms of clinical labor? As historian Ruth Schwartz Cowan points out in *More Work for Mother: The Ironies of Household Technology from the Open Hearth to the Microwave*, every new domestic technology produced new forms of women's work. The forms of unintended, unforeseen, and unacknowledged labor laid out in Firebaugh's observations remind us that many of these new forms of labor tended to be gendered female. In medical communications, women were tacitly understood to shoulder the labor of the frothy optimism put forth by male physicians like Baldwin and Brobeck—though male physicians soon found themselves complaining about these multiplying forms of work as well.[38]

As the unanswered telephone engendered new forms of medical labor within the physician's office and household, it also enabled new kinds of medical workers.[39] In 1921, A. J. De Long argued in the *Journal of the Indiana State Medical Association* that the introduction of the telephone into medicine produced a new form of twenty-four-hour accountability too vital to be left to chance:

> Especially during evenings and on Sundays there are times when a doctor's telephones are unattended. Ordinarily a patient might call intermittently for several hours but could receive no answer until either the doctor or his family returned. At times this obvious lapse in service might be annoying, sometimes it would be alarming, and again it might prove fatal.[40]

De Long was not a physician but an entrepreneur. Wheelchair-bound at the age of seventeen—most likely as a result of poliomyelitis—he transformed his boyish enthusiasm for tinkering with electronic communications technology into a new form of economic life. His first step toward independence as a self-described "invalid and practically a shut in" involved selling magazine subscriptions by telephone. "I was confined to my home, with a telephone already installed, and greatly in need of something to do. What circumstances could be more favorable for the establishment of an Exchange? It was not only convenient for me to always be on the job but rather a necessity." With a basic switchboard, his Doctors' Information Exchange could easily redirect calls when a physician reported themselves to be at a country club or any other place accessible by telephone, or could redirect callers to another doctor who had agreed to cover their practice.[41]

For De Long the telephone was an adaptive technology that made it possible to build a new business model around an emerging need. He shared this model freely, laying out for others the equipment and personnel that would be necessary to set up similar exchanges in other towns and cities in Indiana. Other doctors' exchanges emerged in Vienna, Berlin, San Francisco, and Los Angeles, but De Long's advocacy was crucial to the dissemination of exchanges across the American Midwest. "Every community has its shut-ins and elderly people," he noted. "In this city there are persons who will gladly take up the work should I decide to devote my time entirely to other interests."[42] Within a few years, the Doctors' Information Exchange had become self-sufficient, and De Long was profiled in *Boys' Life* magazine as a "Wheelchair Business Boy." DeLong urged *Boys' Life* readers to consider his story, "in the hope that other boys who find themselves limited physically may be encouraged" to build social and financial foundations using communications technologies.[43] De Long's story of the medical telephone as a route to economic independence forms an important counterstrain to the laments of Drs. Schaefer and Hildreth. If able-bodied physicians worried that this "device of the Evil One" reduced their own independence as practitioners by chaining them to the accelerating tempo of the medical marketplace, a disabled entrepreneur like De Long could see the bargain in a very different light.

As telephone exchanges continued to spread in the 1920s and 1930s, the private life of the physician became an increasingly public affair. "The reports given by members of the Exchange make an interesting record," De Long noted. "You soon learn a doctor's friends, his clubs, his politics, his religion, his favorite sports, and his other interests. You know where he sits in the theater and the kind of shows he likes best."[44] In spite of their discomfort with having their own movements tracked, physicians found they could not stop using these services once they began. "It is difficult to see how one could get along without such an exchange," one doctor wrote in the *California State Journal of Medicine* "after he has once become acquainted with it."[45]

By the mid-twentieth century the telephone was not just an essential attribute of the physician's office and home, it was also a symbol of the constant availability of the modern doctor. Images of the physician answering a telephone in the middle of the night became part of the iconography of the doctor as an accessible authority figure in mid-twentieth-century American life. This mid-century advertisement, one of several heroic images of physicians in the R. J. Reynolds company's highly successful "More Doctors Smoke Camels Than Any Other Cigarette!" campaign (fig. 1.2), emphasized that "24 hours a day your doctor is 'on duty' . . . guarding health . . . protecting and prolonging life." The irony of the tobacco industry's efforts to promote the physician as a modern hero would not escape later critics, but these images deliberately invoked the telephone as a symbol of modernity and twenty-four-hour accessibility.

The necessity of the telephone in medical practice in the early twentieth century did not cross over to all households. Telephone advertisements of the day depicted white, middle-class consumers, and while middle-class Black professionals and households took up the early technology as well, they did not do so in the same numbers or at the same pace as middle-class whites. By the late 1930s, however, telephone use was common enough among readers of the *Baltimore*

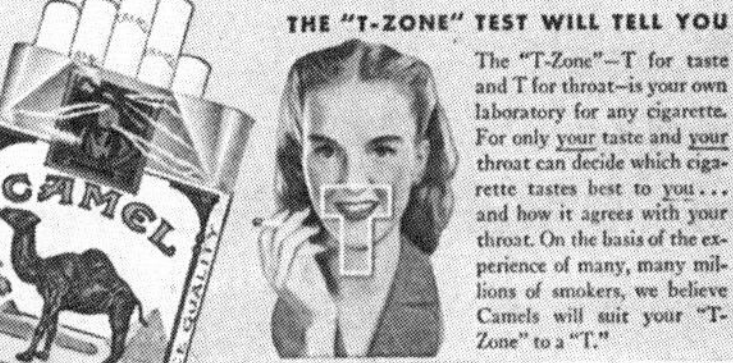

FIGURE 1.2. "I'll Be Right Over!" Advertisement for Camel cigarettes emphasizing the twenty-four-hour accessibility of the physician by telephone, part of Camel's long-lived advertising campaign promoting the American physician as a professional who was always available when needed (but who enjoyed Camel cigarettes during their limited leisure hours). Courtesy of Stanford Research into the Impact of Tobacco Advertising.

Afro-American that Howard University professor Dr. Algernon Jackson devoted part of his regular "Afro Health Talk" column to plead for mercy from the excessive telephone calls that the average *Afro-American* reading doctor received from patients who "evidently think he either sleeps with clothes and boots on or jumps out of bed and slides down a pole into them as the fireman does."[46] Household access to telephones among Black families nonetheless lagged substantially behind white households throughout the twentieth century—an early indicator of lingering divides in access to electronic medical communications in later generations.

The ubiquity of the telephone in medical practice, and the need to have someone take and respond to calls at all hours, introduced new liabilities for the practicing physician. In 1941, the *American Journal of Public Health* noted: "One could not conduct a modern business, a practice of medicine, or a public health program without a telephone . . . upon that much all will agree."[47] Legal cases began to allege that a physician's failure to answer a phone, let alone the failure to own one, was a form of culpable neglect. By the late 1940s medical journals urged doctors to have a staff person or "robot telephone" answer their calls when absent.[48] As the British Medical Association explained in 1947, physicians had to "solve the problem of the unattended telephone."[49]

By the 1950s, a public debate over "telephone neglect" made clear that the communications device had become omnipresent as an infrastructure of medical practice. *The Lancet* and *New England Journal of Medicine* closely followed a malpractice accusation in London regarding an unanswered telephone. A mother had complained to the General Medical Services Board that she waited all day for a doctor to see her young daughter with tonsillitis, but the doctor never answered his phone. After an inquiry, the members of the board found that there was no clear statute violated by a physician who failed to answer the telephone, but all the same found such behavior to be deeply irresponsible.[50] While the British minister of health stopped short of requiring that all physicians be available by telephone at all times, he added that "it seemed a matter of common sense" that a general practitioner was "not in a proper position to fulfill his obligations to visit his patients if he did not make reasonable arrangements for them to communicate with him by telephone."[51]

In 1960, a British court issued the first ruling that declared "culpable neglect of telephone" as a novel form of medical malpractice.[52] Similar policies were soon adopted in the United States. Not only was the telephone, like the X-ray or the routine blood test, a technology assumed to be accessible to all physicians, it had become a standard of practice such that the *absence* of this technology could now be considered a form of malpractice and neglect. The culpability of a physician who failed to answer their phone signaled the integration of the technology into the infrastructure of healthcare. The telephone's early promise as a laboratory reagent or diagnostic technology may have gone largely unfulfilled, but as a communications device it gradually became an essential tool for patients and doctors. Once this transformation was complete, the new technology effectively disappeared. By the mid-twentieth century the telephone became an assumed feature, its presence no longer newsworthy.

* * *

Coverage of telephones in medical journals after 1950 focused on how they served as platforms for other medical technologies.[53] The telephone itself was disappearing as a new technology. Now the doctor was on call, reachable at any hour, and culpable when he wasn't. These things were simply accepted features of the modern world. But the telephone's capabilities continued to inform and fuel the development of newer medical communications devices.

Add a magnetic data storage device, for example, and a telephone became a Dictaphone. By 1951, the Indiana University Medical Center announced a novel system that would allow physicians to pick up any telephone handset and dictate the patient's medical record, to be transcribed into text "using the dictating phone in much the same way he would use an ordinary telephone in talking to the referring doctor." A stenographer would then transcribe the dictation from a magnetic disk into the hospital's central data repositories.[54] By 1957, Toronto's Hospital for Sick Children had fully integrated a Dictaphone system with its telephone switchboard, using three tele-voice writers with voice-operated relays.[55] It was clear to all parties that the Dictaphone, not the telephone switchboard, was the newsworthy innovation at this point. Like the telephone-answering service, however, this labor-saving device served both to create a new kind of medical labor and to further separate that labor from the clinic or clinician and patient. The dictating physician did not see the stenographer anymore—just the telephone.

At this point the *network*, rather than the telephone itself, spurred researchers to imagine electronic medicine in a newly wired world. Early tape recorders weren't the only new devices to tap into the power of the telephone network as an electronic grid for storing and sharing medical records. Early fax machines took on medical roles as well. In the 1940s, a group of physicians began sending digital images of chest X-rays between the Jewish Hospital of Philadelphia and the Ventnor Clinic in Atlantic City, New Jersey, using commercial telephone lines.[56] Based on earlier findings by Argentinian physicians that can be traced to the late 1920s, "telephone facsimiles" of X-ray images could be found in several rural hospitals by the early 1960s. These systems formed the basis for the picture archiving and communication systems (PACS) found in all electronic medical records today.[57] Here the medical fax machine, not the telephone itself, was understood to be the innovative medical technology.[58]

Any device that recorded any aspect of a patient's health in analog or digital form could be hooked into the telephone network to make a new telemedical technology. In 1956 another "ingenious new device," called a telephone electrocardiograph, transmitted a series of electrocardiograms from the Kansas State Tuberculosis Sanatorium in Norton to a cardiologist more than 300 miles away in Emporia.[59] They could be read just fine on the other end.[60] By the late 1960s telephone electrocardiography could be found within the sprawling complexes of large urban hospitals where cardiologists were plentiful but not always right at hand, and across large rural spaces in which heart specialists were few and far between.[61] Obstetrics and gynecology, pediatrics, and neurology soon joined cardiology in linking modes of physiological monitoring, from fetal electrocardiograms to electroencephalograms, to telephone wires for remote transmission.[62] All of these devices required telephone networks to work. But it was the new black boxes that were seen as medical devices. The telephone itself was now infrastructure, not innovation.

As interstate, long-distance telephone networks spread across the country, the opportunities for transmitting medical information also grew. In 1963 Atlantic City hosted a dramatic demonstration of a three-channel system that linked the second-by-second peripheral nervous, cardiovascular, and brain wave data of a patient in the VA Hospital in Boston, beamed live to the American Medical Association Convention Hall on the Jersey shore. If doctors on one end of the long-distance call could manage the real-time care of a patient hundreds of miles away, soon perhaps they could do so from the comfort of their own home. Evocative of the techno-utopian devices found in the new animated sitcom, *The Jetsons*—which first aired a year earlier with an episode featuring telemedical care of six-year-old Elroy Jetson—the telemedography set seemed on the cusp of becoming a real consumer technology. A small, portable telemedography set would soon be available for the physician in private practice, the AMA presenters announced, "establishing his bioelectric consultant no farther away than the nearest telephone plug."[63] The implications were thrilling, but also threatening. On the one hand, for the isolated rural practitioner, expert consultations could soon be obtained immediately from urban specialists, while their patient was still in their office. On the other hand, cardiologists in Cleveland might soon find their local expertise questioned and their secure practices challenged by newfound competition with counterparts in New York, Chicago, and Boston.

As mainframe computers found their way into hospitals and clinics in the 1950s and 1960s—a subject discussed in greater detail in later chapters of this book—the newly introduced push-button telephone became both a metaphor and a concrete tool for integrating networked computer systems that might become part of everyday medicine. As Dr. Scott I. Allen and Michael Otten from the Division of Computer Research and Technology of the National Institutes of Health noted in 1969, the standard twelve-button telephone could also be outfitted with a punch-card reader and serve as "a computer input-output terminal."[64] As with the Dictaphone, the technological marvel in this case was not the telephone itself but the Dataphone that connected the computers to the telephone grid and used the telephone grid to connect doctors to computers. Telephones were used for talking. But the telephone *network* could do more than connect people to people, it could now connect machines to other machines. Unseen and unsung, the telephone grid had become the basis for launching other futures of healthcare information technology.

Magnetic tape, fax machines, X-rays, ECGs, television screens, computer systems: all of these devices and more converged with the telephone network over the course of the twentieth century. The telephone itself had long since stopped being a newsworthy medical technology, but it could nonetheless be iteratively rediscovered as a platform for new inventions—some of which changed the way medicine was practiced, and some of which fell away into obscurity.

* * *

Most dreams of telephone medicine faded nearly as soon as they were spun. Recall the disappointment of Alexander Graham Bell's sometime partner and sometime antagonist, the ear specialist Clarence John Blake, who lamented that the "web of wires" had not become a site for a new specialty of telephone medicine, in part because of "the very delicacy of the telephone . . . and its almost fatal propensity—if such an expression may be used—to pick up sounds that did not belong to it."[65] Yet as the web of wires continued to extend into hospital, clinic, and home, it brought along new understandings of an electronic network as both an abstract concept and a material thing. As the modern hospital became an increasingly specialized organism in the early twentieth century, telephone wires formed its rapidly branching nervous system.[66] At the same time, nineteenth-century concerns about the "almost fatal propensity" of the telephone to distort clinical judgment remained keenly felt into the late twentieth century. Even if the telephone was no longer celebrated as a new medical technology, it remained a source of anxiety. Telephone errors could ruin a physician's practice—or threaten a patient's life.

As late as 1972, phone diagnosis was still described as a "practical but perilous method": too useful to abolish but too dangerous to be entirely comfortable with. While the telephone network played a key role as a prototype for later networks of biomedical communication, it also prompted a set of questions about the problem of "being there" in medicine. Which patients could be treated at a distance? Which needed to be seen in person? And how to tell the two apart? To be accepted as part of everyday practice, telephone medicine required scripts and protocols for the physician and patient to follow.

"Like an extra arm to the physician is that wonderful but pesky invention—the telephone," the American Medical Association counseled its members in a 1954 guide on the proper use of telephones in medical practice. "It is an instrument, however, that requires special handling."[67] Part public relations manual, part technical publication, the AMA telephone manual emphasized comprehensive rules of etiquette by which receptionists, nurses, and physicians could use the telephone to help rather than hinder the business and practice of medicine.[68] Many hospitals followed the Mount Sinai Hospital of Cleveland in hiring consultants from the local Bell Telephone Company to instruct nurses, interns, and physicians in proper use of the telephone to maximize the benefits and minimize the harms the telephone brought to practice (fig. 1.3).[69] These risks were financial and legal as well as clinical. Tactics to minimize the risks of telephone medicine were not neutral or natural: they needed to be learned, and often they needed to be explicitly taught.

As the reports of telephone-related malpractice began in the 1960s and multiplied over the 1970s, hospitals and group practices struggled

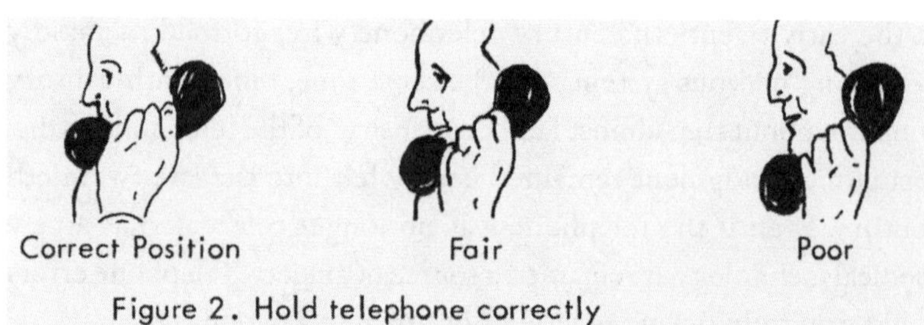

FIGURE 1.3. Telephone etiquette manual for healthcare workers, showing the correct positioning of the telephone for optimal communication. Courtesy of Academy of General Dentistry.

to contain the legal risks of the communications device (fig. 1.4). "It's all very seductive in the beginning" to conduct medical practice by telephone, Stu Chapman wrote in *Legal Aspects of Medical Practice*, "and it may lure you with a false sense of security, convincing you that its technology can help you decide what ails your patients and give you a big hand in prescribing for them. But doctors who never figured that their dependence on the telephone would land them in court have discovered how fickle such a relationship can be for their fortunes and practices."[70]

Chapman described the telephone as an addictive substance—like amphetamine, which first seemed to enhance performance but over time inculcated a dependence the casual user often did not recognize until far too late. The expanded reach of the telephone allowed physicians to broaden the scope of their practices well beyond their individual means, which left them all the more dependent on the device as a way of checking in on patients they might never even have met. As one physician complained after an evening telephoning forty-four separate prescriptions into the pharmacy for patients he did not know, "when you get into a situation like this, you're talking to patients you've never seen before. You wouldn't know them if you saw them on the street. You've never seen them in the office."[71]

The facelessness of telephone communication effaced both doctor and patient. This could lead to some form of dehumanization even

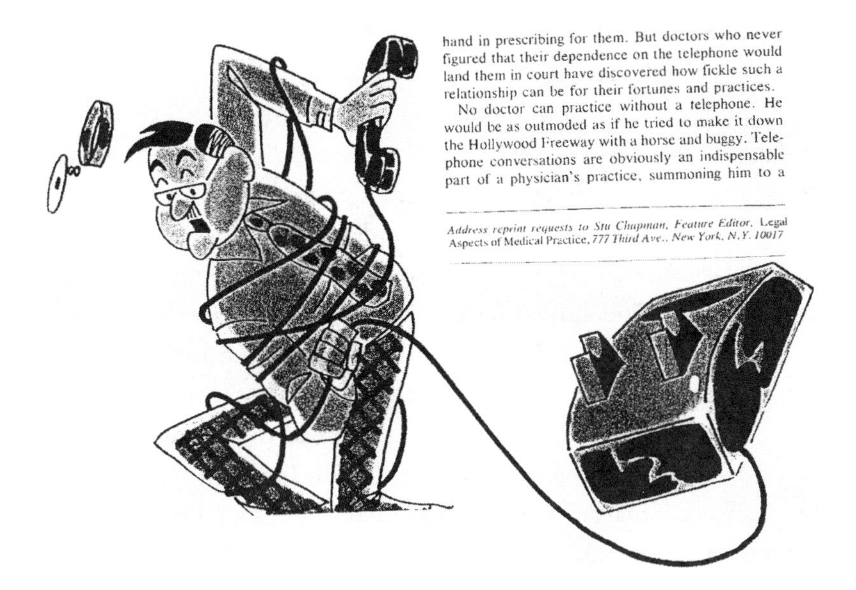

hand in prescribing for them. But doctors who never figured that their dependence on the telephone would land them in court have discovered how fickle such a relationship can be for their fortunes and practices.

No doctor can practice without a telephone. He would be as outmoded as if he tried to make it down the Hollywood Freeway with a horse and buggy. Telephone conversations are obviously an indispensable part of a physician's practice, summoning him to a

Address reprint requests to Stu Chapman, Feature Editor, Legal Aspects of Medical Practice, 777 Third Ave., New York, N.Y. 10017

FIGURE 1.4. As the telephone became an expected feature of medical practice, it also introduced new forms of malpractice into American medicine, represented metaphorically in this illustration from Legal Aspects of Medical Practice as a physical as well as legal entanglement. Courtesy of American College of Legal Medicine.

in the best cases and to unforgiveable lapses in clinical judgment in the worst. Chapman cited a New Hampshire physician group whose telephone practice was found culpable of negligence in the case of a patient who suffered a potentially preventable stroke. When Walter Dwyer called the Concord Clinic in February 1974, he related to a nurse that his wife Jeanne, known to have high blood pressure, had developed a bad headache with nausea and vomiting. The doctor who called him back had never met Jeanne, and half-heartedly prescribed an antinausea pill by phone. When the headache returned the next day, the husband again called the clinic, and a second doctor called in a narcotic painkiller to treat the symptoms. After the symptoms returned a third time, Walter drove Jeanne to the clinic and demanded an examination. She had been bleeding into her own skull for days.[72] This story underscored the risks the telephone posed to patients as well as physicians, if they could not convey the urgency or severity of their symptoms over the wire.

As a new wave of self-help books for American medical consumers flourished in the late 1970s and early 1980s, a subset of these books taught patient and family members how to communicate their medical needs more effectively by phone. In the early 1980s, Cornell pediatrician Jeffrey L. Brown published a pair of books on the topic: *Telephone Medicine* (for physicians) and *The Complete Parents' Guide to Telephone Medicine* (for patients and parents).[73] Telephone encounters, Brown claimed, now took more than 20 percent of a practicing pediatrician's time. In his opinion this reflected major shifts in the economics of the American household in the second half of the twentieth century. Working mothers had less time during the day to bring their children in to see the doctor. Inflation and stagnant wages "forced many families to seek free telephone advice rather than pay for an office visit."[74] Somewhere in this blend of patient empowerment and demand for more cost-effective care, Brown saw a neglected aspect of medical communications.

Brown was careful to point out that his manual was not a substitute for seeking the advice of a physician. Instead, *The Complete Parents' Guide to Telephone Medicine* coached family members to frame problems more efficiently so that a receptionist could better triage them for the physician's attention. Brown calculated that most pediatricians received four to six calls per hour, seven days a week, and relied on their receptionists to help sort out the urgent matters from the routine. Yet the receptionist herself (always coded female) had no special training in diagnosis. Brown's advice helped patients make themselves more digestible as data for triage algorithms by paramedical personnel. When speaking with physicians, he advised patients and parents to introduce themselves properly, tell the doctor the *real* reason for the telephone call, give the most important information first, be brief, and understand what was said to them.

Brown's book sold several editions precisely because these scripts were *not* obvious. "How-to" guides like *The Complete Parents' Guide to Telephone Medicine* make visible some of the social protocols through which medical use of the telephone was gradually scripted to the point that doctors, receptionists, and patients eventually needed no instruction to know how to conform to their roles. "This sounds so obvious as not to deserve mention," Brown himself concluded, "but it is surprising how many parents have such difficulty in organizing their thoughts that the doctor honestly cannot tell the real reason for the call."[75]

In Brown's view, the telephone not only made the pediatric patient available to the physician at a distance, it also required the parents to become both medical historian and physical examiner inside their own homes. More generally, the telephone brought a new form of labor into the business of being a patient. The savvy telephone patient knew how to present only the most relevant signs and symptoms of disease, in the correct sequence, to most effectively aid the receptionist or medical assistant in the act of triage. Brown's book enabled the parent to probe in a formalized sequence the proper observation of common symptomatic complaints like fever, sore throat, earache, vomiting, diarrhea, and rashes, so that the information could be described over the telephone in a manner intelligible to medical receptionists, nurses, and physicians. For cuts and lacerations, a doctor should be called immediately if "your child has a gaping cut which is longer than one-quarter inch" or if "you are unable to stop the bleeding after 20 minutes."[76] Telephone medicine required the anxious parent to learn the skills of a triage nurse.

Telephone triage required patients to code their own experiences in a way that could be efficiently processed by the algorithms used on the other end of the line. It also required parents to develop their own sense of whether their child was *really sick* in the sense that required the immediate presence of trained medical personnel—the meaning of "sick" that healthcare professionals use to denote the patients they fear may not last the night. "While we have tried to make the advice in this manual as accurate as possible," Brown advised, "*you* are the only one who is actually looking at your child when a problem arises."[77]

* * *

The Complete Parents' Guide to Telephone Medicine was more than an instruction manual. In these pages the telephone stood at the crossroads of the changing power structures by which patients related to doctors. The telephone stood as a metaphor for the paradox of medical consumerism, supposedly granting more power to patients and families, on the one hand, while at the same time reducing their unique voices to algorithmic sequences.

Other guidebooks, meant for nurses, likewise dissected the telephone script to examine the even more notoriously problematic power dynamic between nurse and doctor. Jo Simms and Reba McGear's 1988 book, *Telephone Triage and Management: A Nursing Process Approach*, focused deliberately on the new challenges and opportunities the telephone presented for nurses seeking more autonomous clinical roles in the primary care setting. As nurse-educators based in the Pacific Northwest—Simms was active within the Group Health Cooperative of Puget Sound, an important laboratory for managed care and health maintenance organization (HMO) design—the two saw "telephone management as vital to more kinds of nursing than ever before," and packaged the book with an audiocassette of telephone encounters to help train these new "telephone nurses."[78]

As historian of nursing Margarete Sandelowski has described, most stories we tell about medical technologies focus on heroic tools wielded by (largely male) physicians and overlook the far more pervasive role of (largely female) nurses in the integration of everyday technologies of care. These everyday uses, however, can create powerful forms of identification between nurses and the technologies they use. While the thermometer in nineteenth-century medicine was initially a precision tool used by male physicians, in the course of a decade the thermometer became so strongly associated with nursing technique that some nurses were referred to as "thermometer nurses."[79]

So too with "telephone nurses." As the telephone became a technology of nursing, its new uses were drawn from a formal field of nursing practice and theory that differed substantially from the medical profession's obsession with the "doctor/patient relationship." Nursing theorists attended to different elements of the telephonic encounter than physician-scientists, who were more focused toward basic sciences and clinical research. Instead of a disease-oriented schema that moved from the patient's history and physical examination to laboratory and diagnostic studies, the theory of nursing process emphasized human interactions of assessment, planning, implementation, and evaluation.[80]

Nurses understood the advantages and limitations of the telephone as a means to gather pertinent data for diagnostic and therapeutic algorithms, to provide reassurance to anxious parents and patients, and to triage the need for urgent care visits. Telephone triage allowed patients to access healthcare from a position of relative empowerment: in their own home, sitting on their own couch rather than on a sterile exam table wearing a johnny open at the back. Yet the telephone also restricted the sensory world of a typical clinical encounter to a single dimension. The nurse could not see, smell, touch, or observe the patient. The nurse's voice on an anonymous line was not necessarily comforting to patients in need of reassurance—although with effort it *could* be. To be effective, telephone nursing required the establishment of authentic telephone presence through a deliberate and self-conscious process called "developing a telephone personality." Like the makeup television news anchors apply just to appear normal on the camera, telephone nurses were encouraged to amplify their emotional presence so that a real human connection might come across the wire.

Telephone Triage also served as a field guide to the different kinds of telephone personalities that *patients* developed as well. The effective telephone nurse needed to be able to apply different interpersonal as well as clinical algorithms to the care of callers. Clients could be histrionic, manipulative, angry, passive, anxious, depressed, suicidal, or stoic, and each of these states called for a different telephonic script. The "cry wolf" caller, for example, presented a vexing problem. The patient might have real chronic or acute medical problems, but also might simply be lonely or looking for attention. Nurses also had to contend with the obscene caller, who "violates the nurse's trust in the unknown caller and exploits her vulnerability in being available to the public."[81] The cry-wolf caller was misunderstood and deserved guidance; the obscene caller was hostile and needed to be discouraged from their sociopathic tendencies. Dealing with the amplified affect of these encounters while also managing real medical emergencies could be exhausting.

The burden of practice for the telephone nurse was substantial and burnout was common (fig. 1.5). The telephone nurse encountered more patients in a given day than a face-to-face nurse, often meeting with raw, unbridled anger and aggression from anonymous contacts, without freedom to move from her desk or cubicle. "For many reasons," Simms and McGear conclude, "telephone communication can be stressful to the nurse, and especially so because the client is neither

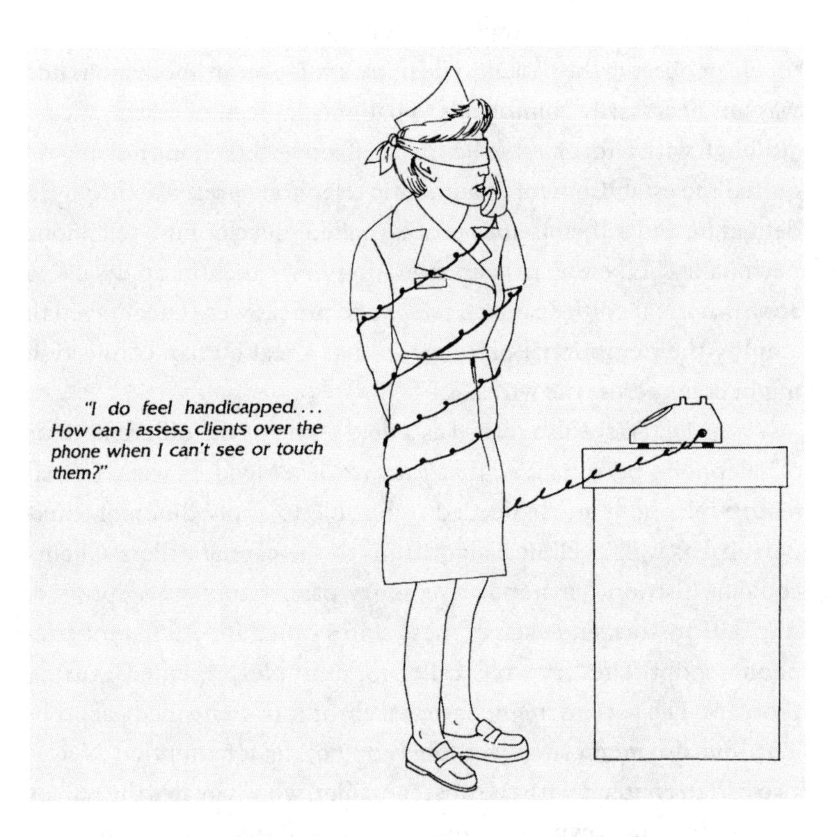

FIGURE 1.5. Burnout and the "telephone nurse": new entanglements for nurses constrained by the limitations of telephone practice. Courtesy of Elsevier.

visible to the nurse nor present in her environment. She can compensate for these handicaps by developing her listening skills and her intuition. She can also use effective communications skills, take care of her emotional well-being, and reduce stress, all of which promote her vital connection with the client." Telephone nurses were advised to take frequent breaks, to consider taking up running, meditation, self-affirmation, group therapy, massage circles, or other forms of mutual support to counter the new and repetitive demands of telephone clinics.[82] Like Mary Firebaugh and her many bells, telephone nurses used many strategies to manage the unanticipated strain that telephones brought into their practice.

Both of these "how-to" guides—Brown's popular book for patients and parents, on the one hand, and Simms and McGear's technical handbook for nurses, on the other—make visible the scripting of roles for telephone practice in late twentieth-century medicine. Both of these books promised a form of empowerment to their users: the former, by amplifying the patient's voice; the latter, by establishing a greater space for autonomous nursing practice. Yet both also made visible the limits of the telephone as a tool of empowerment. To be empowered by the telephone, parent and patient and nurse all needed to willingly code themselves into a set of algorithms, logical "if-then" chains written by others who held more power.[83] These algorithms could only go so far, and what limited forms of agency they provided were quickly curtailed in cases that demanded the in-person attention of a (typically male) physician.

* * *

Mid-century fears around the telephone in medicine were never fully resolved. In the first years of the twenty-first century, a report of the American College of Physicians urged doctors to reconsider the merits of the telephone. Noting that one out of every four encounters between doctors and patients now took place over the telephone, the authors claimed that the telephone offered the healthcare provider an untapped opportunity. Using the phone more self-consciously, doctors could diagnose disease earlier, manage chronic conditions better, lower costs, and expand access to care to more people. Yet the telephone continued to pose unique challenges to doctors and patients. Used incorrectly, it could lead to dissatisfaction, loss of human connection, and bodily harm. The ACP report, a "practical, evidence-based guide for the internist," replicated for physicians a literature that had already been developed a generation earlier for nurses, parents, and patients, as well as for physicians.[84]

As this chapter has shown, different generations of physicians applied different sets of hopes and fears to this communication technology. To the recently minted MD of the 1880s, the telephone was a new and experimental device, and both enthusiasm and skepticism around the technology were largely limited to speculative fiction and laboratory trials. To the young physician of the 1920s, the telephone was a symbol of connected modernity, linking American physicians and their patients. The hopes and fears for the technology had taken on more serious stakes. On the one hand, the telephone promised near-universal access to specialized medicine; on the other hand, it threatened to place those specialists in unwelcome competition with one another and to reduce medical judgment to a detailed triage algorithm. For medical students and interns of the 1960s, the telephone had become a "convenient nuisance": an important part of daily practice, to be sure, but hardly anything to get excited about anymore, unless it could be hooked up to a newer device like a Tele-ECG or a Dataphone.[85] The first century of the telephone in American medicine traces a broad arc from radical innovation and exuberant uptake to widespread amnesia and banality.

Forgetting is an active as well as a passive process. The eventual erasure of the telephone from the pages of medical journals speaks volumes on the powerful role it had taken on in structuring norms and patterns of medical practice. Yet even as it receded into the infrastructure of medical practice, in the late twentieth century the telephone continued to be rediscovered anew as a technology of empowerment, giving patients an amplified voice in clinical interactions, giving nurses a space of their own to communicate their value as autonomous providers. As a tool of empowerment, however, the telephone could not escape the broader power differential between doctor and nurse, doctor and patient, in the charged landscape of American healthcare. As the telephone empowered individual patients, parents, or nurses, it further encoded power relations in what categories were accepted as part of the world-as-it-is.

Like all new media, the telephone as a medical technology was introduced with revolutionary promises, but ultimately realized a far more evolutionary change in concert with the distinct set of opportunities it offered to different sorts of users—doctors, nurses, patients, parents, and entrepreneurs—to speculate on new forms of practice. As a site for speculation, the new communications platform of the telephone allowed a wheelchair-bound young man like A. J. De Long to imagine a new form of career as a physicians' answering service provider, at the same time that it provoked contemporary physicians to imagine a dystopic future of alienation from their own labor. These perceived benefits and costs of telephones did not play out evenly across all parts of American society.

How, then, does the present-day reader properly account for the similarities and the differences between the use of telephone, thermometer, and stethoscope in the late nineteenth century and the funding-round pitch for the CliniCloud in the early twenty-first century? Both of these stories, not just the older one, are inexorably situated in a set of historical contingencies. True, the optimism behind the telephone as a medical technology in the 1880s did not live up to initial expectations. But neither did the performance of the CliniCloud. In the beginning of 2017, the CliniCloud Facebook page had more than ten thousand "likes," thousands of units had been sold to consumers, *The Lancet* had published a review citing the utility of the technology, and the US Food and Drug Administration was moving toward a Digital Health Innovation Plan that could speed approval and draw attention to further CliniCloud peripherals, like a blood pressure cuff and pulse oximeter. But by March of that year, Clini-

Cloud founder Andrew Lin wrote an essay complaining that slow adoption to date had hampered the uptake of the smartphone-based remote patient-monitoring tool.[86]

By 2018, the CliniCloud's social media outreach had ceased entirely, and the product was no longer available in Best Buy, or anywhere else: another new type of media grown old and soon forgotten.

2
THE WIRELESS BODY

Nearly half the US adult population will pass out at some point in their lives. Doctors call this "syncope," and it is bread-and-butter practice for any emergency room or urgent care clinic.[1] While most cases are benign—a symptom of dehydration or mistimed medication—syncope can also be a sign of something gone terribly wrong. It may be a symptom of a heart attack, a blood clot in the lungs, an embolus to the arteries supplying the brain, or a life-threatening arrhythmia. After a series of tests ruling out the worst, most patients go home without incident. Many also go home with a Holter monitor.

The Holter monitor is a portable box that records the electrical activity of the heart. It has become such a common object in clinical medicine that few pause to consider its origins. As new Wi-Fi– and cloud-enabled devices, smartphone apps, and other "wearables" claim to revolutionize the world of preventive healthcare, it is easy to overlook this older instrument of medical surveillance.[2] But in 1949, when Norman "Jeff" Holter first dreamed of a wearable monitor that would broadcast the electrical activity of patients' hearts as they went about their daily business, he was animated by the newest wireless technology of his day: FM radio. Just as the telephone technologies promised to annihilate time and distance, radio technologies promised to overcome the opacity of the body and make physiological processes transparent, for all to see.

What hidden disease might lurk in the daily fluctuations of the seemingly normal heart? "None of us," Holter warned a few years later, "knows what transient changes, if any, may occur in the electrocardiogram of a healthy Cabinet member during the course of an all-day, smoke-filled conference on some international crisis." But what if a wearable device could tell us? Using Holter's monitor, the functioning of the ordinary human heart could be continuously tracked, minute by minute. Holter's ambition wasn't limited to heart monitors. He envisioned "a more general project of *broadcasting physiological data*"—collecting data from people going about their everyday lives and transmitting it to analytic systems to be interpreted.[3]

The first Holter monitor was a bulky FM transmitter hooked up to electrodes on the chest that could broadcast heart tracings with a range of about one city block (fig. 2.1). When worn by people with known heart lesions, the radioelectrocardiogram could detect abnormalities just as well as a conventional ECG machine.[4] Over the next few years, Holter retooled the device into a smaller form that could fit into a suitcase or satchel. Soon, he believed, anyone would be able to wear one of his devices as they went about the business of a normal day.

Holter's early successes in broadcasting bodily data inspired other applications. In private correspondence, the conservative public opinion researcher Gerald Skibbins joked with Holter that his machine could lead to forms of scientific workplace management that CEOs of the day could only dream of. Firms could monitor the efficiency of blue-collar workers as they performed their daily tasks on the shop floor. To manage the managers, Skibbins suggested a more exclusive club that would give a member of the executive class "the comfortable frame of mind that his variations from the norm are going to be detected at a time when he can do something about maintaining his efficiency." The joke cut two ways. Skibbins, like Holter, was deeply skeptical of the expansion of the surveillance state proposed by Cold War liberals. At the same time he was attracted to the ways that similar technologies could economize the management of the private-sector workplace. "I am quite sure," Skibbins concluded,

FIGURE 2.1. Holter's original wearable health technology: the radioelectrocardiograph, c. 1957. Courtesy of New York Academy of Sciences.

"that ideas such as these which we take in jest today will be taken seriously by somebody tomorrow."[5]

He was not wrong. Wearable technologies that transmit physiological information in real time are now commonplace. Your smartphone or smartwatch can transmit your sedentary quotient, sleep duration and quality, and pulse rate to a central server for collection and data mining by a series of unknown parties. One out of every five US employers that offered health insurance in 2018 collected wireless physiological information from the wearable devices of their employees.[6] Many firms now use Fitbits and Amazon Halos and other wearable sensors not just to monitor the health of their workers but also to optimize their daily work routines—and dock their pay for bathroom breaks. Wireless physiological tracking is simultaneously an opportunity for new interventions in public health and for newly invasive forms of workplace surveillance.[7]

Yet as new as the current state of omnipresent physiological surveillance may seem, more than a half-century ago radio devices were already generating similar hopes and fears for making bodily data trackable to the outside world. With the right transmitter and the right receiver, wireless technology could access the inner workings of the human body, translate those functions into electronic form, and then broadcast them outside of the body. Radio promised healthcare providers, public health agencies, employers, and anyone else who might be interested the ability to detect disease earlier and to make the inner workings of the body legible from the outside.

The Holter monitor is one of many artifacts from an earlier era of wireless medicine we have learned not to see precisely because its impact has been so profound. Radio pills, brain wave recorders, and heart monitors may have started as the stuff of science fiction, but between the late 1940s and the early 1960s, these technologies laid the foundation for the modern medical electronics industry. At the same time, increasing preoccupation with secrecy, surveillance, and the broadcasting of hidden messages colored how both inventors and users interpreted and applied these new technologies. The Cold War may be over, but these anxieties—about where one's information goes, who has access to it, and how it might be used—remain.

* * *

Jeff Holter was fascinated by the unseen. One of his earliest projects was to find photographic evidence of square raindrops. Another was to build devices that could detect evidence of Russian atomic blasts halfway across the world using only the supplies available at the hardware store. Most of these projects used ordinary objects to detect extraordinary things. To make visible the invisible.

Holter represents an unusual and understudied kind of American scientist. His research foundation, based in Helena, Montana, was initially funded by donations from family (his father was a successful hardware merchant) and from his own personal accounts. The lab was a ratty, deserted, second-floor space situated above a grocery store. Even after he could afford an assistant, the men still, reportedly, "peed in the sink" for lack of a bathroom. Still, Holter was selective about his funding. He terminated a $250,000 grant from the federal government because he bristled at having to account for his spending. His research was freewheeling, and he wouldn't be hampered by having to justify his expenses. "Most cretins can tell you that independent non-goal-directed research doesn't know its thoughts tomorrow, much less of all next month. We might want a roll of scotch tape or we might want a polarizing microscope."[8]

Holter was both a child of privilege and a self-styled rugged outcast. He sought support from the federal government even as he was deeply suspicious of it. He fiercely defended the "open society" of basic science, but was concerned over threats to recognition of his own discoveries. His lab embraced a small-scale ethos while also connecting to powerful networks of "Big Science," from the Manhattan Project to the War on Cancer. Holter was simultaneously well connected and self-isolated, peripheral and central. He infused an idealistic zeal for basic research and celebration of the individual tinkerer with a deep suspicion of large corporations and big government.[9]

Holter left Helena as a young man to study physics and medicine, and then trained to be a cardiologist. But patient care quickly seemed like drudgery to him. He quit clinical practice after a few years to return to his father's hardware store, which he could also use as a laboratory supply source. Occasionally, prominent scientists in coastal cities called him a "hardware store owner" instead of a cardiologist or the head of an independent scientific laboratory. While he seemed to take this in stride, he actively sought recognition for his work and the work of the research foundation that bore his name.[10] He distributed his early publications to eminent figures he admired, like the cardiologist Paul Dudley White. Holter and White's ongoing correspondence soon led to entry into the complex world of federal research funding. It was here, among his other investigations into the detectability of atomic explosions and the possibility of producing square water droplets, that Holter's Helena laboratory began to focus on the role of radio communications in medicine.

Holter's interest in radio as a medical technology began during his undergraduate studies at the University of California, Los Angeles, when his research advisor asked him to reproduce a classic physiological experiment devised by Luigi Galvani in 1771 using an electrified frog leg. Galvani demonstrated that an electrical wire, placed at the exposed nerve of a recently severed frog leg, will cause the muscles to contract and make the leg kick.[11] Working with the experimental psychologist Joseph Gengerelli, Holter was able to put his own spin on the classic experiment. In 1941, the pair demonstrated that Galvani's experiment could be repeated using *wireless* electrical signals rather than a direct wire.[12]

If a severed frog leg could be made to move wirelessly, like the rudder of a model airplane, what other kinds of biological activity might also be amenable to radio "remote control"? Holter and Gengerelli turned to animal behavior. Soon they were fitting miniature radio receivers into the skulls of lab rats to test this hypothesis. While they couldn't directly steer their rats through a maze like so many small, furry, radio-controlled cars, their experiments showed that wireless neural stimulation could alter animal behavior.[13] The idea of "remote-controlled rats" captured the interest of the national press, and in 1949, *Life* magazine published a photo-essay on the experiment. But Holter and Gengerelli were already thinking one step further, swapping the positions of the radio transmitter and receiver. "If we could do something *to* the nerve," Holter mused, "what might we get *from* the nerve at a distance?"[14]

Physiologists had long struggled to build a physical science of living bodies without rendering their subjects unliving bodies in the process. Most classic physiological experiments, from Galvani's electrified frog legs in eighteenth-century Bologna to Claude Bernard's vivisected rabbits in nineteenth-century Paris, required the use of traumatically severed animals whose death became part of the experiment. The phenomena observed in these animal subjects could hardly be everyday experiences for these frogs, rabbits, or dogs—let alone for human subjects. The Russian physiologist Ivan Pavlov won the Nobel Prize in 1904 in part for developing a technique of "chronic experiments" with dogs that could be monitored for months and years through surgically placed tubes as they went through their daily activities. But even as Pavlov's techniques offered a way to better approximate the study of everyday life, they still were only possible in laboratory animals.[15]

In Germany, Bernard's contemporary Carl Ludwig developed an approach to monitor and record bodily function in human patients without breaking the skin. Ludwig's universal physiological sensor, the kymograph, used cuffs and other mechanical connections to transmit living processes like blood pressure and breathing cycles to a lever arm connected to a needle holding a pen against a rotating drum of paper. Ludwig's device recorded tracings that could be studied by physiologists and clinicians—the first of many mechanisms to make internal biological processes visible without vivisection. The kymograph became a platform for other ways to inscribe human physiology on paper, including Willem Einthoven's development of electrocardiographic (ECG) heart tracings in 1895 and Hans Berger's development of electroencephalographic (EEG) brain wave tracings in 1924.[16] Some of these devices, as we saw in chapter 1, used telephones.

While these techniques were less invasive than vivisection, they still required the subject to be directly attached to a cumbersome, stationary apparatus. Einthoven's first electrocardiogram, for example, required dunking all four limbs in buckets of saltwater for several minutes. Wired sensors were useful for brief analysis but not for long-term evaluation. It was only slightly less impractical to ask a person to carry out their everyday life with wires constantly trailing behind them than it was to ask them to do so with a surgical cannula sticking out of their neck.

Radio cut the cords. Wireless devices offered the chance to study the human body doing what human bodies do: performing normal exercise, undergoing the normal stress of work and play, moving through the normal rhythms of life at home. Where Pavlov's chronic experiments provided a means of tracing life-rhythms unperturbed by the stresses of vivisection, Holter and Gengerelli's experiments offered researchers a tool for "transmitting physiological data by radio" to study everyday life on an even more natural basis.[17] If radio links could be established with rat brains, why not human brains? Yes, they knew this sounded like science fiction. But then again, didn't any new science start out as fiction?[18]

* * *

The Luxembourgian American inventor and publishing impresario Hugo Gernsback moved to New York in 1904 with plans for electrical innovations he hoped might find a better future in the more favorable patent regime of the United States. After selling his first patent to the Packard Motor Car Company, Gernsback used the proceeds to set up the Electro Importing Company, a mail-order radio technology firm that formed the basis of his publishing empire. Gernsback's launch of the magazine *Amazing Stories* in 1926 is celebrated as the foundational moment in American science fiction, and his name is celebrated as the "Hugo" of the Hugo Awards—the top literary honor for writers in the field.

Gernsback began writing advertisements to inspire amateur technologists to purchase his do-it-yourself Telimco wireless kits and consider what else *they* might invent with the components. Gernsback had a passion for the speculative fiction of Mark Twain, Jules Verne, and H. G. Wells, but his own stories read mostly as recipes for making futuristic gizmos using parts available at any hardware store. Most subscribers to his many publications—*Modern Electronics*, *Electrical Experimenter*, *Radio News*—were speculative tinkerers like him. They were more interested in generating plausible new uses for existing radio technologies than in ensuring the technologies themselves could be successfully built.[19]

In the May 1919 issue of *Electrical Experimenter*, Gernsback described the basic specifications for a new "Coming Invention" called the Thought Recorder. The human brain, he reasoned, was increasingly understood as an electrical system: nerves like wires, brain like a telephone switchboard.[20] If "thoughts are of an electrical nature, having probably a very short wave length," he wrote, they should be detectable using the right kind of radio kit with the right kind of oscilloscope.[21] If this seemed far-fetched, he reminded the reader, "fifty years ago the recording of the human voice would have appeared just as fantastic as the recording of thought appears today."[22] Nikola Tesla, a frequent correspondent to Gernsback's magazines, agreed that with the right instruments, "the continuous play of thoughts might be rendered visible, recorded, and at will reproduced."[23] Of all the possible uses for a thought-reading device, Gernsback's illustrators settled on the most banal, depicting the Thought Recorder as an office technology enabling a secretary to take dictation without having to listen to any words being spoken (fig. 2.2).

The Thought Recorder is one of hundreds of plausible technologies that Gernsback and Tesla imagined but never built. As Gernsback reminded his readers, it was not always clear at any given moment which elements of the seemingly fantastical were truly incredible and which might be become available as consumer technologies in another decade or so. Perhaps this is why Gernsback's illustrators depicted a near future in which today's speculative objects had become tomorrow's workaday office devices. Yet just a few years after Gernsback's paper on radio broadcast of thought waves, *Scientific American* carried a report of ships at sea using radio to broadcast the heart sounds of patients on board to physicians in a distant port for analysis.[24] Radio broadcasting of brain waves might soon follow suit.

Many radio enthusiasts in the early decades of the twentieth century believed wireless media could transmit psychic data. While the possibility of manipulating the transmission of brain waves generated

The Thought Recorder is an Instrument Recording Thoughts Directly by Electrical Means, On a Moving Paper Tape. Our Illustration Shows What a Future Business Office Will Look Like When the Invention, Which as Yet Only Exists in the Imagination, Has Been Perfected. By Pushing the Button A, the Tape is Started and Stopt Automatically So That Only Thoughts That Are Wanted Are Recorded.

FIGURE 2.2. Hugo Gernsback's "thought recorder" depicted in the most mundane setting imaginable: the office.

fears for privacy and thought control, it also fueled speculative experiments. The British radio pioneer Oliver Lodge, for example, persuaded the British Broadcasting Corporation to conduct a public trial of radio-telepathy in 1927. During the trial, members of the British Society for Psychical Research attempted to transmit a series of images from the BBC studio to radio listeners through radio-telepathy. Though the 25,000 responses from listeners around the world showed almost no correlation with the images in the BBC studio, Lodge and other radio engineers continued to promote the concept that human thought could be transmitted through radio waves.[25] Their spirits were raised again in 1929 when the German psychiatrist Hans Berger published his successful recording of electrical brain waves using a novel technique he called the *Elektrenkephalogramm*, and buoyed still further in 1934 when British electrophysiologist and Nobel Laureate Edgar Douglas Adrian confirmed that this new technique of electroencephalography (EEG) established the physical existence of brain waves as demonstrable physical artifacts of the "electrical brain."[26]

Continuing EEG research in the 1930s and 1940s was supported not by large-scale corporate R&D laboratories but through smaller sites for "craft science" and tinkerers.[27] The Holter lab was one of them. Holter's original radioelectroencephalogram (REEG) consisted of two large boxes strapped to the chest that connected electrodes on the scalp to an amplifier and FM radio antenna. Though bulky, the cases could fit into a suit jacket pocket or be attached to a subject with a combination of straps and tape.[28] At scientific meetings, Holter displayed photographs of these devices in use. Figure 2.3 shows "brain waves being broadcast by a cycling subject" to a nearby cathode-ray tube display. A follow-up version might act as a preventive headband for people with epilepsy "to sound an alarm to the bearer shortly before an epileptic seizure."[29]

As newspapers across the country carried reports of this new machine that could broadcast brain waves, Holter began to receive strange letters from readers who thought he had built a radio thought

FIGURE 2.3. "Brain wave being broadcast by cycling subject," in the laboratory of the Holter Research Foundation. Courtesy of Montana Historical Society.

broadcast/receiver straight out of a Hugo Gernsback magazine. "For approximately 26 years," one woman wrote Holter, "my brain has been electrified by remote control through the use of short wave. Within the last two years I have become so electrified that now I can also hear messages transmitted to me as well as send brain messages."[30] Another young woman wrote on behalf of her brother, who regularly heard voices in his own head when no other people were nearby. Might these voices be another form of radio transmission?[31] As news of his research spread, Holter fielded questions from people experiencing what psychiatrists would call auditory hallucinations and delusions of reference. All of these letter writers had been told that their experience of thought control by radio was a sign of psychosis and therefore evidence of their own lack of reason. News of Holter's new thought-broadcasting machine presented a new opportunity to affirm their own sanity by producing a credible physical explanation for their experiences.

The Cold War context explicit in Holter's work on remote sensors to detect Russian nuclear weapons testing was implicit in these remote sensors to detect brain waves. Americans who were worried about brainwashing and mind control projected several layers of paranoia onto Holter's new device. In 1950, the *Miami Daily News* had carried the first report of "brainwashing" techniques associated with the Communist Party first in China and then in Korea. Could Holter have found the mechanism by which thoughts were implanted into American brains by sinister Communist agents?[32] After reading an article in a Cleveland newspaper, referring to a "Tiny Radio That Puts Brains on Air," one concerned citizen wrote to Holter to ask if the Russians already had developed a similar technology. He claimed to be "a victim of that radio station somewhere on East Coast maybe Russian Consulate in Washington, D.C." What had first appeared to be hallucinations were now revealed to be thought broadcasts from a device like Holter's prototype. "I wish you write to President Truman or Edgar J. Hoover of F.B.I.," he concluded, "and tell them that such a thing is possible as I have written three times now to each of them and they do not ever acknowledge the letter."[33] Holter replied that he needn't worry: if anything of the sort was possible, the Federal Communications Commission would have picked up on it already.[34] But letters kept coming.

Another man wrote Holter from Long Island in 1951, asking if there was any evidence "of such or similar experiment in which the instrument can 'send' to the brain as well as pick up its electrical impulses?"[35] A year later this same man would write Holter from a hospital for the criminally insane in Wisconsin. He had been engaged in legal hearings disputing his diagnosis of psychosis and insisting that the voices he heard in his head were directly transmitted by radio from the central administrative building of the state asylum. When told by the chief psychiatrist that his "accusations were based on a 'physical impossibility,'" the patient countered with a copy of the *Milwaukee Journal* reporting Holter's demonstration of the broadcast of brain waves. His letter to Holter was a direct appeal for expert witness on his own medico-legal case for discharge, providing "any information concerning the 'means' and 'medium' questioned here."[36]

It is no coincidence that the invention of wireless physiological broadcasting during the Cold War was immediately infused with fears of surveillance and manipulation. Holter was sympathetic, but frustrated. "We seem to get quite a correspondence from paranoid types as a result of our work in radioelectroencephalography," he wrote to the director of the Wisconsin mental hospital, adding that he had a personal interest in the topic, as "unfortunately, my first wife has had acute attacks of severe paranoia during which time she sounded almost exactly like the letter written by [your patient]." All the same, he gradually pulled back from the REEG project and from correspondence with those eager to discuss its hidden meanings. "I have not replied to [your patient's] letter," he concluded, "and will not do so unless you indicate that it would be of any value to him."[37]

* * *

Frustrated that his radio-EEG device was of more interest to paranoid schizophrenics than practicing neurologists, Holter turned from minds to hearts. In his earliest work on the role of radio in brain physiology, Holter speculated that electrical rhythms of the heart could be broadcast using similar principles. By September 1949, he described a prototype radioelectrocardiogram (RECG) in the *Rocky Mountain Medical Journal* and the following year presented an improved version to the American Association for the Advancement of Science. By that time other researchers had begun to publish their own investigations on radio-ECGs as well. When a cardioelectrophysiologist based in Athens presented a prototype "Télé-électrocardiographie" device at the World Congress of Cardiology in Paris, Holter worried he might lose recognition for his discovery.

The following summer Paul Dudley White passed through Helena while vacationing in the nearby Yellowstone and Glacier National Parks. He was impressed by the RECG as a potential clinical device, with the Holter Research Foundation as an independent physiological laboratory, and with Holter himself as a fellow cardiologist and independent inventor. White happened to be in the process of finalizing the 1951 edition of his seminal cardiology textbook, *Heart Disease*, and offered to include Holter's device as an important new prospect for the telemetry of the human heart.[38] At White's urging, Holter also applied for federal funding through the National Heart Institute (NHI) to explore practical implications of radioelectrocardiography for the long-term management of chronic heart disease.[39]

Holter shared his vision of broadcasting physiological data with colleagues in a variety of fields, leading others to speculate about new futures for wireless medical devices. Recall Skibbins's correspondence with Holter on this very point. Why not strap a Holter box onto every employee? Why not "Holter Diagnosis Days" and more exclusive "Holter Satchel Clubs" to monitor efficiency and optimize performance?

> DIAGNOSIS DAY AT GENERAL ELECTRIC—this is a day when all key General Electric employees will carry around a Holter satchel and a central radioelectrocardiography recorder will be monitored

> by some of America's most famous medical analysts who within a few days will report on any signs of cracking up under the system.
>
> THE HOLTER SATCHEL CLUB IN NEW YORK—this unique institution is open to membership only to those who have under their control more than 10,000 employees and entitles each member to receive for his $10,000 per annum fee a monthly Holter Satchel Day or perhaps a weekly Holter Satchel Day for the $15,000 rate, and he can carry this life-saving item around with him for subsequent analysis with the comfortable mental frame of mind that his variations from the norm are going to be detected at a time when he can do something about maintaining his efficiency.[40]

Skibbins's joke contained a seed of strategy by way of satire. His chilling vision foreshadowed the power that new wearable technologies had, not only in producing new forms of labor management, but in reproducing clear class distinctions (white-collar versus blue-collar, management versus labor) in the design and use of wearable devices. Other industry leaders laid out more concrete plans for wireless monitoring of their workforces. In the fall of 1958, a professor at the Yale School of Forestry reached out to Holter to explore how his device might open up new horizons in understanding the physiology of labor in the logging industry and arranged field trials in nearby Montana forests to test the use of the radio-ECG in scientific labor management.[41]

By this time, Holter's lab had outgrown its space above the grocery store. With the funds from his NHI grant, the foundation moved its laboratories to the old Great Northern Depot of Helena. Part of the new facility was rented out to local physicians who agreed to test out new prototypes on their patients, with an immediate focus on making the devices smaller.[42] With transistors, the Holter monitor shrank from several pounds to just three ounces. An improved battery promised several days of continuous use, while a portable carrying case allowed recording throughout the day.[43] By June 1961, the

Holter Radioelectrocardiograph could be "carried by a subject free to go where he wished as long as he took the 'ECG brief case' and left it in his general environment."[44]

"How does it feel, Ginger," Holter wrote Gengerelli in June 1961, "to be a father of Radioelectrocardiography, which now after fourteen years is practically sweeping the country?"[45] *Businessweek* and the *Wall Street Journal* projected a growth market for Holter's devices. With plans in place for a new Electrocardiocorder and Electrocardio-caster, Holter now sought to license these technologies, along with the original Radioelectrocardiograph, to a growing medical electronics industry that could help move his device from a local prototype to a nationally and internationally available consumer good.

As radiotelemetry showed promise as a growing field for clinical practice, it developed a potentially lucrative market as well. Visiting New York to attend the new International Conference and Exhibition on Medical Electronics, Holter was encouraged to see that other physiologists and cardiologists were building their own devices for radio-ECGs.[46] Many of these units were still quite heavy: the vacuum-tube equipment developed by one team at the University of Nebraska weighed more than two pounds.[47] But these do-it-yourself (DIY) physician-inventors, and the manufacturers who joined them for the event, were confident that transistorized circuits would continue to yield smaller and smaller devices. As the size of radiotransmitters continued to shrink, with one "already made small enough for a man to swallow," the editors of the *Journal of the American Medical Association (JAMA)* speculated on a medical future in which these wireless devices could be, literally, consumed.[48]

* * *

In the spring of 1961, in the Grand Ballroom of the Waldorf Astoria Hotel in Midtown Manhattan, recently inaugurated President John F. Kennedy approached the great challenge of balancing "the need for far greater public information" with "the need for far greater official secrecy" that characterized the developing Cold War.[49] Only a few weeks later, the same ballroom was full with some 1,600 physicians, engineers, and would-be medical electronics manufacturers discussing the role of secrecy and openness in the flow of medical information. Holter would later claim the International Conference and Exhibition on Medical Electronics—the first event of its kind in the United States—as the origin point of the American medical electronics industry. As he scanned the rooms during panels and presentations, he found that the number of manufacturers' representatives often exceeded the number of scientists. "After my paper," he recalls, "I was swamped by these scouts but won't go into the many headaches involved in trying to select a company."[50] Holter was excited but ambivalent. Here, finally, was the prospect of securing a licensing deal to make his inventions available to more physicians and patients. Yet he remained wary of large, well-funded R&D units like those headed by conference host Vladimir Zworykin, whose work was supported by both the Radio Corporation of America (RCA) and the Rockefeller Institute for Medical Research.

Like Hugo Gernsback, Zworykin immigrated to the United States early in the twentieth century and established a career at the intersection of invention, entrepreneurship, and radio technology.[51] Unlike Gernsback, however, Zworykin climbed the corporate ladder to become vice president for research at RCA, one of the twenty-five highest-grossing corporations in the world when *Fortune* created the first "Fortune 500" list in 1955. At RCA, Zworykin had a well-funded R&D team, along with an institutional disposition to develop products that could not only be built but also marketed and sold. When he retired from RCA a year later to found the Center for Medical Electronics at the Rockefeller Institute, he maintained close ties with RCA's research laboratories and the upper echelons of the corporation.

The first technology developed through Zworykin's new venture was a radio pill (fig. 2.4). As the *Chicago Tribune* reported in 1957, the Center for Medical Electronics had successfully demonstrated "a pill containing a radio station so small it can be swallowed to permit it to broadcast reports on the health of your insides." The gastroenterologist John T. Farrar quipped that it was likely "the first time anyone

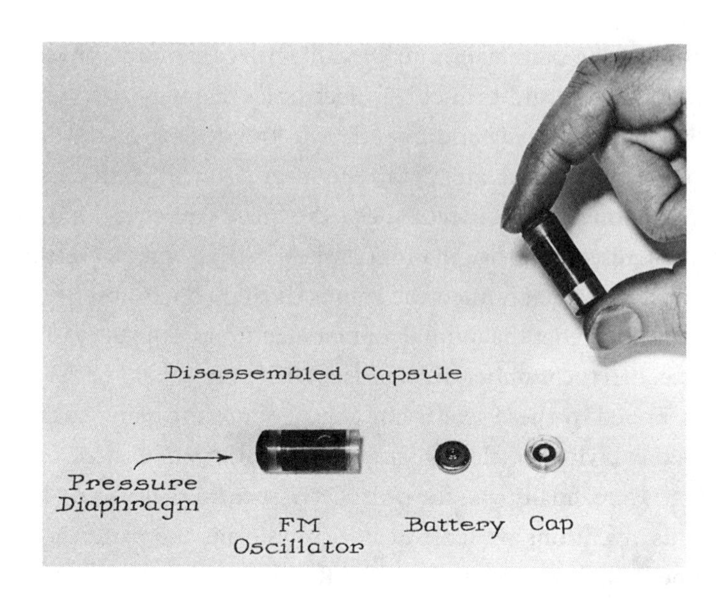

FIGURE 2.4. The RCA radio pill, demonstrated by Dr. John Farrar on the body of Vladimir Zworykin. Courtesy of Hagley Museum and Library.

has passed a radio transmitter down an entire gastrointestinal tract."[52] Similar reports, however, soon came in from Sweden and East Germany.[53] Where Holter had been concerned that he might lose recognition for his discovery of the RECG, Zworykin seemed more concerned that international work on the radio pill was fragmented and not circulating fast enough to make meaningful progress. This, then, was the origin of the International Conference and Exhibition on Medical Electronics that brought Holter and many others to New York.

These researchers had converged on a similar platform from different starting points. R. Stuart Mackay, professor of medical physics and clinical professor in the space sciences laboratory at the University of California, Berkeley, first became interested in the idea of "telemetering from within the body" in the early 1950s. For Mackay, the problem began with the bladder, not the gut. Working with undergraduate volunteers at Berkeley, he succeeded in transmitting real-time physiological data on pressure within the bladder from a small radiotransmitter inserted via the urethra. While visiting at the Karolinska Institute in Sweden, Mackay collaborated with Bertil Jacobson to produce what they believed to be the first version of an intrahuman radiotransmitter, which they called by the clunkier term "endoradiosonde."[54] By 1960, the Swedish group had adapted their capsule to produce a visual tracing of its passage through the abdomen. Another team in Japan, led by Akihito Uchiyama, adapted a glass electrode to transmit blood pH via radiotelemetry.[55]

As research on radio pills accelerated, Zworykin hoped that academic teams in different countries would collaborate to make the transmitters smaller and more efficient, and to ease the transfer of technology to medical industry. "What does miniaturization mean for medicine?" Zworykin asked a group of radio pill researchers gathered in 1960. "It means, first of all, that we can expect the introduction into the various organs of electronic probes so small that they will not interfere substantially with the functioning of the organ in question, and yet can transmit significant information regarding its operation."[56] Tiny radio devices could broadcast details of human physiology without interfering with its workings, thereby resolving the technical limitations that had vexed so many medical researchers for so long. By 1961 a team in Heidelberg had produced a minuscule capsule that could broadcast information about temperature, chlorine ions (to study gastric acid production), and pH, and could be fitted with strings and magnets so that it could hover in place to measure physiological function of a single organ over time. Meanwhile the Rockefeller group cut the size of their pill by 30 percent by doing away with batteries altogether.[57]

Shrinking the radio pill expanded both the possible uses and the possible markets for this new device.[58] RCA did not patent or market its first radio pill. Instead it displayed the device much as automobile manufacturers display concept cars: not as products to be bought and sold but as platforms to encourage future speculation. Zworykin's radio pill succeeded as a gambit to build collaborations in the emerging medical electronics sector and to attract long-term partnerships with pharmaceutical and medical device firms inspired by the recent success of the cardiac pacemaker.[59] Shortly before the New York conference, Zworykin met with leaders in medical electronics at RCA and executives from the pharmaceutical giant Smith, Kline & French (now GlaxoSmithKline) to discuss this terrain. Smith, Kline & French was one of the top five pharmaceutical firms in sales and net profits, and one of the top two in terms of marketing activity and sales force, but "as far as prestige (both among physicians and the pharmaceutical industry) and research activities," Zworykin considered "this company [to be] by far the most outstanding."[60] RCA executives had heard that SK&F was looking to diversify its holdings in the medical device industry as well, aiming to develop new markets in medical electronics.[61]

The radio pill was not the end but the means. This "simple building block," Zworykin hoped, would serve as a platform for the proliferation of medical telemetry devices in general medicine.[62] When he shared the stage at the Waldorf Astoria with Holter, Zworykin also presented a prototype for a "Miniaturized Hospital Telemetering System." Noting that the recording of vital signs like temperature, pulse, and respiration was often left to chance in hospitals, he proposed a "simple system of telemetry which can be conveniently and economically used by all patients."[63] Pulse, temperature, and respiration could be broadcast by FM radio technology from bedside to the nursing station to maintain constant surveillance of the patient's physiological processes at a distance.[64] Just as Holter's presentation of his radio-ECG found eager industrial partners at the 1961 New York conference, Zworykin's prototype hospital telemetry unit attracted attention from several pharmaceutical and device firms.

Smith, Kline & French soon asked RCA to build a version of this device that would be called a "diagnostic chair": a tool to simultaneously record and transmit blood pressure, ECG, and respiration data from the patient "without causing alarm as the present technics do."[65] The two firms and the Rockefeller Institute speculated on general radiotelemetry devices for clinical and consumer purposes, radio-transmitted blood pressure, radio-transmitted ECG, and other miniature devices designed to burrow into the inside of the opaque body and render its inner functioning transparent. They were not alone. Like the representatives of several electronics manufacturers that buttonholed Holter after his talk, other pharmaceutical and medical device firms were also interested in translating wireless medical devices from the desks of independent inventors into patent-protected and trademarked consumer goods. Radiotelemetry had become a speculative field for large corporations as well. But this didn't rule out the ability of individual radio enthusiasts to continue to imagine their own DIY versions of wireless medicine.

* * *

Holter and Zworykin's wireless prototypes were also taken up outside of the medical electronics industry, especially by those seeking to build a role for biomedical telemetry in public health. What could be learned by broadcasting the functioning of not one but thousands of hearts at a time? Here the vision of wireless medicine was conceived not in terms of growth markets for individual patient use but as a tool for screening whole populations at a time. Dr. Cesar Caceres, the chief of the new Instrumentation Field Station of the Division of Chronic Diseases of the United States Public Health Service, became a leading figure in this effort.

Caceres first moved to Washington, DC, when his father was appointed the Honduran diplomat to the United States; he studied medicine at Georgetown and trained in both cardiology and medical electronics at George Washington University before entering the US Public Health Service.[66] His interests in medical electronics and chronic disease led him to the 1961 conference on medical electronics in New York, where he met Zworykin, whom he considered a giant in the field. Later in life he recalled meeting Zworykin at that conference as a seminal moment for his own career in biomedical telemetry and medical informatics.

But where Zworykin emphasized the use of these devices by individual doctors and patients, Caceres saw a broader role for radio technology in preventive medicine. A technologist and tinkerer himself, Caceres was convinced that physicians with some electronics know-how should help build their own new electronic communications devices that could serve public health. There was emerging evidence from the field of cardiology suggesting that telemetry could do just that. In 1962 the *New England Journal of Medicine* described results from more than 1,000 patients' experiences with radiotelemetry.[67] A 1964 review in *JAMA* established the importance of this technology in cardiac intensive care units. As these units became part of the infrastructure of American hospital care, the continuous remote surveillance of the ailing heart played a key role in reducing the lethality of heart attacks and enabling newer and more effective forms of cardiac surgery.[68]

The term "telemetry" had long referred to the tracking of inanimate objects, especially the missiles, rockets, satellites, and other projectiles that were the early basis for the Cold War sciences of cybernetics and systems analysis.[69] But as the Space Race heated up, competing Soviet and American efforts involved the telemetry of living beings as well. In a celebrated flight in 1957, the canine cosmonaut Laika was connected to radiotelemetry leads that conveyed electrocardiogram, temperature, respiration, carotid artery movements, and other life functions.[70] Likewise radiotelemetry of animal and human vital functions became crucial to NASA's early Mercury program. "Hidden under the silver space suit of Commander Alan B. Shepard, Jr.," Rita Chow reported in the *American Journal of Nursing* shortly after the Mercury 1 mission, "were telemetering devices, such as electrocardiograph leads adhering to his chest, a rectal thermometer securely in place, and a respirometer hovering over his mouth." As above, so below. Chow told her audience of nurses, "Just as telemetering equipment recorded the astronaut's vital signs, so we are seeing multiple patient monitoring systems recording physiological data in some hospitals today and in many tomorrow."[71]

In theory, radiotelemetry devices could be built by anyone who understood the basics of electronics. Even as Holter began to trademark and license his own devices, radiotelemetry attracted a cohort of amateur medical technologists. As Capt. Frederick W. Fascenelli described in an article titled "Electrocardiography by Do-It-Yourself Radiotelemetry," DIY approaches to radio-ECG could make the vital technology more affordable and more accessible. For radiotelemetry to have its maximal impact, he argued, hospitals should not depend on purchasing expensive devices from medical electronics firms like SK&F or RCA, but instead enable their own, in-house technicians to find affordable means to produce their own devices.

While commercially available radiotelemetry systems cost $500 to $1,500 per channel, the tools to build a single-channel radioelectrocardiography system could be found at the corner RadioShack for less than $15. Fascenelli distributed blueprints for "a simple one-channel electrocardiogram radiotelemetry system that can be built by any interested person in four hours."[72] This included a circuit diagram, breadboard, and a parts list with prices included. Twenty cents of the project budget represented two silver dimes that could serve as electrode leads. Readers wrote to the *Journal* to suggest modifications and improvements in the circuit diagrams. It became clear Fascenelli was speaking not just to researchers, but to a broader community of do-it-yourself medical technologists, not unlike Gernsback's early radio enthusiasts.[73]

Yes, medical electronics represented a growing space for corporate speculation, but the field was still open to anyone with an idea and a soldering iron. Why pay the markup to a for-profit medical device manufacturer when you could build the tool you needed yourself? This open-source model of DIY radiotelemetry spoke directly to Caceres's hopes to scale up the use of this technology. Widely available wireless medical devices could spawn an entirely new field of public health research into the epidemiology of "silent diseases." The Holter Radioelectrocardiograph didn't need to be a brand-name product; it could be a generic instrument.[74] Likewise, as the RCA Radio Pill became a generic endoradiosonde or 'sonde, it turned into a universal machine for translating physiology into information. By 1965, newer radiotransmitters with integrated circuits could serve as miniature Geiger counters or Doppler flow detectors, or they could detect bleeding through a chemical reaction with hemoglobin. Any aspect of bodily function might be made transducible and transmissible for broader study. "The number of variables that might profitably be telemetered by these methods," Stuart Mackay concluded, "is limited only by the imagination."[75]

Biomedical telemetry reimagined physiology as a science of pure information and recast preventive medicine as a field of big data. With wireless devices, physiological information could be transduced into electronic form inside the body, transmitted across distances ranging from a few feet to hundreds of miles, picked up by a distant receiver, and processed and stored as meaningful information. In the hands of physiologists like Mackay at Berkeley, guided by the cybernetic theory of Norbert Weiner and the information theory of Claude Shannon, the radio pill became an infinitely modifiable platform to convert more and more of the functioning of the human body into analyzable data.[76] In the hands of public health officials like Caceres, biomedical telemetry promised to transform the study of preventive medicine into a field of transparent and continuous surveillance—

especially for the larger time frames and data sets necessary to study the epidemiology of chronic conditions like heart disease.

The expansion of personal health data in public health research also created new problems. As an employee of the US Public Health Service, Caceres favored the adoption of telemetry technologies as part of an expanding surveillance state. Yet he agreed with Gerald Skibbins that there was risk as well as potential in the widespread collection of physiological data. Caceres urged caution. Where would all this data be stored? How would it be protected? What analytic power could make sense of the reams and reams of tracings these new devices collected? As electronic transmissions of physiological processes became transmuted into biomedical data, these challenges of storage and analysis offered a glimpse of the challenges to come. A new crisis of data deluge was on the horizon.

* * *

The sheer scale of data that telemetry systems produced soon became onerous. As Holter sought to license and market new radio technologies for collecting bodily data, he realized that these devices also brought new problems related to "the very voluminous data acquired." Each Holter monitor could store twenty-four hours' worth of ECGs. At an average heart rate of 80 beats per minute, a single device being worn by a single patient for a single day would generate more than 100,000 ECGs to be analyzed. The regular use of just one Holter monitor in any given month could quickly exceed the analytic capacity of an entire small cardiology practice.[77]

New magnetic storage and digital computing techniques, available by 1960, offered a solution. The first data analysis prototype produced by the Holter Research Foundation, the Audio-Visual Superimposed ECG Presentation (AVSEP), enabled twenty-four hours of continuous ECG recording to be evaluated in less than twenty minutes. Basically, this technique involved playing back the recording at twenty times normal speed, as a visual image on an oscilloscope screen and as an acoustic pattern on loudspeaker that sounded like a background static "growl." As the cardiologist scanned the screen and listened to the underlying rhythm, any signal of deviation, either seen as a "spike" or heard as a "blip," could be revisited more closely with the playback speed slowed down for closer inspection. Accused of adding a depth and volume of data to electrocardiography that cardiologists were simply not yet ready for, Holter countered that "it is in the nature of science sometimes to find oneself working on the roof before the second floor is finished, and this cannot be helped."[78]

The automation of data monitoring helped make this information legible to nonphysician users as well. As Caceres observed, while medical monitoring systems were originally developed by teams of physicians and engineers, in the future their "use will be primarily in the hands of nurses and nonelectrical technicians" who needed simpler interfaces, alerts, and alarms to know when the telemetry of bodily signals indicated danger or required an immediate response. As remote telemetry transformed vital information into electronic data, it also allowed this information to be displayed in real time across visual displays in nurses' stations equipped with alarms that would beep whenever a parameter deviated.[79]

The automation of telemetry technology changed *who* had access to this vital medical information. For generations the ECG was a cryptic tracing legible to physicians only after years of training. Now, automated alarm systems attached to telemetry monitors allowed more nurses, aides, and technicians to perceive these hidden signs of disease and act on them. Caceres developed other diagnostic algorithms to automatically detect arrhythmias and signs of heart strain. These devices performed a double move, making the inner functions of the patient's body more transparently visible to physicians, while also making the art of diagnosis more available to nonphysicians.

While Caceres admitted that use of radiotelemetry chiefly took place in intensive care units, he was hopeful that monitors might soon be built into the monitoring of every patient in the hospital.[80] Eventually, every hospital bed could become a relay station in a field of continuous surveillance. "If one thinks of each patient or bed as a radio station and each monitoring apparatus as a receiver," he concluded, "one may observe physiological function by tuning from one 'station' to another without any interference."[81] This vision of continuous surveillance extended from intensive care unit to the hospital floor, from hospital floor to outpatient clinic, and from outpatient clinic to consumer technologies in everyday homes.

Storage was another problem. Each device added to these surveillance systems contributed another pile of data to be analyzed. As the scale of data collection grew, magnetic storage devices took on increasing importance in the expansion of biomedical telemetry. By 1957, it was possible to record twenty-four hours of radio-ECG on a reel just 5 inches in diameter.[82] In 1958, Holter trademarked a new device that could dispense with radio entirely and record findings directly onto an even smaller, integrated tape drive built into the electrode unit.[83] In this model, which he called the Electrocardiocorder, magnetic tape replaced radio. "If one thinks it through," Holter wrote Gengerelli, a few years later, "it appears that the Electrocardiocorder should be the really routine instrument of the future in the whole field of continuous electrocardiographic monitoring together with rapid data presentations."[84] Why bother transmitting and receiving data through radio, Holter asked, if the wearable device could simply store it in magnetic form to be analyzed later?

The Holter monitor still in use today is a recording device, not a broadcasting device.[85] Zworykin, too, became more interested in storage than the transduction or transmission of physiological data. By 1960 he had dropped his radio pill projects for medical computing and electronic records systems. When, in 1975, *JAMA* highlighted the work of Holter and Caceres in a brief review of telemetry as a crucial technology for reversing the rising epidemic of coronary heart disease, the role of radio was omitted entirely.[86] Paradoxically, exactly at the moment when telemetry became commonplace in the hospital and recording devices like the Holter monitor became staples in outpatient clinical workups, the wireless imagination went fallow. As telemetry systems became part of the infrastructure of the hospital, they no longer animated the ambitions of corporate researchers like Zworykin or independent investigators like Holter or public health researchers like Caceres, as each continued to seek new means to transform medical information into analyzable data. In turn, as the patient became more mobile through smaller and smaller iterations of the Holter monitor, their data were made available and accessible to more parties than previously anticipated.

* * *

In the early decades of the Cold War, wireless devices occupied a key role in the technological imagination of American medicine. Radiotelemetry could broadcast the hidden processes taking place inside the human body for all to see. Yet if these radio devices were new, the dream of telemetering signals from inside the body related to something much, much older. "The sickly sweet odor of diabetes, the moist, slight cough of tuberculosis, the faint pulse of shock, all belong to telemetry," cardiologist Robert P. Grant argued in the foreword to Caceres's 1965 textbook *Biomedical Telemetry*, "in each instance an arc between signal source and receiver is made."[87]

Clinical medicine had been concerned for centuries with separating signals from noise. Whether scanning, smelling, tasting, auscultating, or percussing, physicians had long used their own senses as telemetering devices to detect subtle signals of disease. Several tools designed to transcend the limits of those senses—the stethoscope, the thermometer, and the electrocardiogram—also functioned as telemetering devices. All represented ways to amplify faint signals from inside the body and make them legible to the outside world. What *was* new, however, in the Cold War origins of biomedical telemetry, was a particular speculative fiction of transparency. Now information could be silently transmitted across the borders of the human body, enabling continuous, remote surveillance of processes within. While many of these fears were projected onto broader fears of the surveillance state—both Soviet and American—in hindsight it appears that many of the more insidious uses of wireless surveillance have taken root in the private workplace. Evidence of this, today, is as close as your nearest Amazon warehouse.[88]

Some sixty years after Zworykin's initial demonstrations, the radio pill has once again become a platform for expanding healthcare through Wi-Fi–enabled smart technologies. In November 2017, the FDA approved the first "digital pill"—a capsule that, once consumed, tracks the consumption of other medications. Subsequent cloud-enabled pills were being developed for specific cancer chemotherapies, each embedded in a logic of precision medicine, each supported by speculative capital in the hopes of receiving a federally sanctioned monopoly over a new medical market.[89] Yet at the same time, an open-access movement in smart medical technologies has in recent years used Wi-Fi technologies to produce new DIY medical devices. This is most visible in the community of amateur technologists and persons with Type I diabetes who have used Wi-Fi connections to produce their own "electronic pancreas" by hacking the wireless connections on their continuous glucose monitors to communicate with the wireless connections on their portable insulin pumps. These new modes of wireless medicine today recapitulate earlier conflicts over the role of corporate R&D versus amateur research in the field of medical electronics, visible to Holter and Zworykin more than half a century ago. Then, as now, wireless medicine was simultaneously a site for the speculative interests of an emerging medical electronics industry and for the speculative imagination of amateur medical technologists.

While it might be tempting to read a figure like Zworykin as a visionary ahead of his time, it is far more instructive to read him, as well as Gernsback, Holter, Caceres, and others mentioned in this chapter, as figures very much of their time. They built pieces of their own present into devices that continue to carry their names today. The infrastructure of electronic patient monitoring and surveillance that took shape in the Cold War moment, and expanded to the monitors, alarms, and alerts common to every hospital floor today, still bears the marks of that earlier era's concerns with transmission and reception, signal and noise, surveillance and secrecy. Likewise the marketing exclusivities and trade secrets established in the medical device industry in the postwar decades continue to prioritize innovation over access in the field of medical electronics today.

These problems have not been solved by the passage of time or by the emergence of newer technologies. Increases in data storage capacity and microprocessor speed have made it easier to transmit, store, and analyze physiological information, but they have also compounded problems of a growing data deluge.[90] While new modes of data encryption initially appeared to offer some privacy protections for users of wearable medical devices, recent years have seen a host of concerns about the hacking of wearable medical devices, from pacemakers to insulin pumps.[91] The fears that first coalesced around Holter's brain wave–broadcasting device are reactivated with each new demonstration of the vulnerability of wireless medical devices to new forms of medical hacking. The wireless body is a source of limitless potential interventions to improve individual and public health, but it is also a space of surveillance and control—and loss of control.[92]

It is easy to forget the role that earlier radio technologies have played in building the structures of physiological surveillance we now inhabit. To the extent that the use of a device like the Holter monitor has become commonplace, it disappears among other medical processes that have become routine. Radio was crucial to introducing telemetry into medical practice. Once telemetry devices became everyday tools, however, they ceased to be recognized as radio technologies. While the dream of the wirelessly monitored body continued to animate many technological projects in medicine in the decades that followed, it increasingly shared this space with another medical use of radio, one that made the physician knowable instead of the patient.

3
THE ELECTRONIC LEASH

Dr. Eliot Archer woke up, reached for his watch and pager, and set out for his office in downtown San Jose, California. It was a typical day for this general practitioner and surgeon: clinic hours in the morning, house calls in the afternoon, and then hospital rounds on his sicker patients before returning home. Like most American physicians in 1966, Archer ran an independent private practice, on call twenty-four hours a day for the needs of his patient population. Unlike most American physicians, however, Dr. Archer was a fictional creation of the Motorola Corporation.

Archer's "typical day" was orchestrated by the communications staff of the *Motorola Newsgram* corporate magazine to launch the Motorola Pageboy he kept under his pillow—and to highlight the centrality of radiocommunications in professional life (fig. 3.1). From the personal pager to the radiophone in his car to the radio-dispatched ambulance and the nurse call system at each patient's bedside, Archer's day was made routine by radio—a technology so useful that its very presence had become almost invisible in the everyday life of the physician. "Our fictitious Dr. Archer," the magazine admitted, "thought his day was pretty dull."[1] Dr. Archer's ignorance of the enormity of his dependence on radio communications is itself a sign of the success and importance of this wireless medium in modern medicine. The story behind his pager is also the story of how a profession that

A Day in the Life of Dr. Archer

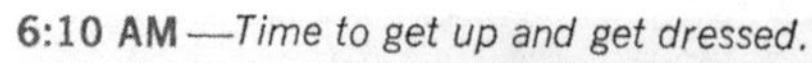

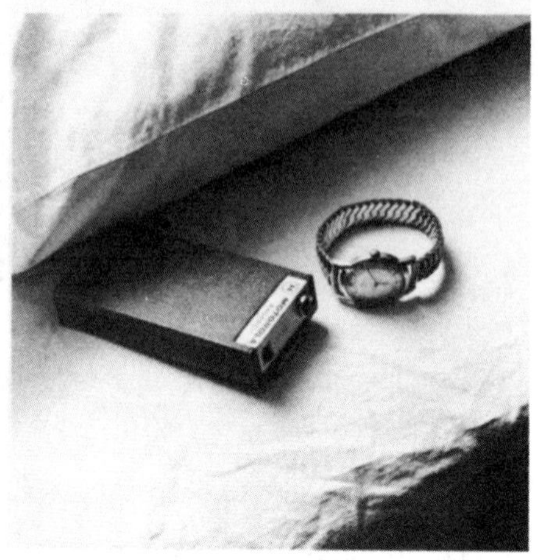

FIGURE 3.1. "A Day in the Life of Dr. Archer." *Motorola Newsgram* 23, no. 3 (1966): 10–15, at 15. Motorola, Inc. Legacy Archives Collection. Motorola Solutions Archives, Chicago. Reproduced with permission.

prized the autonomy of the individual practitioner came to tether itself to an invisible electronic grid.

Archer may have been a fictional physician, but the role that radio played in the transformation of medical life in the real city of San Jose could be well documented. Both major San Jose hospitals had installed Motorola radio nurse call systems. The hospital featured in the "Dr. Archer" photoshoots had installed a radio-based Motorola Total Hospital Communications system linking Motorola televisions in patient rooms to Motorola electronic nursing stations and an electronic physician register that used Motorola pagers to track which physicians were in the hospital at any given time. In this tech-forward city nestled in the corridor soon to become known as Silicon Valley, Motorola radio ambulance communications had already "helped reduce what formerly were emergencies to routine problems." Inside and outside the hospital, these paging systems transformed the terror of emergency into the cool, coordinated efficiency of urgent response. "Though he has never fully realized it," the *Motorola News-*

gram concluded, Dr. Archer's "safety, his comfort, and his well-being are dependent to a large extent on radio communications."[2]

The physician's pager represents one of the most successful schemes so far to adopt a new telecommunications technology in the daily practice of medicine. If the white coat and stethoscope became key symbols of American medicine in the nineteenth century, the pager became newly iconic for the practicing physician of the late twentieth century. As an everyday technology, however, its importance has been generally overlooked.[3] Part business history (as seen in the Motorola communications discussed above), part telecommunications history (all cities are now at least as "smart" as Dr. Archer's San Jose), the history of the physician's pager is also a history of professional transformation and the mediation of medical work. The pager promised free range to private practitioners like Dr. Archer, who could conduct their business from their automobiles or from the golf course. But the pager also promised new tools of surveillance—like the Motorola Radio-Register—through which medical institutions could coordinate and control the labor of healthcare workers as well.

Although some uses of AM (amplitude modulation) radio to contact medical personnel can be found in the early twentieth century, the pager was largely a byproduct of the development of more powerful FM (frequency modulation) radio in the 1950s.[4] As this new technology showed promise in locating specific individuals and patching them into call networks, pagers first took on significance as extensions of telephone-answering services, then as tools of rationalizing care in emergency services through larger institutions (police, fire, rescue, medical emergencies) and then later for personalized delivery of services in homes (locksmiths, drug dealers, sex workers, and physicians in private practice). Two-way radio devices were initially sold to physicians in the 1950s and 1960s as a means to enjoy greater liberty and autonomy while "on call." Yet by the 1970s, physicians routinely complained that their pager—also known as the "beeper" or the more expletive "bleeper"—had become a yoke by which hospitals bound them to expectations of constant service.[5]

Pagers did not create paging.[6] The cultural authority of the physician has for some time derived from being an expert to turn to in time of crisis. Physicians had been paged for centuries before they carried the electronic devices we know as pagers. As Mercutio lies bleeding by Tybalt's blade in *Romeo and Juliet*, he cries out to a stunned Romeo: "Where is my page?—go, villain, fetch a surgeon."[7] Just as the term "computer" formerly referred to a person and not to a machine, the "page" here was a person, a servant who could be sent to summon medical care when urgently needed.[8] By the turn of the twentieth century, the page had become an action, not a person. In New York's Mount Sinai Hospital, attending physicians were paged by a steam whistle to summon them for emergent needs. When this was criticized as a nuisance, one house officer recalled, it was replaced with "an enormous gong, the number of strokes of which indicated the arriving M.D."[9] As gongs were replaced by electronic public address (PA) systems, individual physicians could be paged by microphones and loudspeakers, prerecorded tones, flashing lights, and dedicated Morse code signals.[10] By 1929, one of the largest paging systems in the country was installed in the hospital of the University of Pennsylvania, in which light boards with more than a hundred numbers were visible from every hallway in the twelve-story building.[11]

These early electronic paging systems were a means for hospitals to rationalize the labor of a physician workforce whose members prized their own autonomy. Yet a 1948 survey of hospital administrators warned that fewer than one out of every three American hospitals had been able to install an electronic paging system—and fewer still had figured out how to use one effectively. Most hospitals simply used the telephone operator to seek out the physician in question using a "hunt and call" method. Physicians were good at evading these calls. Flashing lights and numbers could easily be ignored, and, as one administrator grumbled, "half of the doctors don't know what their number is in the first place."[12] A few hospital administrators were optimistic that portable radio devices might be devised to make each physician available at all times. "When that system goes into operation," they concluded, "it is plain, all the objections to existing paging systems will vanish and doctors will sign in as eagerly as a youngster with a new electric train."[13]

Doctors signed in, eagerly. These new devices, with names like Pageboy and Bellboy, were marketed to them as electronic servants that would enhance the physician's ability to enjoy a life of leisure outside of the hospital and clinic. In practice, however, doctors soon began to complain that the pager had made them into the servants, instead.

* * *

Radio entered medical practice as an extension of the telephone. By the mid-twentieth century, widespread adoption of telephones in professional life had created new expectations for practicing physicians to be accessible to patients. As the increasing use of telephones to summon "on call" physicians led to a sense of twenty-four-hour availability, the telephone helped give rise to a new industry of physicians' answering services that would take messages and track down the physicians so that they could receive them. As we saw with the career of A. J. De Long in chapter 1, the limitations of a communications technology could also become the basis for a thriving business.

Sherman Amsden founded the New York Doctors' Telephone Service after a frustrating night in 1922 in which he tried without success to reach his own physician by telephone during a family emergency. Amsden's firm, soon renamed Telanserphone, logged 20,000 calls within the first year of operations. Telanserphone quickly expanded to include plumbers, electricians, and other urgent service industries in New York City.[14] By the early 1940s, Amsden could boast more than fifty operators throughout the city, thousands of clients, and millions of calls logged.[15] But the physician in motion remained an elusive quarry. "Too many calls were piling up on absentee doctors," one local paper reported. "A means must be devised to reach them hurriedly, instead of waiting for them to call 'Telanserphone' at their leisure to check on the calls their office had received."[16] What could a patient or worried family member do when the matter to be discussed was too urgent to wait for the doctor's timely return to home or office?

Amsden imagined a market for wireless devices that might further extend the availability of the on-call physician. His radio-paging service, Aircall, supplied each physician with a two-tube radio receiver, weighing just 6 ounces, tuned to a radio station transmitting over a twenty-five-mile radius from the top of the Pierre Hotel at the base of Central Park. Each doctor was also given a three-digit code that corresponded to a length of 16 mm tape with a voice recording stored in Telanserphone's headquarters next to a loop of chain that wound along a sound head of the sort used by movie projectors. Any time that Telanserphone received a call for a particular physician, the operator would take a message and hang that physician's "code stick" on the loop, which would repeatedly broadcast their specific code until the physician called in.[17]

Advertisements for Aircall emphasized the leisure time users could enjoy because of the Aircall connection: time spent at weddings, baseball games, or on the beach. *Popular Science* featured a story of a "well known New York doctor" at a recent wedding, who "unobtrusively removed from his pocket a small plastic box, held it up to his ear for a moment, then got up, tiptoed out of the church and rushed to a hospital in time to deliver the baby of a patient."[18] Aircall sold selective radio-paging devices to physicians and other professionals as a personal extension to their telephone-answering services. They broadcast individual signals in an encrypted form over a discrete urban space, allowing users to remain accessible to their patients while at their personal leisure outside of the hospital, home, or clinic. It was an added service that individual physicians purchased in order to enjoy new liberties. Although the Aircall service was used by nonmedical professionals as well, physicians were central figures in popular press coverage of the new technology.[19]

Aircall's closest competitor, Royalcall, pursued a different marketing strategy for its prototype physician paging systems. Instead of appealing to individual private practitioners, Royalcall targeted its radio-pagers toward hospital administrators seeking greater control over their physician workforce. Royalcall was named after the electronics equipment manufacturer Harry Royal, but the heart of the venture was a partnership between hospital acoustics consultant Charles E. Neergaard and citizens band (CB) radio innovator Al Gross. Neergaard traced the origins of his career as a sound engineer to a negative medical experience he had as a young adult. Hospitalized for two months in early 1939 due to an illness with a slow recovery, he found himself increasingly distracted by the constant pages for physicians on the hospital loudspeakers, and vowed, as he put it, "to banish all audible paging systems."[20]

Gross was a serial inventor and radio enthusiast. He built his first shortwave rig from scratch in 1934 at the age of seventeen and by 1939 had designed his own handheld, battery-operated transceiver.[21] During World War II, he developed aircraft-to-ground radio systems for relaying messages to and from Allied spies via ground-to-aircraft radio communication, and in 1945 created a peacetime market for this technology as citizens band (CB) radio.[22] Much like his contemporary, Norman "Jeff" Holter, Gross was better at imagining new radio technologies than he was at marketing them or developing their consumer potential. By the time CB radio found extensive use in the fields of trucking, shipping, emergency rescue, fire, ambulance, and police work, Gross had already moved on to other endeavors. He sold the Citizens' Radio Corporation in 1949, and was deep in a new radio venture, Royalcall.

Royalcall used a specialized CB radio at the hospital switchboard that could send encrypted signals that would only buzz the desired receiver, allowing more than 800 different users to be reached privately through a single system.[23] Neergaard and Gross emphasized its silence and selectivity: the buzz of the Royalcall receiver was brief but memorable. "Like a shrill mosquito," Neergaard claimed, "it cannot be ignored; but unlike the mosquito, it is very short in its audible range."[24] The physician wearing a Royalcall could not miss the signal, but others nearby would not be able to hear it.

In 1952, Royalcall conducted a successful test of the system at the Long Island Jewish Hospital, using thirty receivers and a 50-watt central transmitter. The firm was invited to demonstrate its prototype system at the Pentagon before "a large group of officials representing the Army, Navy, Air Force, V.A., Public Health Service and State Department."[25] Gross and Neergaard also approached a series of hospital directors to gauge the potential market, culminating with an interactive demonstration booth at the 54th meetings of the American Hospital Association (AHA). But the AHA meetings were a bust, and subsequent pitches for the paging system failed to produce any buzz among hospital administrators.[26]

Royalcall did arouse the interest of at least one potential buyer. As the firm folded, the Motorola Corporation swiftly moved to purchase the rights to the device.[27] Where Aircall and Royalcall probed the established limits of radio-paging outside and inside the hospital in the 1940s and 1950s, Motorola played a dominant, even definitional role in the widespread adoption of the personal pager by American physicians in the 1960s and 1970s. This company, which in 1955 introduced the first consumer product to be branded as a "Pager," still controlled an 80 percent market share of the $2 billion pager market at the turn of the twenty-first century.[28] Where Aircall started by marketing its services to independent practitioners, and where Royalcall failed at marketing its services to hospital administrators, Motorola would succeed by combining the two markets.

* * *

The brand name "Motorola" was the result of a company redefining its identity around the concept of wireless mobility. When the Galvin Manufacturing Corporation branched out from battery manufacturing to introduce the first mass-produced car radio in 1930 (a device named the Motor-ola), the company renamed itself around its flagship product. A decade later, Motorola engineer Donald H. Mitchell proposed a handheld, two-way radio for military use, which he called the Handie-Talkie. Between 1942 and 1945, Motorola supplied 387,958 Handie-Talkies to the US Armed Forces.[29] After the war, the Handie-Talkie became a central product in Motorola's lineup.

Motorola was quick to promote new uses for mobile two-way radio systems in civic life, especially for emergency services (fire, police, park services, and so forth) whose "battles" against natural disasters could be analogized to the war effort, and whose valiant operators required the ability to communicate and coordinate while remaining agile and unencumbered. By 1947, highway police in thirty-four states carried Motorola two-way radio units for emergency communications, along with scores of fire departments, electrical utilities, cab companies, trucking and shipping companies—and rural physicians.[30] Over the next five years, domestic Handie-Talkie sales increased fourfold: nearly every issue of the *Motorola Newsgram* carried accounts of people whose lives were saved by Motorola mobile communications.

The in-house magazine featured many stories of medical emergencies that featured two-way radio as an agent of public health and public good.[31] One North Carolina physician, who installed Motorola two-way radios in his Jeep and his clinic buildings in 1949, found it led to "one of the greatest gains during the years of his practice."[32] A colleague in South Carolina recalled that, prior to the advent of two-way radio, he would spend the bulk of his day in transit, "out of touch completely" with "no way of knowing how his other patients were faring." After he installed a Motorola two-way radio system in his car in 1954, "no matter where he goes on a call, he is in constant touch with his home or office and is in a position to answer emergency calls immediately." His son, also a physician, credited the two-way radio with saving the life of a local boy wounded in a hunting accident, who would have perished without the Motorola link.[33]

The 1955 photo-essay detailing the life of Dr. John Otten of Chicago's Presbyterian Hospital, however, stands out. This was the first issue of the *Motorola Newsgram* to feature the new device Motorola called the Radio Pocket Pager, especially geared toward the on-call physician (fig. 3.2).[34] Smaller, sleeker, with transistors instead of vacuum tubes, it was the first product to be branded as a "Pager."[35] While the original Handie-Talkie was the size of a carton of cigarettes and weighed 5 pounds, Dr. Otten's new Radio Pocket Pager was the size

FIGURE 3.2. Dr. John Otten of Chicago's busy Presbyterian Hospital uses a Motorola Handie-Talkie radio-pager while relaxing in a locker room after performing an operation. "Transistorized Handie-Talkie Radio Pocket Pager: New Transistorized Wireless Paging System Provides Truly Private Plant-Wide Calling," *Motorola Newsgram* (November–December 1955): 28–29. Motorola, Inc. Legacy Archives Collection. Motorola Solutions Archives, Chicago. Reproduced with permission.

of a "king-size package of cigarettes" and weighed only 10 ounces. Unlike AM radio receivers, its FM receiver was immune to interference from X-ray machines and other hospital equipment. The central page operator at the hospital switchboard could buzz the desired doctor using a silent and selective paging mechanism similar to the Royalcall. The discreet pager would "make greater privacy possible, preserve silence during sleeping hours, and conceal emergency situations which might be upsetting to patients," and would also relieve "paging insensitivity" among physicians who had learned to disregard their name being called on the loudspeakers.[36]

The Radio Pocket Pager attracted immediate publicity. "These two-way radios will keep doctors and nurses in constant touch with their patients," a 1956 Universal newsreel announced over footage depicting doctors and nurses in an urban hospital. "Each staff member has a wavelength assigned to him which is signaled by a beep on the receiver, called by an operator at the central switchboard . . . paging Dr. X!" Before-and-after footage showed doctors, patients, and nurses whose schedules were liberated by the new technology and the added mobility it provided—including an odd scene in which a new mother was pleased to hear the "distant cry of her brand-new baby" through radio link from the hospital nursery to her hospital room.[37]

When Boston's Beth Israel Hospital adopted the Radio Pocket Pager, the *Boston Globe* quipped that "young Dr. Kildare is no longer called; he's beeped on a pocket radio."[38] New York's Mount Sinai Hospital also issued Radio Pocket Pagers to 175 house officers and nurses. Hospital operators had complained that the growing campus, which now stretched over three Manhattan city blocks, was too big for PA systems to effectively locate the "'house physician on the run' who hasn't the time to keep one eye on the call board, who prefers instead to keep both eyes on the patient."[39] Thanks to the new Radio Pocket Pager, one hospital administrator gushed, "the whole transaction takes only seconds, and those little transistor radio receivers give us instantaneous 100 percent coverage over 1,077,000 square feet and more of building area."[40]

These systems could only broadcast inside spaces in which every floor of the building had been individually wrapped in metal coils to locally transmit the signals. Beth Israel alone required a half-mile of aerial wire, and coverage could still be patchy in areas that were not carefully wrapped. Ira Eliasoph, one of the first cohort of Mount Sinai medical residents to receive Motorola Handie-Talkie pagers in 1961, recalls being "at times annoyed because the eye clinic in the outpatient building was a dead spot."[41] As Mort Weinberg, a marketing executive in Motorola's Paging Division, remembered, "The cost of putting in the wires and maintaining the wires was horrendous, as I recall . . . it cost you a small fortune. And then someone would go up and repair the roof and cut the line." In turn, only those institutions that could sustain regular in-house maintenance of their systems pursued these early radio-paging systems. The hospital, Weinberg continued, was a "natural market."[42]

In Motorola's early marketing of two-way radios, medical uses formed only one of many demonstrations of this potentially lifesaving technology (fig. 3.3). The hospital was depicted as a public institution crucial to public health. In these marketing efforts, the doctor was merely a helpful figure in promoting the virtues of products more generally geared toward clients in transportation, fire, police, and industrial sectors. But as two-way radios shrank to pocket-sized devices that required extensive local infrastructure to function, the hospital began to emerge as a "natural market" in its own right.[43] Within just a few years, Motorola Radio Pagers could be found in fifty-nine hospitals across the country, from academic medical centers like Mount Sinai in New York City and Beth Israel in Boston to community hospitals in Tennessee, Wisconsin, and Arizona, and federal hospitals serving veterans and active duty military personnel, as well the Alaska Native Health Service hospital in Anchorage.[44] As the market for hospital communications broadened, Motorola doubled down on its investment in the field through R&D, acquisitions, and litigation.

* * *

Motorola created its own Hospital Communications Division shortly after its 1959 acquisition of the Dahlberg Company, which specialized in communications products related to health and healthcare. Two of Dahlberg's principal product lines, the Miracle-Ear (the first fully transistorized hearing aid able to fit inside the ear) and the Pillow Speaker (a radio speaker, placed under a hospital pillow, that allowed direct communication with ward nurses), especially complemented Motorola's new emphasis on healthcare communications.[45] When the Dahlberg Company was sold back to its founder five years later, Motorola retained the hospital radio communication systems. Under the new name RADIO W-E-L-L, this technology became the centerpiece

FIGURE 3.3. Advertisement promoting Motorola transistorized Handie-Talkie radio-pager, c. 1957. Motorola, Inc. Legacy Archives Collection. Motorola Solutions Archives, Chicago. Reproduced with permission.

to the new Motorola Hospital Communications System. The system allowed for a combination of patient entertainment (four channels of broadcast television and a host of radio stations) as well as in-house links though closed-circuit television and RADIO W-E-L-L. This last part, a two-way radio connection to the nursing station, ensured that "nurse and patient speak quietly, confidentially, [so that] the nurse can evaluate the patient's needs without leaving her station."[46]

Motorola theorized the interconnections of physician, nurse, and patient through radio as system of "Total Hospital Communication" that linked stationary patients with ward-bound nurses and highly mobile doctors (fig. 3.4).[47] As the new division expanded its product suite over the next decade, RADIO W-E-L-L remained a central feature alongside older two-way paging systems and radio-pagers, and newer ventures using closed-circuit television systems to transmit patient orders, X-ray images, and patient communications. Combining

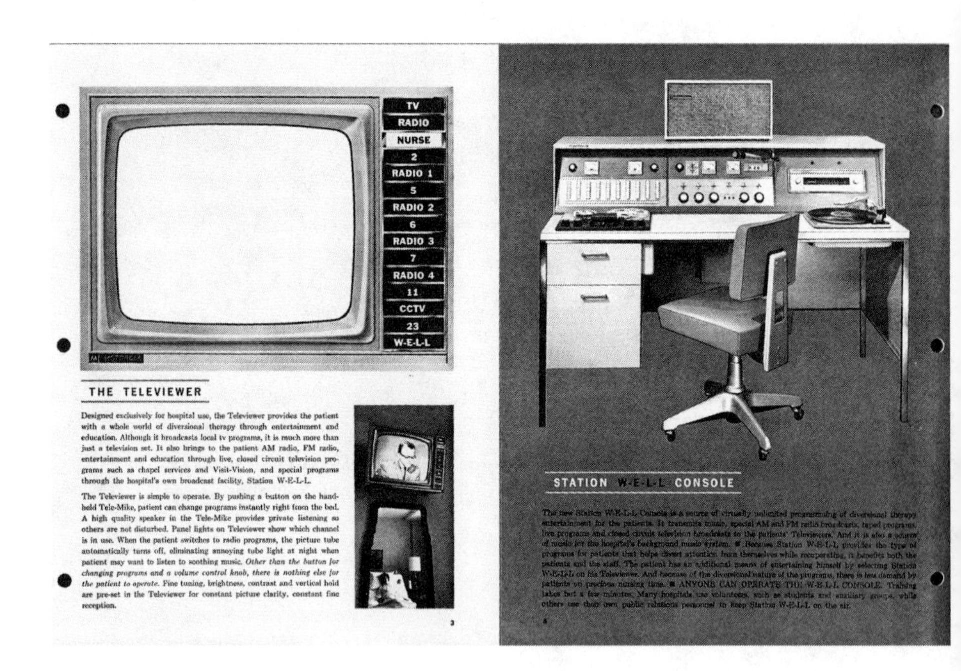

THE TELEVIEWER

Designed exclusively for hospital use, the Televiewer provides the patient with a whole world of diversional therapy through entertainment and education. Although it broadcasts local tv programs, it is much more than just a television set. It also brings to the patient AM radio, FM radio, entertainment and education through live, closed circuit television programs such as chapel services and Visit-Vision, and special programs through the hospital's own broadcast facility, Station W-E-L-L.

The Televiewer is simple to operate. By pushing a button on the hand-held Tele-Mike, patient can change programs instantly right from the bed. A high quality speaker in the Tele-Mike provides private listening so others are not disturbed. Panel lights on Televiewer show which channel is in use. When the patient switches to radio programs, the picture tube automatically turns off, eliminating annoying tube light at night when patient may want to listen to soothing music. *Other than the button for changing programs and a volume control knob, there is nothing else for the patient to operate.* Fine tuning, brightness, contrast and vertical hold are pre-set in the Televiewer for constant picture clarity, constant fine reception.

3

STATION W-E-L-L CONSOLE

The new Station W-E-L-L Console is a source of virtually unlimited programming of diversional therapy entertainment for the patients. It transmits music, special AM and FM radio broadcasts, taped programs, live programs and closed circuit television broadcasts to the patients' Televiewers. And it is also a source of music for the hospital's background music system. ■ Because Station W-E-L-L provides the type of programs for patients that helps divert attention from themselves while recuperating, it benefits both the patients and the staff. The patient has an additional means of entertaining himself by selecting Station W-E-L-L on his Televiewer. And because of the diversional nature of the programs, there is less demand by patients on precious nursing time. ■ ANYONE CAN OPERATE THE W-E-L-L CONSOLE. Training takes but a few minutes. Many hospitals use volunteers, such as students and auxiliary groups, while others use their own public relations personnel to keep Station W-E-L-L on the air.

4

FIGURE 3.4. "The Televiewer" and "Station W-E-L-L Console" in *Total Hospital Communications*, c. 1962. Motorola, Inc. Legacy Archives Collection. Motorola Solutions Archives, Chicago. Reproduced with permission.

Motorola's Radio Pager with the nursing radio console system originally developed by Dahlberg, Motorola Hospital Communications Systems pamphlets could now boast of "the most modern, most comprehensive hospital communications system ever developed by a single manufacturer."[48]

Motorola's system used different devices to plug patients, nurses, and doctors into a comprehensive grid of Total Hospital Communications depending on their relative need for mobility. Patients were static entities, typically limited to their beds. Nurses were mobile agents within a given hospital unit or floor, but their activity was clustered in tighter or looser orbits around the central locus of the nursing station on their particular ward. Physicians were highly mobile entities who could range over a variety of places inside the hospital. The innovation of the pager linked the mobile physician to the same communications grid that less-mobile nurses and immobile patients were more tightly tethered to.

Professional distinctions, labor functions, and gendered expectations were directly built into the different products marketed for wireless connectivity.[49] Motorola's promotional pamphlet contrasting the Automatic Nurse Call device for nurses alongside the Radio Paging system for physicians illustrates a stationary woman at a console in both systems (fig. 3.5). The Automatic Nurse Call offered connectivity to the nurse at *her* station; the physician's pager promised to find the physician wherever *he* might be. In the Motorola Total Hospital Communications system, the doctor-on-the-move was now as constantly and, in their terms, "automatically" findable as the ward nurse. Doctors and nurses both needed to be accessible at any given moment. But the kind of value ascribed to physicians, called from afar to provide expert guidance at a critical moment, did not extend to a nurse, whose value was bound to the moment-to-moment care of the patient's body. Later on, when a line of pagers was designated for nursing staff as well as physician staff, Motorola took pains to distinguish the sleek, silver, hi-fi portable systems carried by male doctors from the chunkier, pastel-colored, ward-bound radios intended for female nurses.

MOTOROLA

MOTOROLA

RADIO PAGING

The Motorola Radio Pager offers many benefits to the hospital and the staff. It is based on low frequency radio waves that radiate everywhere in and immediately around the hospital. The system is custom engineered so there are no "dead spots" anywhere from the basement to the roof. No matter where the person being paged happens to be, his pager will give him the message.

This modern method of paging eliminates the confusion of watching for light signals or trying to hear every message on the public address speakers. Because only the person being paged receives the alerting tone in his pager, the pager does the listening for the user. And it virtually does away with the problem of missed messages.

Since this method of paging is personal, it drastically cuts down the amount of paging over the public address speakers, thus reducing the noise caused by public paging. Further, it takes a substantial load off the switchboard because the voice message eliminates the need to call the operator to find out what the message is.

Weighing only 5 ounces, the Radio Pager is slightly larger than a package of cigarettes, and fits snugly in a pocket or on a belt. It is convenient to carry, easy to use. To hear the voice message, the user waits until the alerting tone stops, then presses a button. The voice message is clear and distinct, and is repeated twice for positive message understanding.

A choice of several methods can be used to transmit the voice message to the person being paged. A console with microphone and push-buttons (shown above) and a telephone desk set are available. Or the paging system can be connected to the telephone system so that any telephone can be used to originate a page.

The soft alerting tone in the receiver is compelling. Only the person being paged hears the tone, gets the message. The receiver is fully transistorized, operates with tiny batteries that have a low operating cost of about a penny a day.

Because the Motorola Radio Pager is efficient, delivers a personal voice message, and saves the hospital and the staff precious time, it is the finest paging method available to the hospital.

10

THE AUTOMATIC NURSE CALL

Here is the most practical, most advanced Nurse Call System ever developed for the hospital. Based on modern electronic principles, it is fully automatic, very easy to use. Here are some of the benefits it provides both the nurse and patient.

22B

The Nurse Control Station

EASY TO USE. The patient presses the "Nurse" button on the Tele-Mike. Instantly, the "Call" light (A) goes on and a buzzer sounds at the Nurse Control Station. To talk with the patient, the nurse merely picks up the handset (B). There are no switches or buttons to operate. It is just like using a telephone. The "In Use" panel lights when the system is in use (C).

IT TELLS THE NURSE WHO IS CALLING. The digital readout panel (D) tells the nurse what room and bed is calling. She thereby knows which patient is calling her, and can answer him by his name.

MOST IMPORTANT CALLS ARE ANSWERED FIRST. When a patient is on a priority basis, his call automatically supercedes other regular calls in the system and the "Priority" panel lights up (E). Emergency calls supercede Priority and Regular calls, and the "Emergency" panel lights up (F). This feature allows calls to be automatically switched in order of their urgency.

ELECTRONIC MEMORY STORES CALLS. When several calls are in the system at one time, they are stored in sequence of priority. When the nurse completes a call, the next one automatically registers at the station. Should Priority or Emergency calls come in while the nurse is handling a Regular call, the appropriate panel will light to alert her.

NURSE CAN CALL PATIENT. By simply dialing the room and bed number of the patient (G), and pressing the "Call" button (H), the nurse can call or monitor any patient. This feature allows private nurse-to-patient communications, and also allows her to monitor just the patient through his Tele-Mike, and not the distracting sounds of the whole room.

NURSE SAVES VALUABLE TIME. Because the patient can privately tell the nurse his wants, whether he asks what time it is or requests a service, the nurse can determine his need instantly without going to his room. The nurses time and energy is saved so she can devote it to patients requiring the most care.

5

FIGURE 3.5. "Radio Paging" and "The Automatic Nurse Call" in *Total Hospital Communications*, c. 1962. Motorola, Inc. Legacy Archives Collection. Motorola Solutions Archives, Chicago. Reproduced with permission.

Another key interface between the physician's Radio Pager and the Total Hospital Communications system was a console called the Radio-Register (fig. 3.6), which promised efficiencies in physician labor management previously attainable only in nursing systems. The Radio-Register was effectively a new form of electronic time clock for physicians. On the way into the hospital, they would check out a pager, with an assigned number, and would automatically be "registered" with the central communications switchboard as on duty and available for paging. Before checking out of the hospital, they would sign out by putting their pager back in its electronic berth (where its batteries, presumably, might also recharge).[50] The independence of the private physician had long posed a problem for hospital administrators in their efforts to operate hospitals along principles of scientific management. Radio-pagers offered a powerful new tool to make physicians and their behavior more trackable.

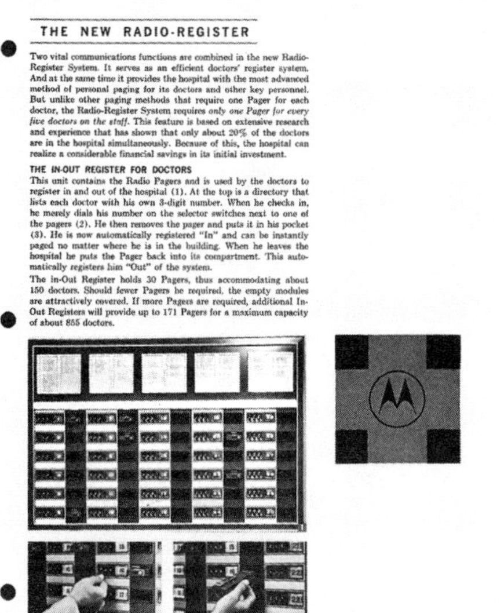

THE NEW RADIO-REGISTER

Two vital communications functions are combined in the new Radio-Register System. It serves as an efficient doctors' register system. And at the same time it provides the hospital with the most advanced method of personal paging for its doctors and other key personnel. But unlike other paging methods that require one Pager for each doctor, the Radio-Register System requires *only one Pager for every five doctors on the staff*. This feature is based on extensive research and experience that has shown that only about 20% of the doctors are in the hospital simultaneously. Because of this, the hospital can realize a considerable financial savings in its initial investment.

THE IN-OUT REGISTER FOR DOCTORS

This unit contains the Radio Pagers and is used by the doctors to register in and out of the hospital (1). At the top is a directory that lists each doctor with his own 3-digit number. When he checks in, he merely dials his number on the selector switches next to one of the pagers (2). He then removes the pager and puts it in his pocket (3). He is now automatically registered "In" and can be instantly paged no matter where he is in the building. When he leaves the hospital he puts the Pager back into its compartment. This automatically registers him "Out" of the system.

The In-Out Register holds 30 Pagers, thus accommodating about 150 doctors. Should fewer Pagers be required, the empty modules are attractively covered. If more Pagers are required, additional In-Out Registers will provide up to 171 Pagers for a maximum capacity of about 855 doctors.

11

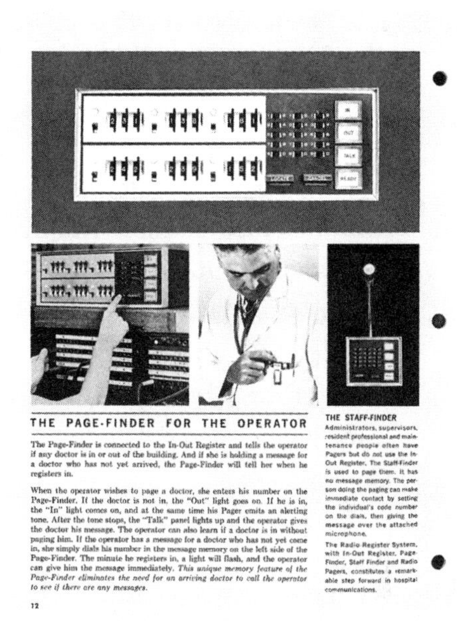

THE PAGE-FINDER FOR THE OPERATOR

The Page-Finder is connected to the In-Out Register and tells the operator if any doctor is in or out of the building. And if she is holding a message for a doctor who has not yet arrived, the Page-Finder will tell her when he registers in.

When the operator wishes to page a doctor, she enters his number on the Page-Finder. If the doctor is not in, the "Out" light goes on. If he is in, the "In" light comes on, and at the same time his Pager emits an alerting tone. After the tone stops, the "Talk" panel lights up and the operator gives the doctor his message. The operator can also learn if a doctor is in without paging him. If the operator has a message for a doctor who has not yet come in, she simply dials his number in the message memory on the left side of the Page-Finder. The minute he registers in, a light will flash, and the operator can give him the message immediately. *This unique memory feature of the Page-Finder eliminates the need for an arriving doctor to call the operator to see if there are any messages.*

THE STAFF-FINDER

Administrators, supervisors, resident professional and maintenance people often have Pagers but do *not* use the In-Out Register. The Staff-Finder is used to page them. It has no message memory. The person doing the paging can make immediate contact by setting the individual's code number on the dials, then giving the message over the attached microphone.

The Radio-Register System, with In-Out Register, Page-Finder, Staff-Finder and Radio Pagers, constitutes a remarkable step forward in hospital communications.

12

FIGURE 3.6. "The New Radio-Register" and "The Page-Finder for the Operator" in *Total Hospital Communications*, c. 1962. Motorola, Inc. Legacy Archives Collection. Motorola Solutions Archives, Chicago. Reproduced with permission.

To the hospital physician, Total Hospital Communications promised freedom for the practitioner to move inside the walls of the institution with confidence that they would always be accessible and accountable to the grid. To the hospital administrator, Motorola products allowed a more seamless management of physician as well as nursing labor—while building highly gendered distinctions between different classes of healthcare providers into the design of the devices themselves. As physicians became accustomed to using pagers inside hospital walls, Motorola engineers searched for ways to promote the use of pagers outside the hospital, as well.

* * *

Enter Dr. Archer, invented in 1966 to showcase the chromed transistor radio kept beside his bed, which worked both inside and outside of the hospital. This device, the Pageboy, was also Motorola's first pager to find widespread success as a general consumer product. In the late 1950s and early 1960s, using very high frequency (VHF) radio, Motorola engineers were able to combine the selectivity imagined in the Royalcall with the wider urban range of the Aircall. Motorola merged these two markets, promoting the pager as a tool that would simultaneously extend the autonomy of the practicing physician and the ability of hospital administrators to surveil and control their physician workforces as well. As a device that used a piece of the telecommunications spectrum, however, the Pageboy came under the regulatory purview of the Federal Communications Commission (FCC).

This placed Motorola on a collision course with the telecommunications behemoth American Telephone & Telegraph (AT&T), its subsidiary Bell systems, and their primary supplier, Western Electric, which had also been working on a VHF device, called the Bellboy, to extend the reach of the Bell telephone system into "radio calls."[51] Before it was able to bring the Bellboy to market, however, AT&T was successfully sued by Motorola and several other device firms, which alleged that the telephone monopoly's actions in the wireless space represented an anticompetitive practice.[52] Under the terms of a consent decree, AT&T was barred from selling its own radio-pager and obliged to license its radiocommunications-related patents, free of royalties.[53] Motorola was now able to access more than 7,500 unexpired AT&T, Bell Labs, and Western Electric patents and sell its own products direct to consumers as "Pageboy" systems.[54] In an added windfall for Motorola, AT&T had already petitioned the FCC for the channels needed to operate Bellboy devices. Where hospitals had been the first markets for Motorola paging systems, the new linkage with Bell systems allowed Motorola the opportunity to consolidate first citywide, then regionwide, and ultimately national wireless connectivity through paging.

Dr. Archer, the imagined early beneficiary of the Motorola Pageboy, was cast in the city of San Jose for good reason. Release of the first Pageboy was delayed until 1965 in part because Motorola was cultivating a relationship with the city of San Jose to showcase the South Bay metropolis as a model city enabled by wireless connectivity.[55] When the 1965 Motorola Annual Report announced the launch of the Pageboy to Motorola investors, it described San Jose as "a city that captures the feel of a nation on the move."[56] Dr. Archer and his Pageboy were suspended in an urban infrastructure whose seamless efficiency at every level—industrial production, domestic consumption, emergency services, education, security, and governance—was made possible by radio. Most of the radio stations of San Jose, the *Motorola Annual Report* concluded, were invisible to residents. These radio stations were not accessible on the FM or AM dial, but present in their everyday lives as a matrix of two-way radio systems operated by public utilities, local industry, and independent practitioners like Dr. Archer.[57] The Pageboy's integrated circuit linked them all. The Pageboy was a device you could wake up with, go to sleep with, and take everywhere in between. "It was hardy," Mort Weinberg, who worked in Motorola's paging division at the time, recalled. "It was really—we were beginning to learn that you had to make pagers that could drop . . . it was a maintenance man's dream." More than a decade later, thousands of first-generation Pageboys were still in use.[58]

Motorola also made a bid to sell pagers directly through a chain of Chicago-area stores called "Page One." As Ed Bales, who worked in Motorola's paging division, recalled, "The idea was to encourage people to use pagers, to buy pagers, buy Motorola pagers . . . and then what we would do is sign them up on the Illinois Bell or radio common carrier system, depending on what they chose to do."[59] Page One stores and other efforts to market Pageboys to doctors outside the hospital allowed Motorola to focus on the physician as independent business owner rather than as a hospital staff member. In Bales's terms, the main job of the Page One storefronts was to "develop the market." The salesman would evoke, for a prospective physician client, the problem of missing out on leisure time while on call—social gatherings, fishing trips, golf games. "Wouldn't it be nice to be able to go anywhere, you know, when you're on call?"[60] The key to selling Pageboys to private physicians was to emphasize the added freedoms they would gain in the bargain. Who wouldn't like to spend that extra hour out on the links, or knee-deep in a favorite fishing hole, rather than holed up by the telephone in home or office? Motorola dominated the growing pager market among physicians and many other professionals using pagers, for whom wireless mobility promised a combination of freedom, productivity, and efficiency.

Over the course of the 1970s, however, the wirelessly connected physician also became the symbol of the risks that came along with the constant availability that radio-paging provided. The pager became one of many tools of control in what sociologist Erving Goffman referred to as the "total institution" of the hospital. While Goffman's use of the term was initially meant to describe how the closed social system of mental hospitals, like prisons, came to have a degrading effect on the identities of all who inhabited them, many medical trainees began to feel that the same applied to all hospitals, and the constant beep of the pager became associated with the sense of autonomy lost instead of freedom gained.[61] In the 1978 best-selling exposé of medical training, *The House of God*, published by psychiatrist Stephen Bergman under the pen name Samuel Shem, the narrator Dr. Roy Basch repeatedly refers to his pager as the "Grim Beeper."

The House of God was a thin disguise for Boston's Beth Israel Hospital, where Bergman trained. The Grim Beeper was almost certainly a Motorola product.[62] One of the key axioms in Basch's dehumanizing descent into physician burnout was the realization that "the patient can always hurt you more," and the Grim Beeper was a key instrument in that torture. On the first day of his internship as a newly minted MD, amid extensive pangs of self-doubt, Dr. Basch is pulled back from a brief feeling of competence and self-sufficiency by the insistent buzzing of his pager:

> It didn't last. I got more and more tired, more and more caught up in the multitudinous bowel runs and lab tests . . . I hadn't had time for breakfast, lunch, or dinner, and there was still more work to do. I

> hadn't even had time for the toilet, for each time I'd gone in, the grim beeper had routed me out. I felt discouraged, worn.[63]

Physician resistance to paging predated the invention of the pager. Hospital administrators in the 1920s had complained that "there are always some persons who do not listen for their calls or look at the annunciator as they pass through the corridors."[64] But resistance to the pager was a more complex affair. The notification board, the loudspeaker, the time clock: these managerial tools could be left behind when one left the workplace. But the pager, as an instrument of control, stayed with you—in your pocket, on your belt, under your pillow. "A lot of people resented paging," Motorola's Mort Weinberg admitted, "because it was an electronic leash around their necks."

As the bleep of the pager became increasingly associated with negative affect by many who wore them, some physicians began turning their pagers off, even though this move conflicted with the moral and legal imperatives for physicians to provide constant accessibility to their patients and colleagues. As sociologist Eviatar Zerubavel noted in a study of doctors' and nurses' paging practices, presented at the American Sociological Association in 1978, the first chapter of the code of medical ethics adopted by the American Medical Association in 1847 began with the injunction that a physician must be "ever ready to obey the calls of the sick"—in other words, that *ever-availability* was a fundamental professional obligation of the physician.

Physicians had long been expected to be continuously responsible and accountable for their patients' well-being: always on the job, expected to be ready for duty whenever needed. But "during the last two decades," Zerubavel observed, "a special one-way radio system has been introduced into hospital life." Now that physicians carried beepers wherever they go, "practically as well as symbolically, they can be reached *at any time* . . . they have no periods of time which are utterly private and not vulnerable to intrusion."[65] The technology of the radio-pager had extended the expectation of access into every moment of the physician's life.

Zerubavel studied the effects of technological change and institutional organization on the experience of time. His 1979 book, *Patterns of Time in Hospital Life*, traced the impact of radio-paging on how physicians experienced the temporal distortions that came with expectations of constant service and accountability. Many physicians training in the 1970s agreed that the introduction of the pager changed the lived experience of tempo and temporality within the space of the hospital. By the late 1970s, Zerubavel noted, physicians increasingly invoked concepts of "private time" to push back against the expectations of continuous availability and accountability. Physicians in his study "also admit—though unofficially—that the practice of turning one's beeper off is not uncommon among them."[66] Certainly the denizens of Shem's *House of God*, both real and fictional, had also figured this out by 1978. Toward the end of the novel, one of Dr. Basch's co-interns walks past him in the hospital with a jubilant expression, and asks him, "Hey man, notice anything different? . . . My beeper, man, it's off. They can't get me now."[67]

Yet one could not always just turn off the Grim Beeper, since each 'bleep' carried an aura of urgency and emergency. Many physicians-in-training in the 1970s recall the sense of pride they felt upon being issued their first pager as interns, and the deep sense of responsibility that the object was imbued with. One intern at the University of Chicago in the mid-1970s recalled the day he received his first pager, a voice-pager whose "bleep" would cue him to press a button to receive a message of some urgency. Yet this very set of temporal expectations led to a popular prank among his class of medical interns called "barking," in which one would page a fellow intern or resident using the telephone system and, in the place of urgently needed information, simply bark into the telephone. The unwitting mark would hear the bleep, fumble quickly for their pager, and press the button, only to hear barking emerge from the speaker.[68]

By the end of the 1970s Pageboy and Bellboy beepers were actively promoted to doctors and other drug dealers as nuisances, but convenient and necessary nuisances all the same. New York Telephone advertised the irritation the new Bellboy II induced, as an attractive

FIGURE 3.7. "You can depend on it to make a nuisance of itself": advertisement for the Bellboy Beeper (manufactured by Motorola).

feature. "You can depend on it," the advertisement reads, "to make a nuisance of itself" (fig. 3.7).

* * *

The physician's pager became the electronic leash. It was neither the first nor the last telecommunications device whose superficial promises of freedom through connectivity concealed hidden costs of constant accountability. Already by the first decades of the twentieth century, wireless connectivity presented a Mephistophelean bargain that cut into the freedoms and leisure of the practicing physician.

By the first decades of the twenty-first century, the smartphone presented the same bargain to the general population. Perhaps you, too, remember purchasing your first smartphone as a tool to expand your own autonomy and mobility—and now feel that it is hard for you to spend an hour, let alone a day, without checking it. Somewhere in between the landline and the smartphone, the pager became an original site for articulating the dilemma of the "electronic leash" that now extends well beyond physicians.[69]

The pager may not have produced paging, but it did transform the meaning of the page from a person (servant) to an action (service) to a consumer good (the pager) that still contained nominal and functional relationships to the original meaning of the page as a servant. It is not an accident that product names like "Pageboy" and "Bellboy" made explicit reference to this lineage between human servants and solid-state electronics—another piece of the history of technology loaded with several levels of servant/service meaning.[70] As a new object whose utility was folded into older technologies and practices of locating physicians in times of urgency, the pager at first only made sense in certain places (within a specially coil-wrapped building for physicians, mechanics, store detectives, police, and firemen), then within the context of localized broadcast systems (police, fire, ambulance), then in urban and suburban environments more generally. Originally promoted as a servant *to* the doctor, the electronic leash also tethered the physician in deeper servitude to the growing influence of large institutions in the practice of American healthcare in the late twentieth century. Promises of increased autonomy and independence gave way to practices of increased surveillance and loss of control of one's own time.

As the pager moved into different environments, it took on different meanings. Within a single hospital system, a nurse paging system might be viable for a floor, while physician pagers could work throughout a hospital, city, or region. By the 1970s, however, physicians were not the only users, and still larger ranks of professionals began to sport pagers in the 1970s and 1980s, including sales representatives and law enforcement officers, as well as sex workers and drug dealers. "It used to be doctors carried pagers, and therefore everybody wanted to carry pagers to look like a doctor," Motorola's Jim Wright noted in 1988, "but it is now picking up teenagers, which is a real problem we have, because pagers are being banned in high schools, because there is an assumption that every teenager that has a pager is a drug dealer."[71] The Pageboy had originally been a marker of a middle-class professional. But the same Motorola device was understood to mean something different in the hands of a drug dealer compared to a physician, even if the shared use of the technology underscored key similarities between the two trades.

By the end of the 1980s the pager was embraced as an icon of hip-hop culture as well. Sir Mix-A-Lot released a track called "Beepers" in 1989, while The Tribe Called Quest led off the first track of its platinum-winning 1991 *Low End Theory* with an ode to the pager as a marker of maturity: "back in the day when I was a teenager / before I had status and before I had a pager." One of the last tracks, "Skypager" is fully devoted to the role of the pager in the hip-hop lifestyle—including its discontents. "People tend to think that a pager's foul," Tribe front man Q-Tip drawls, "well it really is 'cause it makes me scowl." The burden, as well as the benefits, of the pageable lifestyle was perhaps best captured a few years later in the opening line of Notorious B.I.G.'s "Warning" which begins after an irritating string of electronic beeps, "Who the fuck is this? / Pagin' me at 5:46 in the morning / crack of dawn an' / Now I'm yawnin' / wipe the cold out my eye / See who's this pagin' me, and why?"[72]

Whose interests did the pager serve? Q-Tip, Biggie, and Dr. Basch agreed it could be hard to tell. This glittering device sold to professionals to convey mobility, accessibility, and authority quickly became a gilded cage. Still, as pagers came and went as icons of the hip-hop life, and were subsequently ditched by drug dealers for disposable "burner" phones and other modes of telecommunications, they stayed relevant in the medical world much longer. "The drug dealers have all gone to cell phones," an analyst at Cisco Systems told *American Medical News*, but "the doctors really haven't."[73] Long after car phones, then mobile phones, then routine use of Short Message

Service (SMS) texting and smartphones become prominent, the pager remained a key element of dedicated medical communications well into the twenty-first century. When Motorola finally sold its paging division to Google in May 2012, 85 percent of American hospitals were still using Motorola paging systems.[74]

Many of us have, like Dr. Elliot Archer, overlooked the role of the pager as a medical technology precisely because its introduction as a small, everyday device was so very successful. For more than a half-century, the pageability of the on-call physician was never in question. This ethos of constant accessibility, embodied in the "bleeping box," inadvertently became a symbol of the physician as evocative and potent as the white coat or the stethoscope. While the physician was pageable before the pager—whether by page boy, steam whistle, gong, loudspeaker, telephone, or flashing lights—the materialization of that function into a wireless Pageboy created new geographies of access, new forms of accountability, and even new bodily habits. A new diagnosis, "phantom pager syndrome," has subsequently been described in the medical literature: that is, the feeling of a pager buzzing on one's belt even though no device is currently attached to one's person. This unique form of pathology did not exist prior to the pager—but subsequently it has spread well beyond the medical profession to anyone who might use their mobile phone in "vibrate" mode.[75]

In part the pager persevered as long as it did because it materially embodied a set of values that connected current trainees in medicine to older traditions of accountability and service central to the self-perception of medicine as an honorable vocation. Yet the automation of these humanistic obligations has not produced more humanistic physicians. To the contrary: the externalization of accountability in the form of the pager (and in subsequent electronic media of constant availability) may have further accelerated the alienation of healthcare workers from control of their own labor and the rising tide of physician burnout.[76]

While it is better for physicians to curse at the Grim Beeper than at their patients, in many cases they have now learned to do both.

4
THE AMPLIFIED DOCTOR

In the second half of the twentieth century, television became a site of intense speculation over the power of new media to reshape social relations. This was certainly the case in 1966, when Reba Benschoter approached the podium at the New York Academy of Sciences to talk about the role of television in the future of medical care. A crowd of luminaries had gathered—advertising executives taxied over from Madison Ave, celebrity physicians flew in from across the country, and overseas dignitaries beamed in via satellite. The anticipation was palpable. New communication technologies in healthcare, organizers argued, promised a biomedical revolution every bit as important as antibiotics, anesthetics, or organ transplants. "Just as seven out of ten prescriptions written today are for items unknown to medicine before World War II," the conference brochure read, "seven out of ten possibilities for communicating the knowledge behind those prescriptions were not available before 1946."[1] Television had transformed American public life in the two decades since World War II, and Benschoter was about to reveal how it could transform clinical practice as well.

Benschoter had spent much of the decade exploring the potential of medical television. Since the late 1940s, closed-circuit television had occupied a growing role in the rethinking of medical education, especially in the field of surgery. Executives at electronics corporations like RCA and pharmaceutical corporations like Smith, Kline & French saw medical education as a natural market for color and high-definition television. They convinced many medical educators that the "liveness" of closed-circuit television could replace the observation gallery of the old surgical amphitheater with new spaces for medical education and could enhance the ability to connect the hospital and the classroom.[2] Benschoter's work at University of Nebraska Medical Center (UNMC) and at national medical organizations was part of a broader effort to use television to revolutionize the way doctors learned. Her accomplishments at UNMC included founding a Department of Biomedical Communications, establishing the first journal in the field, and joining the Council on Medical Television for the National Institutes of Health.[3]

Benschoter's experience led her to propose other, more ambitious applications for audiovisual communications in medicine. Television networks already served as a communications infrastructure for emergency alert systems. Why couldn't they also serve as an infrastructure for the delivery of medical care? By allowing doctors to see their patients at a distance, she suggested, a new form of television medicine—soon to be named "telemedicine"—was poised to radically change the very nature of clinical practice, biomedical research, and medical institutions themselves. All you needed were two televisions, two cameras, and some way to link them into a circuit.

In the late 1940s and early 1950s, most Americans experienced television as a broadcast medium dominated by the "big three" major networks: ABC, CBS, and NBC. Yet by the mid-1960s, newer, more interactive modes of community antenna television (CATV) and cable television promised a more dynamic interplay between viewers and creators. Interactive closed-circuit television offered instantaneous multimedia connection to conversations that had previously operated on sound alone, adding a full palette of visual cues to the aural connection of the phone call or two-way radio. As the "live" feel of two-way television opened up new possibilities for virtual presence in private business and public service, medicine became an especially important site for imagining the benefits of networked multimedia.[4]

Television seemed to offer limitless possibilities for reshaping healthcare. With interactive television circuits, one could conduct a physical exam, review an X-ray or a microscopic blood specimen, even provide psychotherapy in real time, though hundreds or even thousands of miles might separate physician and patient.[5] As with telephone medicine and radio medicine, however, many doctors responded to this new medium of care with concerns over what might be lost in the exchange. The prospect of television medicine brought new risks of artifact and error into diagnosis and treatment, and new forms of distance and distrust into doctor–patient communications. Other, deeper strains of physician resistance were powerfully connected to a desire to maintain professional control over their own clinical spaces, which the presence of the television camera threatened to violate.

Yet television also offered the possibility of expanding the range, reach, and influence of the individual physician. Just a few years after Benschoter's speculations on interactive television in everyday clinical practice, a young physician took the podium at Washington's Dupont Plaza Hotel with a copy of Marshall McLuhan's best-selling *Understanding Media: The Extensions of Man* to describe how new electronic media would produce what he called "the Amplified Doctor." Dr. Kenneth Bird had just become director of the Massachusetts General Hospital Telediagnostic Field Station, a new laboratory to test technologies to "extend competent medical services to areas that are either too primitive or too sparsely settled to support a full-time doctor."[6] Working with an array of interconnected two-way television devices, a dedicated engineer from CBS, and a staff of nurse practitioners, physicians, and technicians, Bird coined the term "telemedicine" and became the world's most visible advocate for using television medicine to extend the reach of the average doctor. In early 1968 Bird claimed, with a bravado and bluster McLuhan would have admired, that "medicine as we know it became obsolete."[7]

Bird regularly seized on McLuhan's metaphor of television as "the most recent and spectacular electric extension of our central nervous system" to imagine new ways for doctors to make meaningful human connections with distant patients.[8] But his audiences did not all agree on *which* aspects of medicine should be extended through telemedical circuits, and who won and who lost in the exchange. Should television primarily play a role, as Bird suggested, in amplifying the reach and power of physicians? Or should the real extension of television medicine lie in developing new roles for "physician extenders" who, like the nurse practitioners in Bird's Boston teleclinic, were able to take on increasing autonomy through these television links? Expansive visions of interactive televisions in domestic spaces also promised to empower patients to access care outside of the spatial constraints of the hospital or the doctor's waiting room. At the same time, these visions of household medical monitoring carried echoes of the two-way screens that Big Brother in George Orwell's novel *1984* used to peer into private lives of citizens.

Questions of power and presence lay at the heart of all proposals to transform healthcare through interactive television. If physicians abused their professional power behind the closed doors of their examination rooms, perhaps they would not do so in the more open space of the televised encounter. If patients felt disconnected and detached in sterile clinical environments, perhaps the ability to reach their doctor from home might help them open up a bit. The technology we now call telemedicine was imagined and implemented in the midst of wide-ranging popular debates over how television could change social mores and human interactions. As television became a vehicle for reimagining medical encounters, it also became the focus of a series of attempts to combine technological and social-scientific expertise to engineer a better doctor–patient relationship. The shared hubris of telecommunications engineers and social scientists converged in an overly confident belief that a new platform of care could undo the asymmetries of power so deeply ingrained in medical interactions.

The technology of telemedicine on its own seemed ideologically neutral. It could empower the patient or nurse practitioner, or it could extend the reach of the doctor. The question of who gained and who lost power in this exchange was determined not by any fixed attributes of television itself, but by a series of conscious decisions through which telemedicine was conceived and introduced into American medicine.

* * *

Reba Benschoter grew up on a farm outside a small town in Iowa. After college she found a job at a local TV station as a program editor while pursuing a master's degree in psychology. When her husband, Benny, took a new position at a TV station in Omaha, Reba sought to put her graduate degree to work at the Nebraska Psychiatric Institute (NPI). Yet when she introduced herself to the institute's founder, Dr. Cecil Wittson, he offered her a job in biomedical communications instead. Benschoter's master's thesis had focused on the role of television in education, a theme that fit closely into Wittson's own plans to expand the work of neurology and psychiatry beyond the walls of his new institute.

As a psychiatrist directing a field hospital from a US naval base in Sydney, Australia, during World War II, Wittson became convinced that telecommunications had potential to improve the field of mental health. Like many other field operations in the war effort, the mental health services of the armed forces in the Pacific Theater relied on radio communications systems to extend the range of the few mental health caregivers in the military over vast distances.[9] After the war, when Wittson joined the faculty of the University of Nebraska Medical Center in Omaha, he explored the use of telephones to provide a similar extension to mental health services to rural areas in Nebraska. In the mid-1950s, Wittson partnered with three other mental health facilities across the state, in addition to mental hospitals in North and South Dakota, to create an interstate telephone consultation system.[10]

Wittson soon raised $1.5 million to build a new 100-bed Nebraska Psychiatric Institute on the UNMC campus. When the project broke ground in 1955, Wittson announced that the entire edifice would be hard-wired for closed-circuit television.[11] The work of wiring the television circuits, selecting and connecting cameras and screens, and making the system legible to physicians would fall to lead technician

FIGURE 4.1. Van Johnson, engineer, and Reba Benschoter, project director, reviewing the programming schedule for the interactive television link between the Nebraska Psychiatric Institute in Omaha and the Norfolk Mental Hospital. Courtesy of History of Medicine Division, National Library of Medicine, Bethesda, MD.

Van Johnson, one of only a handful of African American professionals employed at UNMC. Johnson joined the Nebraska Psychiatric Institute in 1956 as project engineer to manage its expanding network of media technologies.[12] When Benschoter showed up in Omaha looking for work as a psychologist, Wittson thought her communications background would complement Johnson's technical and engineering skills. "We were an interesting pair," Benschoter recalls, noting the relative absence of women and African Americans from the workforce in academic medicine at the time, and how daunting it could be to make a space, "in the middle of all of these MDs and PhDs," who otherwise tended to be both white and male.[13] With strong backing from Wittson, Johnson and Benschoter collaborated to build the field of biomedical communications at UNMC for the better part of the next four decades (fig. 4.1).

The Nebraska Psychiatric Institute became an experimental space for developing multimedia medicine. After a few months working out the equipment purchases, supply contracts, and wiring logistics, Benschoter and Johnson created a closed-circuit loop between two video cameras, two microphones, and two television sets that connected the new psychiatric institute and the anatomy department across the street. Working closely with Wittson and other psychiatrists on the UNMC faculty, they promoted this new platform for delivering remote lectures and conducting classroom discussions. Within a short while they could demonstrate a detailed neurological exam of a patient in one building by a doctor sitting in front of a television set in another building. This was followed by a flawless demonstration of group psychotherapy over the same closed-circuit television system (fig. 4.2). Like the telephone, the television could be used to diagnose and treat patients at a distance.

Linking two buildings on opposite sides of the same street was useful as a proof of concept, but the real challenge came with longer distances. Wittson, Johnson, and Benschoter applied for federal funding to test the potential for a television circuit between their facility in Omaha and a relatively neglected state mental hospital more than

FIGURE 4.2. Early use of television for group psychotherapy. Courtesy of Special Collections and Archives, McGoogan Health Sciences Library, University of Nebraska Medical Center, Omaha, Nebraska Psychiatric Institute Collection.

a hundred miles away in Norfolk. When a dedicated, closed-circuit microwave TV link between the two facilities went live in 1964, NPI initiated the first large-scale test of interactive television as platform for remote patient care (fig 4.3).[14] By the time Benschoter spoke at the New York Academy of Sciences in 1966, the program had been running for more than two years and had found widespread acceptance among local physicians, psychologists, nurses, residents, students, staff, patients, and their families.

Some of Benschoter's data concerned the role of interactive television in medical education. Grand Rounds from the Department of Psychiatry along with other lectures from the UNMC campus could be beamed into the remote hospital as "telelectures" so that the residents and faculty who happened to be in Norfolk could listen and occasionally ask questions of their distant colleagues. Joint seminars and conferences could also be staged on a regular basis, allowing for even more interactive discussions (fig. 4.4). All of this helped trainees in the satellite institution feel connected to the main campus. But

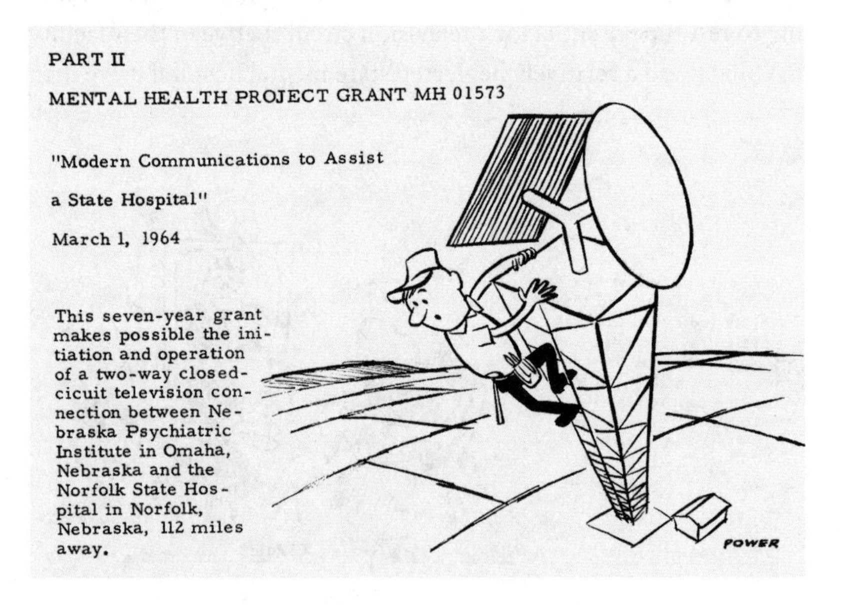

FIGURE 4.3. Modern Communications to Assist a State Hospital. Courtesy of Special Collections and Archives, McGoogan Health Sciences Library, University of Nebraska Medical Center, Omaha, Nebraska Psychiatric Institute Collection.

FIGURE 4.4. Illustration of interactive conferences via the television link between the Nebraska Psychiatry Institute and the Norfolk Mental Hospital.

Benschoter, Johnson, and Wittson's work soon expanded to emphasize the use of interactive television in clinical care, patient quality of life, and ward management as well.

Television offered immediate benefits in the diagnosis and management of complex neuropsychiatric disorders. The underfunded Norfolk state mental hospital could not afford to maintain its own neurologist on staff. Caring for more complicated patients, whose conditions required the input from Omaha specialists during their intermittent visits to Norfolk, was challenging and often dispiriting. Now, however, sitting down with a TV screen and camera in front of him, the neurologist on staff at NPI could observe reasonably detailed neurological examinations and electroencephalograms over the television. Before the TV link, only three staff psychiatrists were stretched across twenty-seven wards at Norfolk, and employees often described themselves as custodians rather than therapists. With interactive television, Benschoter, Johnson, and Wittson were able to add regular therapy rounds on each ward, staffed by psychiatrists back in Omaha (fig. 4.5).[15]

FIGURE 4.5. Illustration of the remote medical supervision of the psychiatry wards at the Norfolk Mental Hospital through interactive television.

Half of the patient population at Norfolk came from the Omaha metropolitan area, and many of them were effectively isolated from their families during the duration of their hospitalization. After Benschoter opened a viewing station at NPI eight hours a week for family members to communicate with their loved ones, the morale of patients and hospital staff, as well as families, improved. One woman from Omaha had not been able to see her brother since he entered the Norfolk hospital several months earlier because of the expensive six-hour bus ride a visit entailed. While she recalled her brother being in "terrible shape" the last time she saw him, she was surprised how much the "TV visit" helped her understand his improvement. She described the video interview as "really wonderful . . . like she was really talking to her brother."[16] The shared visual frame of the television conveyed a sense of presence and connection the telephone alone could not convey (fig. 4.6).

Embracing interactive television as a new medium for medical practice had, in short order, produced real results for physicians, nursing staff, families, and patients. "Today," Benschoter concluded,

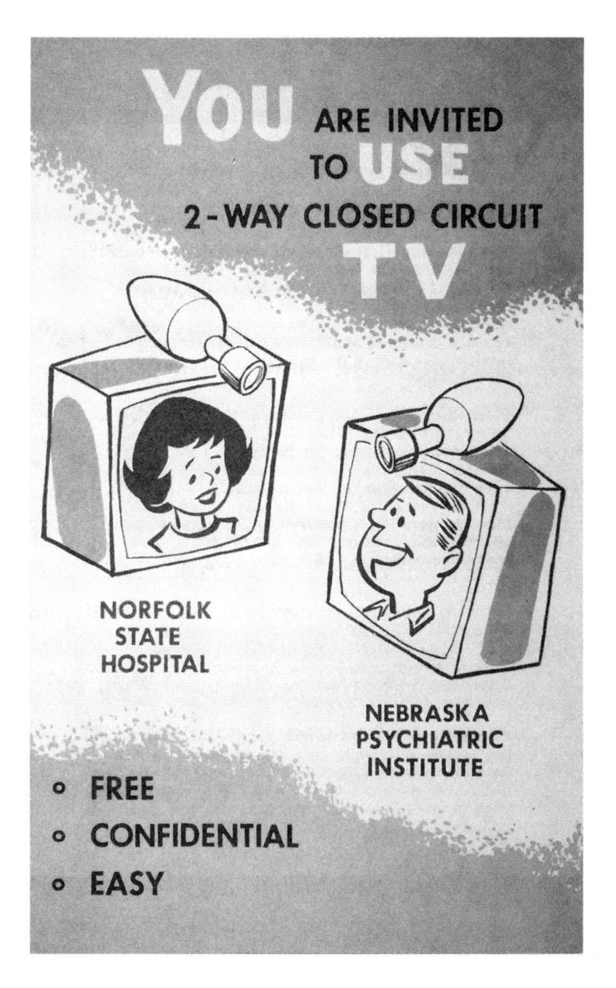

FIGURE 4.6. Pamphlet from the Nebraska Psychiatric Institute promoting the use of two-way television in the care of patients, emphasizing the service as "free, confidential, easy." Courtesy of Special Collections and Archives, McGoogan Health Sciences Library, University of Nebraska Medical Center, Omaha, Nebraska Psychiatric Institute Collection.

"success of the project can be seen in the unlocked doors, the changed staff attitudes, the patient participation in hospital therapy and work programs, and the appearance of the ward itself with its freshly painted furniture and new curtains."[17] A depressed, custodial institution for the mentally ill had become an optimistic site for therapy and recovery. Patients seemed to be getting better during the trial period, as the total inpatient population of the hospital dropped by nearly half, from more than 900 in 1965 to 476 by the end of 1968. Benschoter and the rest of the team at NPI took this as evidence of the broad clinical efficacy of television medicine.

These claims were almost certainly an overreach. In hindsight, it is now clear that the pendulum shift away from mass institutionalization of the mentally ill had started decades earlier, and was in full swing well before Benschoter, Johnson, and Wittson set up the Norfolk television link. A wide variety of social and economic factors drove the mass discharge of patients from state mental institutions at this time—with the not entirely benign effect of shifting the burden of mental healthcare to the nation's jails and prisons, to its police, and to the streets.[18] Yet at the time, it appeared to Benschoter and Wittson that the emptying of Norfolk's beds was a good thing, attributable at least in part to the new television link to Omaha. Television medicine, in their estimate, had not only changed the care of individual patients but also altered the therapeutic function of the mental hospital as a whole. In a moment of optimistic overextension, Cecil Wittson predicted that if the television link from Omaha were expanded further, "Nebraska may well become the first state to do away with state mental hospitals."[19]

Years later Benschoter would refer to the Nebraska television project as the first successful demonstration of a new field of telemedicine. When she presented her findings to the New York Academy of Sciences in 1966, however, the term had not yet been coined. To understand the origins of the term "telemedicine" and its role in the framing of electronic medicine in the late 1960s and early 1970s, we must turn from Omaha to Boston, from rural networks to urban practice.

* * *

On the morning of October 4, 1960, a murmuration of starlings flew into the engine of an Electra turboprop plane taking off from Boston's Logan Airport, causing the plane to stall and collapse in the shallow waters of Winthrop Bay. Several passengers died on impact, but the tragedy of their loss was compounded by many more casualties who survived the crash only to die afterward because of delays in medical attention. As radio broadcasts relayed news of the disaster, hundreds of spectators took to the roads to witness the unfolding event, unintentionally blocking the passage of emergency vehicles on key roadways. Ambulances from Boston's Massachusetts General Hospital (MGH), less than three miles away, were unable to arrive in time to forestall many otherwise preventable deaths. In the aftermath of the crash, the director of the Massachusetts Port Authority reached out to the president of MGH, John Knowles, to build a "miniature hospital" at Gate 23 at Logan Airport to respond to future air emergencies. Kenneth Bird, then a young faculty member in the Department of Medicine, was named its first director.[20]

When the Logan Medical Station opened its doors at Gate 23 three years later, Bird found the logistics of running a clinic located a few miles away from his main hospital to be challenging—especially in terms of communications. Along with fellow internist Joseph Miller, Bird staffed the clinic during peak commuting hours, 8–10 a.m. and 4–6 p.m., and was on call by telephone and pager for the nurses who staffed the clinic twenty-four hours a day. "One of our first patients," he recalled, "was an elderly woman who had injured her hip. The medical station nurse telephoned me a full report of the woman's symptoms, but I just couldn't be sure of the diagnosis."

> Despite all our efforts I had to face the fact that a verbal report just wasn't enough in a situation like this. "If only I could see the patient" I thought. There was no choice but to bring the woman by ambulance to the hospital to determine how extensively she was hurt. But of course I could see patients at a distance. If I could see a space launch a thousand miles away in Florida, and hear an astronaut's heartbeat a thousand miles up in space, then there was no reason why a patient a few miles away couldn't be seen and his vital signs checked, while a nurse led him through a physical examination.[21]

The phone could only convey so much. But experiences with closed-circuit television in medical education allowed Bird to imagine—like Wittson, Johnson, and Benschoter—that he could be *more* present with a patient several miles away, if only the right kind of visual connection could be made. Adding sight to sound promised the remote physician increased ability to diagnose, treat, and manage disease. Sometimes a doctor needed to see a lesion directly, to scrutinize a patient's face while discussing sensitive matters. Just as the Omaha project augmented remote psychiatric care through new television links in rural settings, the Boston project augmented remote urgent care through new television links in urban settings.

Though the distance between MGH and Logan was a tiny fraction of the distance between Omaha and Norfolk, the task of getting a TV signal between the hospital and airport proved in some ways to be more challenging. Privacy fears loomed, as did concerns that unencrypted signals might be picked up by household TV sets or might interfere with other forms of telecommunications found in a heavily populated urban area. Along with engineers from Columbia Broadcasting Service (CBS) and the local television channel WGBH, Bird struggled to build a dedicated link that would be dependable and secure and able to meet the higher standards of privacy needed to protect sensitive healthcare information.

With the help of a three-year grant from the US Public Health Service, Bird was eventually able to set up a line-of-sight microwave transmission pathway between the two sites. Closed-circuit TVs equipped with a range of cameras for long shots and close-ups to aid physical examination were installed at Gate 23 (fig. 4.7), along with other specialized cameras able to transmit X-ray images, electrocardiograms, and videomicroscopy of blood smears and other simple laboratory procedures. All of this information was beamed from Logan Airport to a tower at MGH, where it fed into a wire connecting to a tiny alcove studio tucked off the Emergency Department (fig 4.8). When the telediagnostic clinic opened in April 1968 to great fanfare, the *Boston Globe* reported that "the doctors are never more than a few feet away from their patients, even though the latter are at the airport and

FIGURE 4.7. Telemedicine was arguably most effective as a means of enabling the autonomous practice of nurse practitioners in primary healthcare. Pictured here are the nursing staff who directly encountered patients at the MGH Telediagnostic Clinic at Logan International Airport. Courtesy of Massachusetts General Hospital Archives, Boston.

the doctors are at the hospital downtown."[22] A third-year medical student at Harvard Medical School rotating through the clinic described it as a clinic in which "the doctor's stethoscope is three miles long."[23]

This medical student, Michael Crichton, was also an aspiring science-fiction writer. His first book, *The Andromeda Strain*, was already in press; his second book, *Five Patients*, would devote an entire section to the science-fictional world already present in the contemporary practice of the Medical Station at Gate 23.[24] Bird also described his clinic as a form of science fiction in the present. Where Wittson and Benschoter used the television to bring metropolitan psychiatry to a relatively neglected and state hospital in Nebraska, which otherwise lagged in access to diagnostic and therapeutic technologies, Bird described the jet-setting patient population at Logan Airport as "a prototype of one community of the future." Many of

FIGURE 4.8. Dr. Kenneth Bird, who claimed to have coined the term "telemedicine," at the physician's console in an alcove off the emergency room at the Massachusetts General Hospital. Courtesy of Massachusetts General Hospital Archives, Boston.

the 5,000 employees of the airport were already accustomed to using radar, radio communications, and closed-circuit television on a daily basis. "This community," he told a reporter, "already is convinced of the value of applied electronics and thus has set its own stage for acceptance of medical electronics."[25]

Bird coined the term "telemedicine" as "the practice of medicine without the usual physician-patient physical confrontation."[26] He often quoted Marshall McLuhan in grant applications and public talks, especially McLuhan's observations that in the electronically interconnected society of postwar North America, "time has ceased, space has vanished . . . ours is a brand new world of all onceness."[27] To Bird, the simultaneous experience of interactive television created new possibilities of being together, even when apart. Television permitted a "dynamic interaction which allows interpersonal communication across distance to recreate, and even *enhance*, face-to-face communication."[28]

Telepresence in medicine brought peril as well as promise. Securing the privacy of the television link, especially when streaming video of a partially disrobed patient, was an early and frequently repeated concern. Bird, however, seemed more concerned with questions of image fidelity: signal and noise, enhancement and filters. When was video quality good enough to simulate the face-to-face presence of a direct physical examination, and when might it lead to artifact or to false assurances? Even though color TV had played a key role in the development of medical television as an educational tool, Bird and CBS engineer Stanley Krainin preferred the crisper images of black-and-white television as a medium for clinical practice. Except in dermatology and some forms of pathology, Bird and Krainin saw color to be largely incidental to medical diagnosis. Black-and-white television offered better sharpness, contrast, and accuracy, and the marginal benefits of color did not outweigh the risk that "a poor color picture could lead to erroneous patient assessment and diagnosis."[29]

Their laboratory notebooks, preserved in the archives of MGH, comprise a visual corpus of data intended to establish where, exactly, the threshold of "good enough" diagnostic image quality could be set. Consider the photographic prints pasted onto the page shown in figure 4.9, depicting two television screens, which themselves depict images of the eye exams of a model patient at two different camera settings. Bird and Krainin amassed hundreds of photographs like these to form a database of image fidelity in physical diagnosis. If a doctor could see a lesion in the blood vessels of the conjunctivae in person—that is, the red streaks in the "whites" of your eyes—would that same lesion be visible to another doctor looking at that eye on a TV screen several miles away? The pair steadily tested the influence of different permutations of cameras, lenses, and vidco-enhancement algorithms on the ability to distinguish key features on microscopic, radiological, and physical examinations.

Bird and Krainin started a new science of similarity, documenting the equivalence of telepresence and physical presence. Much of the scientific literature on telemedicine since that time is likewise concerned with demonstrating that the services provided by medicine

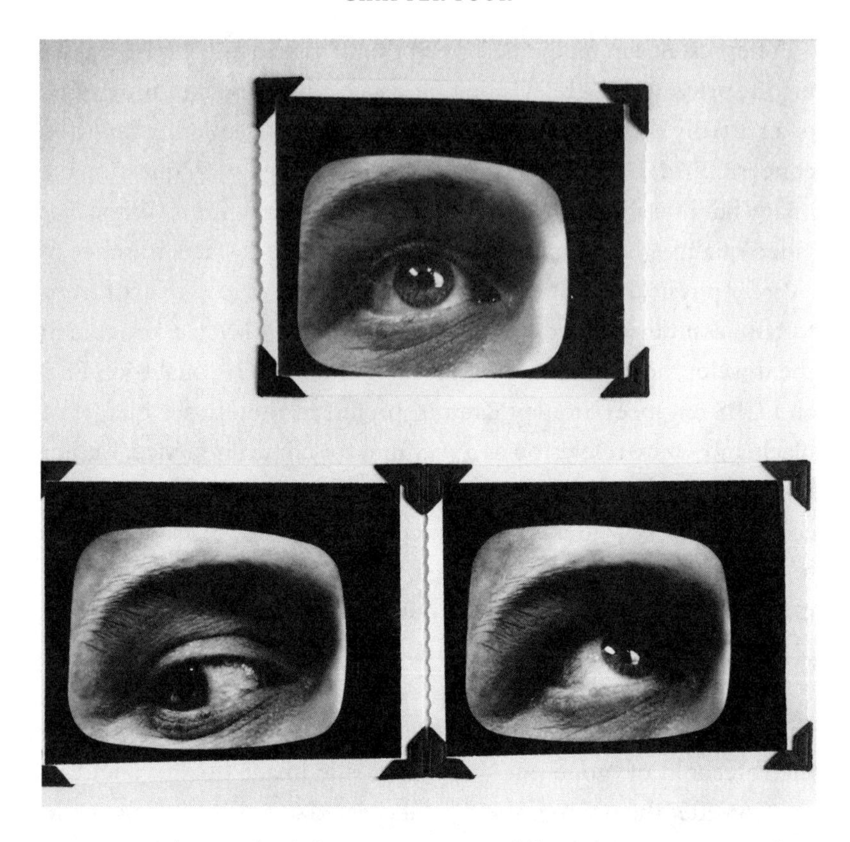

FIGURE 4.9. Photographs of television screens using different camera settings to evaluate differential resolution of the conjunctival vessels: a form of scientific evidence to demonstrate equivalent diagnostic capability. Courtesy of Massachusetts General Hospital Archives, Boston.

at a distance are equivalent, even if not identical, to those provided by flesh-and-blood encounters.[30] Yet diagnostic verisimilitude was only one of several axes of parity that telemedicine needed to establish to convince skeptical physicians, patients, and policymakers. Bird and Krainin's research could demonstrate what was present in physical diagnosis by television, but it could never fully account for what was lost in substituting telepresence for presence. Beyond the technical questions of fidelity and equivalence and diagnostic confidence, the telediagnostic clinic also encountered some degree of skepticism from patients and practitioners alike. "I don't want no Doctor peering / Down my throat from miles away," reads a piece of doggerel typed on Logan Medical Station letterhead. "I don't want nobody listening / To my heart on microwave."[31] Patients described unease over the use of television to broadcast their intimate health information, and wariness that the use of interactive television as a health surveillance technology cast the physician as an all-seeing Big Brother (fig. 4.10).[32]

Echoes of Big Brother notwithstanding, Bird's telemedical field station was roundly considered a success. It helped establish a form of electronic care tolerated and even enjoyed by most patients, and it helped produce publication after publication about the value of medicine by television. Bird in Boston and Benschoter in Omaha began expanding their point-to-point telemedical links into networked systems. By 1968, the Norfolk link had shut down, but Benschoter added three new Omaha-area Veterans Affairs (VA) hospitals to the two-way television system operated out of the Nebraska Psychiatric Institute, soon expanding into a broader Biomedical Communications Center in the University of Nebraska Medical Center. By 1970 this interactive television grid was linked to all UNMC clinical departments, and within a few years to orthopedics, oncology, physical medicine, and rehabilitation services at nearby St. Joseph's Hospital as well.[33] Likewise, in the early 1970s Bird and telemedical nurse practitioner Marie Kerrigan expanded the initial two-point link between MGH and Logan to incorporate the Bedford VA Hospital and other Boston-area facilities in a broader telemedical network. As telemedicine picked up momentum, many hoped that this new platform of medical practice might not only change how people accessed clinical care, but also fundamentally alter the power dynamics between patients and their doctors.

* * *

American physicians perceived a number of threats to the public image, social status, and self-regulation of the medical profession in the 1970s, an observation that perplexed those who felt medicine had more to offer than ever before. More cures existed for more diseases,

FIGURE 4.10. The imagery of telemedicine also evoked dystopian fears of constant surveillance—as illustrated by one *Boston Globe* article on the MGH Telediagnostic Clinic titled "Big Brother, M.D., Is Watching Over You." Courtesy of Massachusetts General Hospital Archives, Boston.

but not enough patients were accessing care. More money was being spent than ever before on the US healthcare system, but fewer people were satisfied with it. The physician-hero so neatly packaged in television shows like *Dr. Kildare* and *Ben Casey* had become a more flawed figure in popular books like the best seller *Intern*, whose anonymous author, "Doctor X," described the blunders and errors made in hospitals under increasing bureaucratic pressure. "Doctor X" was accused by some of aiding in a more general plot "to destroy medicine's meticulously protected public image."[34] Health feminists, radical health activists (including the Black Panthers and the Young Lords), gay rights activists, disability rights advocates, and other civil rights groups criticized the expertise and authority granted to physicians as a key index of structural inequities in American society. Several public scandals, including the 1972 revelation of the racism, exploitation, and deception at the heart of the US Public Health Service Tuskegee Syphilis Study, suggested that physicians were incapable of self-governance and needed to be accountable to a new field of bioethics beyond the control of the medical profession.[35]

These diverse but intersecting critiques of medical authority disagreed in many particulars. But health activists at the time were aligned in their objections to the expert role of the physician as a symbol of white male paternal authority. Power asymmetries were built into the social spaces of the clinic.[36] In the clinic, the physician wore a suit and a white coat and had the knowledge and power. The patient wore a thin paper gown and was encouraged to acquiesce to the "sick role" and follow the doctor's orders. Over time, a groundswell of patient advocacy groups demanded that this model of care change: patients deserved more of a say in their own healthcare decisions. They insisted on newer, more egalitarian models of doctor–patient communication, with a particular twist: the passive role of the patient became redefined as a more active, empowered consumer of healthcare.[37]

Many hoped that the flexibility of telemedical encounters might play a role in further empowering the patient as a consumer of healthcare. The journalist Ben Park, who had helped found the Alternative Media Center at New York University, began to document the transformative role that telemedicine was poised to play in reshaping doctor–patient relations.[38] His initial 1974 report, *An Introduction to Telemedicine: Interactive Television for Delivery of Health*, was followed by other attempts by social scientists to understand how this new medium might transform human relations in medical settings. These included Joel Reich, a sociologist at Washington University's Program in Technology and Human Affairs; Rashid Bashshur, a health services researcher at the University of Michigan; and Maxine Rockoff, who directed the Health Care Technology Division of the US Department of Health, Education, and Welfare (HEW). Each of them viewed Benschoter's work in Omaha and Bird's work in Boston as the first two steps in a new era of telemedical care. Each of them saw potential, in the age of cable television, for patients to demand new agency in healthcare through this new medium.[39]

Not everyone, however, thought that television should empower patients. In Boston, Kenneth Bird and Marie Kerrigan argued that telemedicine should instead empower *physicians* to maintain the traditional doctor–patient relationship in the face of growing social challenges. As they planned the second "electronic health information highway," intended to link MGH with the Bedford VA Hospital, Bird and Kerrigan argued that television "produces an alteration in time and space which actually *expands* the role of the physician." In a well-designed telemedical interface, they stated, "the fundamental doctor-patient relationship is not only preserved, but often it is actually augmented, enhanced, and seemingly more critically focused."[40] These are not accidental terms. Bird's microwave transmitters *augmented* the signal to travel long distances; he developed *image enhancing* filters for TV signal processing; and the ability to shift between wide-angle and long-lens cameras allowed his teleclinic to be *critically focused.* Properly designed, the television frame could engineer more than "good enough" medicine; it could deliver better medicine, with the doctor firmly in charge. "Telemedicine can provide as much *or more*," they insisted, "than the actual physical presence and direct interviewing of the physician."[41]

This expansive language was provocative, but also vague and inconsistent. It suggested that telemedicine could engineer a win-win situation—patients happier with their care, physicians more in control—but did little to resolve existing concerns of what "being there" meant in telepresence, for doctor or patient. Television medicine added more interactive presence than telephone medicine or radio medicine, but still captured only sight and sound. The absence of touch, of smell, and of the global sensibility we use to navigate interpersonal interactions persisted. The sociologist Joel Reich, in a report on the social dynamics of telemedicine that took Bird's clinic as its principal model, tried to catalog all of the things *not* present in televised encounters in order to describe what, exactly, telepresence truly conveyed. Reich's account of telemedicine is a history of the senses: visual and aural were present, yes, but olfactory, gustatory, thermal, and haptic channels were all missing, and that absence was crucial.

"Until such a time as Smell-o-Vision became a reality," Reich half-joked, "with contemporary interactive television the loss of the olfactory channel is complete." Reich compiled a list of roughly fifty diseases for which the use of smell might still play a part in routine diagnosis. The clinical significance of this loss was likely minuscule, as was the loss of the sense of taste—but he insisted they were losses all the same.[42] Nor was it clear that a nurse practitioner, standing in the same room as the patient, could develop an adequate language for describing odors verbally to a physician on the other end of a telemedical circuit. A similar concern related to the relevance of color. Bird and Krainin's early work on visual acuity for telemedical circuits had assumed black-and-white television was more practical for telediagnosis. When color was relevant, as in some dermatological diagnoses or review of hematological slides, practitioners on both ends could refer to numbered codebooks (analogous to the Pantone Color Scale) to convey the right color. Color could be standardized and rendered legible at both ends of the black-and-white television circuit in ways that smell could not.[43]

These losses paled in comparison to the loss of touch. Some forms of haptic information, like the sensation of hot and cold, could be captured using thermometric sensors and transmitted electronically as graphs, charts, or raw numerical data. Yet many clinicians worried that the single variable of temperature did not contain all of the information captured by a physician's hand on a clammy brow. The haptic channel was also a two-way channel: the touch of the physician was both a sense organ and a means of providing communication, reassurance, a form of therapy in itself. Another hand, perhaps that of a nurse practitioner in the same room with the patient and the television camera, might act as a limited prosthesis for some of these functions—but not all. The television physician would have to find a way to establish bonds of trust without the "co-presence ritual" of the physician's touch.[44]

Reich's sensory catalog captured only part of what was absent from telepresence. But physicians and social scientists interested in telemedicine lingered on the loss of the haptic in particular. Bird recognized that the loss of touch in the doctor–patient interaction was significant, but he was optimistic that there might be a sociological, if not a technological solution to remediate this and other things lacking in telemedical encounters. He also thought the loss of touch could be offset by the many other forms of presence made possible in the electronic age. "There are several uses of telemedicine circuitry," he noted, "in which a modification of the normal co-presence ritual may have to be considered eventually." After all, wasn't our own presence in the three-dimensional world itself a constructed aspect of our shared social reality, a set of etiquettes and protocols that could be re-engineered to work over longer distances? Just as deep-sea divers learned to communicate with coded hand gestures in a benthic environment that did not permit oral communication, doctors and patients could figure out new codes for telemedicine. "The primary requirement" to establishing clinical rapport in telepresence, he counseled, "is careful adherence to reasonable behavioral etiquette, maintenance of the normal confidentiality of a face-to-face exchange and an explanation for any off-camera diversion of either audio or video nature."[45] For Bird and others who believed in the promise of telemedicine, anything that the physician might lose in telepresence could be written back in with the proper scripting and staging of the clinical set.

* * *

If telemedicine lacked certain aspects of presence and communication that might build trust between a doctor and patient, it also amplified the doctor as a remote and disembodied authority. At times the very asymmetries and power dynamics that health advocates thought telemedicine would break down actually became more pronounced in these encounters. In a telemedicine trial project at the Lakeview Clinic in Minnesota, for example, nursing staff meetings at the clinic were presided over by external physicians whose enlarged heads were displayed on a big screen at the head of the table. Clinic staff acerbically referred to the super-sized physiognomies of the consulting physician as the "Face of God," and resolved the issue by taking the TV off its pedestal and putting it on the floor, so that clinic staff could look down upon the physician, rather than up. "When the monitors were placed lower," they reported, "and the images were reduced in size, this incipient problem of deification was alleviated."[46]

Putting a television set on the floor was a minor gesture, perhaps, but Ben Park and Maxine Rockoff were fascinated by the idea that distortions of power relations in the televised clinic could be mitigated by changing the physical locations of cameras and television screens. In June 1973 the pair met in New York with Park's longtime friend, celebrated University of Pennsylvania sociologist Erving Goffman, to study the presence and absence of other established social dynamics in videotapes of televised doctor–patient encounters. Rockoff had read Goffman's seminal work, *Presentation of Self in Everyday Life*, and considered him to be "one of the country's leading experts on person-to-person communication."[47] The conversation lasted for six hours and became a flash point for both Park's and Rockoff's thinking on the topic. It also illustrates some of the key limitations of how social science frameworks of the early 1970s approached the relationship between technological and social structures in explaining the dynamics of medical encounters. Like telemedicine itself, these conversations are as notable for what they leave out of the frame as for what they center within it.

Goffman suggested that a key problem for establishing presence in telemedical encounters was that the two figures (doctor and patient) inhabited different backgrounds. "When two people communicate in the same room," Rockoff recorded in her notes of the meeting, "their 'backgrounds' also communicate and they share the same communication space":

> We have developed a 'code' for handling such things as a third person walking into the room, the need to look at reference material of notes, etc. A new code will have to be developed for television interactions. For example, when a person enters one of the rooms involved in a television interaction, the person in the other room is not party to this interruption and feels neglected by the person who turns away from the television set. A 30-second interruption can seem very long from his perspective. We do not know how anxious this will make a patient if the doctor with whom he is interacting temporarily leaves the screen.[48]

This framework was intuitive and appealing, and promised that telemedicine could be used to build new, better codes of relations between doctor and patient. Goffman's "codes" were part of the invisible fabric of everyday life; they tended to be noticed only once disrupted. He suggested two options for dealing with the disruption of shared space between doctor and patient: one could either change the nature of the camerawork to help reproduce the sense of shared space, or one "may choose to take advantage of the technology to develop new forms for interacting."[49]

The spatial disruption of telepresence, Goffman argued, also disrupted the power differentials built into the professionally controlled spaces of hospitals, clinics, and doctor's offices. Telemedicine undid the inequalities of what Goffman called the "right to touch." In the clinic, the doctor may touch the patient, but the patient may not touch the doctor. In telemedical encounters, he suggested, the field was leveled: neither party could touch the other.[50] "The ordinary relationship between physician and patient in the physician's office," Park added, "is extremely asymmetrical . . . the patient waits at the physician's convenience; the patient does what he is told in the sequence he is told to do it; the physician asks questions of the patient, and he does not expect questions in return unless he asks for them; the physician touches the patient and he does not expect to be touched in return." Telemedicine took the patient out of a space the doctor controlled, and into a space *they* controlled. "In the telemedicine setting," Park speculated, "the situation may be restored to considerably greater symmetry *because* the environment, though shared, consists of two spaces."[51]

Telemedicine advocates in the early 1970s seem to have placed as much trust in the application of social science expertise as in the application of technological expertise. But both could be equally reductive, and suffer from similar forms of hubris and overextension. It is challenging now, from the perspective of a twenty-first-century reader looking back, to understand how the reframing of television backgrounds could be imbued with so much subversive potential by health policy and media scholars in the 1970s. This distance reflects some of the limitations and some of the overreaches of applied social science of the time. Goffman's framework is also notable for how many other forms of social disparities in access to care—for example, cost of care, the relative accessibility of physicians, linguistic barriers, and health literacy—it left out.

Goffman's suggestion that telemedicine would undo the steep power asymmetries between doctor and patient stood in stark contrast to Bird's promise that television would amplify the role of the physician, not diminish it. Bird would have been relieved, perhaps, to learn that other sociologists thought telemedicine would do exactly that. Elliott Friedsen, a leading scholar of professional power and social control, argued that any notion that television had the power to restore symmetry in doctor–patient interactions was both overstated and naïve. Telemedicine, he pointed out, could just as easily escalate these power differentials:

> It seems to me that telemedicine makes it possible for everything to be "prepped" in advance so the doctor can simply come in like a surgeon and do purely functional work. I'm not sure I would argue that there is greater symmetry between doctor and patient under the circumstances. Interaction is restricted, unless some new requirement for an elaborate etiquette were introduced. If I am right, this is indeed a selling point to doctors, particularly the super-specialists who like to make use of every minute to perform their esoteric functions, but I am not exactly sure that this is the best thing in the world for human patient care.[52]

Where Goffman presented telemedicine as a disruptive technology with power to level the steep power asymmetry built in the doctor–patient encounter, Friedsen saw another technical means to enhance social control. Both agreed that there was *something* unsettling within telemedical encounters, for better or for worse. Eye contact, for example, was understood by both medical professionals and medical sociologists to be crucial to establishing rapport in difficult clinical situations. Yet the coupling of a television with a camera presented a paradox: one could look either at the screen or at the camera. To look at the camera was to give the appearance of looking into the eyes of the other, but not truly to look at the eyes of the other. To look at the television was to look at the eyes of the other, but to give the appearance of *not* looking at the eyes of the other. To look back and forth from camera to screen and back again was to give the appearance of anxiety and shiftiness.[53] None of these options helped establish trust, sympathy, and rapport.

Social scientists and technicians converged in proposing engineering solutions to the problems of establishing intimacy in social relations. Perhaps, if the camera were placed behind a one-way screen, eye contact could be restored. What other social dilemmas of doctor–patient relations might find similar technical fixes at the intersection of applied social science and communications technology? In his discussion of the Lakeview Clinic "Face of God" problem—where the problem of looking up at the oversized physician's head was in part solved by looking down at the screen instead, Park invoked another reductive social theory, that of "proxemics," recently invented by anthropologist Edward Hall. Hall developed this term in his 1966 book *The Hidden Dimension*, a popular work on the social study of everyday life that made him, like Goffman and Friedsen, a well-known public intellectual if not a household name. Hall described four scalar distances of social interaction: intimate, personal, social, and public, each with a "close phase" and a "far phase" in which ordinal increase in distance denoted a concomitant decrease in familiarity. These could be coded precisely in units of length: social far-phase, at 7–12 feet, was a space of formal business and social interaction; personal far-phase, at 2.5–4 feet, was an "arm's length" distance for discussing sensitive material; while intimate far-phase, at 6–18 inches, caused visual distortions and eye-crossing.[54]

The choice of wide-angle or close-up lens, the distance between camera and subject, encoded a sense of closeness or distance, intimacy or formality into clinical relationships. As an applied social science, proxemics allowed telemedical advocates to envision new ways to physically engineer the desired degree of intimacy in clinical interactions. "At what *apparent* distance," Park mused, "does the physician cease to be a friendly symbol of help and become over-domineering or even a threat?"[55] As with Goffman's overemphasis on scripting codes of interaction, Hall's invocation of proxemics suggested that substantial issues of trust and distrust in medical encounters could be resolved with proper calculations of focal length and set design.

Hall's theory of proxemics is about as reductive as social theory can get: it reduces all human interactions to a single dimension of physical distance. Measure the apparent distance and you measure apparent intimacy. Even if one accepts (as Hall himself admitted) that this system is based entirely on white, middle-class American social values, this theory still explains very little about how intimacy and trust are established within this subgroup. And other forms of social context, like race, class, ethnicity, immigration status, or personal histories of medical mistreatment, had no place in his system. Yet telemedical advocates found in proxemics a model of social interactions that was useful because it was linear, quantifiable, calculable, and suggested that problems in clinical communication were amenable to engineering solutions.

It seems absurd now to argue that apparent social distance was *really* the chief obstacle to overcoming challenges of trust in the clinical encounter for African American patients who just a year earlier had learned of the publicized abuses of the Tuskegee Syphilis Study. Indeed, the idea that a science of proxemics could engineer better clinical relations through emphasis on focal lengths should be read as a broader indictment of the blinkered view and outsized aspirations of medical sociology in the 1960s, and its relative inability to address structural racism in healthcare. The appeal of proxemics was its reductive power—its erasure of historical structures of inequity and complex intersections of race and class, alongside ethnicity, gender, and sexuality. That such an explicitly one-dimensional social theory would be expected to explain much about clinical interactions at all must be read as a firm reminder that medical sociology can be just as reductive as medical technology.

Nonetheless these kinds of overextended promises of utopian social engineering through the design of technological interfaces suffused the early theorization of telemedicine. Park hoped similar techniques might overcome the skepticism many physicians expressed toward substituting telepresence for face-to-face presence. He lumped physician resistance to telemedicine into four major fears: first, that television would "not provide sufficient capacity to 'do the job'"; second that these systems would take too much time to learn to use; third, that "television (probably because of what is seen on home screens) is a second-rate medium (narrow sense) appropriate only to second-rate entertainment and 'kitsch' intellectuality"; and fourth—but most prominent—the standard complaint that "the doctor-patient relationship would be destroyed."[56]

It was never exactly clear which aspect of this relationship would be destroyed by telepresence. Where physicians warned of loss of intimacy, trust, and rapport, Goffman's and Hall's work suggested that problems of changing power dynamics themselves were central to the discomfort many physicians felt with telemedicine. Park suggested that these problems of asymmetry must be resolved before a new "code of behavior for interactive television" could create agreed-upon scripts for doctors and patients. He compiled a long list of awkward interactions in the telemedical milieu, including scenes like this one:

> Two physicians are jointly examining a patient. They arrive at a point where they feel they no longer need to converse with or look at the patient, and turn to talking with each other. They do not excuse themselves. They simply turn abruptly away. During the course of their conversation the image on monitor screens is an extreme close-up of the patient's face. Although the physicians are behaving as if the patient weren't there, the patient is very much there and growing painfully aware of the enforced and awkward presence.[57]

Relating this to Elliott Friedsen, Park was surprised that the sociologist did not see this as in any way unique to the television medium. "What you referred to with some shock," Friedsen replied, "is something that occurs routinely on hospital floors while physicians are making rounds."[58] Where Park saw the television camera as a distorting lens, Friedsen saw it as providing a clear view of behaviors that took place every day behind closed doors. The presence of the camera simply made visible the abuses of power already at work in hospitals and other clinical settings. Doctors had been so accustomed to these displays of power that they treated them casually, even without thought—unnoticed until a television camera made them visible to others.

Without the camera, toxic social relations flourished unseen, unchecked, unnoticed. Hospitals and clinics were designed to give power to the clinical gaze and to strip personhood and agency away from the patient. A similar conversation about technology, absence, and presence has recently developed around cameras and the social spaces of policing in early twenty-first-century United States. The absence of body cameras, dashboard cameras, individual cellphone cameras, or other forms of videorecording allowed racially directed abuses of police power to be routinized in traffic stops over the course of the twentieth century. The increasing proliferation of cameras in the early twenty-first century did not stop these practices, nor did it stop the behaviors of those professionals long accustomed to controlling the social dynamics of these spaces. Yet the presence of these cameras *has* changed the visibility of the problem and allowed a broader scrutiny of these occupational abuses of power in daily life.

The presence of live television cameras in clinics and hospitals exposed these professionally controlled spaces as well. It should come as no surprise that physicians came to resent, and consequently to resist, the presence of the television camera. "Telemedicine is perhaps correctly viewed by many physicians as a threat because they are vulnerable to inspection by others," Park observed. A video record of the doctor's visit "becomes an unwanted form of peer review and conjures up whatever misgivings physicians may have about admitting someone else into what is often perceived as a private preserve."[59] "The point here, however," he concluded, "is that if telemedicine is to help in alleviating some of the problems faced right now in health care, initial resistance on the part of many physicians will have to be overcome prior to the emergence of the 'wired nation.'"[60]

* * *

Reba Benschoter's presentation at the New York Academy of Sciences on that hot summer day set in motion a flurry of speculative development and social theorizing about telepresence and health. Most immediately, this conversation was picked up by a pair of medical futurologists who followed her talk. The first imagined the greater efficiencies of time and money for physicians and patients that could occur once all everyday medical care could take place by television and other telecommunications devices. The other predicted we would soon communicate by television as readily as by telephone, and that by 1990 everyone would carry a "picture gun" and a "personal viewer" in their briefcase or purse that would allow them to record and transmit audiovisual connections wherever they traveled.[61] This technological speculation has also come to pass: anyone who owns a smartphone, tablet, or laptop today can testify to the ubiquity of telepresence technologies. But the sociological speculation—that such technologies could be used to engineer better relationships between doctors and patients—has not been realized.

As interactive television offered a powerful new tool to augment the reach of the physician, it also offered the possibility to empower patients to level the playing field in doctor–patient communications. Suffice to say this development was not embraced enthusiastically by all parties. Doctors, nurses, and patients held different positions of power in the closed spaces of the clinic and hospital before TV cameras made them visible to remote audiences, and they had correspondingly more to lose (or gain) through restaging of clinical interactions in the television frame. The technology of telemedicine did not simply empower doctors or nurses or patients, and it certainly did not grant power in equal measures to all parties. The subversive potential of telemedicine—at least in the 1960s and 1970s—was considerably less than was hoped or feared.

Attempts to reframe power relations in the clinic through telemedicine met substantial resistance even among its supporters. Where Kenneth Bird, from his hospital base, clearly saw telemedicine as a means of augmenting and extending the professional power of the physician, Reba Benschoter hoped that telemedicine would serve more to empower patients and their families. Even among those social scientists who hoped that telepresence would offer new tools to empower patients, many thought it naïve to simply hope that telecommunications engineering would undo the steep power dynamics built into clinical medicine. Indeed, the question of whether telemedicine would disrupt or augment the authority of physicians seemed to boil down to a question of who controlled the design and implementation of these systems.

Like most police in early twenty-first-century America, most physicians in late twentieth-century America did not want to invite cameras or videotape into spaces of power relations where previously their word alone had conferred an authoritative advantage. For a long while, they would continue to be successful at keeping cameras out of their clinics. When documentary filmmakers did manage to bring cameras into medical spaces in the 1960s and 1970s (as Frederick Wiseman did for both *Titicut Follies* [1967] and *Hospital* [1970]), Friedsen's comments held true. What is most fascinating about these films is how physicians' displays of paternalistic behavior do not change even when they know they are being watched.

As he struggled to frame how interactive television might shed light on the nature of presence and absence in American medicine, Ben Park realized most of these studies assumed a "typical" American patient. Much of the discussion of the doctor–patient relation among sociologists, physicians, and policymakers to date had implicitly or explicitly framed the doctor–patient interaction assuming a racially unmarked "typical" American consumer. Not only did this restrictive view lead to bias in the cultural dimensions of telemedical practice, including the "software" assumptions of individual psychology based on "middle-class white Americans of European extraction," but also was written into the physical "hardware" of the telemedical system itself. The known red bias of the standard Plumbicon television camera, for example, which favored light skin tones over dark skin tones, detected inflammation on light skin with a different sensitivity compared to detection of inflammation in darker skin tones.[62]

Though he failed to critique the fundamental hubris at work in suggesting that the one-dimensional studies of "proxemics" could engineer better human relations by television, Park understood that there were limitations to the sociology as well as the technology in play in these early dreams of telemedicine. It was naïve to suggest that telemedicine would empower all users in the same way, when the hardware and software of these systems themselves favored certain users over others. If telemedicine were to spread throughout the American healthcare system, Park warned, discrepancies in health literacy and technological literacy across American racial, ethnic, and socioeconomic divides might themselves exacerbate differential access to a telemedical system designed with white, middle-class bodies, minds, social groups, and behaviors in mind. "If cultural and ethnic differences can, and indeed, often do, make for difficulties in the face-to-face interactions between physicians and patients," he included, "what are the ways in which televised interaction may ameliorate or exacerbate those difficulties? Does the interactive medium have potential for creating *different* difficulties?"[63] More pointedly, he speculated, "inner city" Latino and Black populations, along with American Indians living on underfunded rural reservations, might face different challenges and opportunities in adapting to the new medium of care.

Looking beyond the standard white, middle-class patient who was all too often the assumed user of medical technology, Park joined others who wondered whether the new technology of telemedicine would affect not just interpersonal but also populational divides in American society and disparities in American healthcare. Just as telemedicine could be imagined as a tool to level power asymmetries between doctor and patient, it could easily be imagined as a tool to erase health disparities across divides of race, ethnicity, class, and geography. Then again, it might do exactly the opposite, and serve as a tool to amplify existing power differentials.

As telemedical systems grew beyond single links connecting doctor and patient, to become the nervous tissue of a new projected system for primary care and public health, many began to hope television would improve access to healthcare, but braced themselves for the possibility that it might simply improve access for those who already had it.

5
THE WIRED CLINIC

"Whatever else it may witness," Dr. Roger Egeberg warned in the first weeks of a new decade, "the nineteen-seventies will see the most severe test of America's capacity to deliver decent health care to those who need it."[1] Serving as assistant secretary for health and scientific affairs in the Nixon administration, Egeberg faced a bitter paradox: as medicine did better, health disparities got worse. Perversely, as medical technology offered lifesaving diagnostic, therapeutic, and preventive abilities unimaginable a few decades earlier, the very success of these interventions *widened* the gap in morbidity and mortality between those who had access to care and those who did not. With the recent enactment of Medicare and Medicaid, an increasing number of Americans were now seeking healthcare from an already strained framework of primary healthcare providers. And healthcare costs were soaring—increasing at twice the rate of inflation with each passing year. Before the oil crisis, before the economic crisis, the dawn of the new decade brought with it a self-conscious recognition of a nation in healthcare crisis—a term whose urgency, though born in the 1970s, has not subsided since.[2]

In this critical context, the technological platform of telemedicine took on immediate practical value well beyond Boston and Omaha. Kenneth Bird and Reba Benschoter had shown that if you had two televisions, two cameras, and a way of connecting them, you could connect individual doctors with individual patients across hundreds of miles—at relatively little cost. By 1973, Bird was ready to move beyond his initial two-hospital system to create municipal and regional telemedical networks, with hopes to eventually "mold it into a nationwide interactive health telecommunications system."[3] Where telemedicine links in the 1960s had promised to level power differences between individual doctors and patients, telehealth networks in the 1970s now promised to level population disparities in access to healthcare.

American health policy had placed the hospital at the center of the public health system for decades. The Hill-Burton Act, the largest single piece of federal spending on healthcare to date, disbursed $28 billion between 1948 and 1975 to support more than 10,000 new hospital projects.[4] These funds supported the construction of hospital beds and intensive care units and operating rooms, they funded radiology suites and chemotherapy infusion centers, and they concentrated access to these diagnostic and therapeutic advances to a set of increasingly high-tech medical centers. By the 1970s the limitations of the hospital-centered approach had become all too clear. Primary care physicians objected to resources being funneled into technologically driven medical specialties at the expense of general practice. Minority health activists argued that mainstream hospital-based medicine failed to help, and often actively harmed, urban communities of color. Policymakers from rural areas objected that the concentration of hospital centers and medical specialists in urban areas left their constituents behind. Public health professionals argued that overinvesting in medical technology diverted public funds from addressing more significant economic and social determinants of health. All agreed that the current system provided high-cost, high-tech care to a small fraction of the population, when it should be providing essential care to everybody.[5]

Even elite physicians like John Knowles, former president of Massachusetts General Hospital and head of the Rockefeller Foundation, became vocal critics of the overreliance on technology in American medicine. "We have emphasized high-cost, hospital-based technologies," Knowles reflected in his 1977 book, *Doing Better and Feeling*

Worse, "to the neglect of other services where the benefits are much greater relative to the costs incurred."[6] Yet Knowles and many others still held out hope that investing in the *right* technology—a technology that shifted the context of care and prevention outside of the hospital and empowered communities to take a more active role in their own healthcare delivery—could provide meaningful improvements in public health and help cut growing costs of American healthcare.[7] To Knowles, and to many other would-be reformers of the era, that technology was cable television.

Unlike conventional television, which had concentrated power over content to a handful of nationwide broadcast networks, the new platform of cable TV offered an interconnected system in which every television set could potentially serve as both receiver and transmitter. Cable promised to decentralize the flow of information in American society, and was projected to have an impact on a par with the invention of the printing press. "The stage is being set for a communications revolution," journalist Ralph Lee Smith wrote in 1970, as "every home and office will contain a communications center of a breadth and flexibility to influence every aspect of private and community life."[8] Once the power to create and disseminate television content could be put into the hands of anyone, new flows of information would empower previously marginalized communities. Or at least that was what the backers of cable technologies said. Smith was not convinced.

Information technology promised a means to knit together the divided fabric of American society. It promised to integrate marginalized communities through interactive telecommunications, without needing to engage with the messier roots of structural racism that created and maintained those divisions. Just a few years earlier, the report of President Lyndon Johnson's National Advisory Commission on Civil Disorders (known as the Kerner Commission) had exposed in granular detail the systemic inequalities in education, policing, employment, housing, and welfare that underlay the social divisions within American cities, and called for large-scale federal spending to redress them. Instead of following any of these recommendations, however, the Johnson administration immediately commissioned a new Task Force on Communications Policy to evaluate the role that interactive satellite and cable television might play in healing America's urban crisis.[9] "If it is true," Smith concluded in his 1970 exposé on the topic, "as it seems to be, that cable TV is about to effect a revolution in communications, it must be said that a revolution has rarely been created by persons of less revolutionary intent."[10]

John Knowles may have been an unlikely revolutionary. But revolutionary *technologies* appealed to him. Knowles had been an early supporter of Bird's television experiments while at MGH. He believed the Rockefeller Foundation's leading position in medical and public health philanthropy could broaden applications of telemedicine as a platform for community health.[11] Interactive television, he argued, could extend the benefits of modern medicine to marginalized and underserved populations that otherwise did not have regular access to hospital or clinic sites.[12] After a 1972 Federal Communications Commission (FCC) decision ruled that all future cable TV systems must include interactive features, Knowles imagined telemedicine would only become more common, more affordable, and more practical as a basis for healthcare delivery.[13]

Telemedicine was a technological solution to the healthcare crisis. Cable could connect physicians concentrated in affluent urban and suburban centers to underserved communities of the inner city and rural patients alike. In Washington, DC, health services researcher Maxine Rockoff dedicated research funds toward seven "exploratory broad-band communication experiments" within the new Health Care Technology Division of the federal Department of Health, Education, and Welfare (HEW). By the end of the decade, the federal government sponsored at least fourteen telemedical demonstration programs across the country, each designed to showcase the use of interactive television to counteract a different social disparity in healthcare access. These programs targeted a wide variety of underserved populations, from rural white settlements in remote areas of Vermont and New Hampshire to inner city Latino and Black communities in Harlem and the West Side of Chicago. Other projects explored the role of telemedicine in reducing the steep healthcare disparities found across Inuit and American Indian reservations from Alaska to Arizona, and the rapidly industrializing island population of the "unincorporated territory" of Puerto Rico. Still other pilot programs focused on the internal disparities caused by the American carceral state, as in the growing prison population of the Miami–Dade County Correctional System.[14] The goal of each demonstration program was to provide evidence of feasibility, efficacy, and acceptability: to lay the groundwork for nationwide telemedical networks.

The sheer scope of these demonstration projects shows just how much liberal reformers believed that interactive television could reshape social relations. Each project united a group of medical and public health figures with community activists who hoped this new technology might have the power to produce greater equity in access to healthcare.[15] Yet each project was limited from the start. Telemedicine might connect more patients to more practitioners, but even when these systems worked to expand access to care, they could not deliver the long-term changes they promised. Although many of the urban and rural populations that served as test subjects in these projects hoped that this new technology would improve their own access to healthcare, they had learned to view such projects with skepticism. They had seen many promises come and go.

Yet even cable skeptics acknowledged the allure of a technological fix to the complex social problem of access to care. One 1971 dissent submitted to the Sloan Commission on Cable Communications complained that enthusiasm for telehealth involved a naïve optimism that "given the proper technology our health problems will disappear." Yet even if he knew that "obviously they will not," he still hoped that with the right technology, "we may become better equipped to cope with a few of them."[16] The story of early telehealth networks is a complex chapter in the convergence of healthcare reformers and marginalized communities to search for technological solutions to complex social problems. That these demonstrations failed to resolve healthcare disparities might, in retrospect, seem inevitable. But understanding how the promise of a technological fix continues to hold enduring appeal for many who should know better—this puzzle is every bit as urgent now as it was then.

* * *

When we think of cable today—if we think of it at all—we picture a set of monopolies and near-monopolies that control vast swaths of broadband through integrated telephone, internet, and television bundles. These firms have achieved a high degree of capture over the regulatory agencies assigned to rein in their market power. They have molded much of the structure of telecommunications policy to their liking. Yet in its early days, cable television was hailed as a disruptive, community-based technology, a site of fevered speculation on the possibility of decentralized communications networks. Public debates over the future of cable television in the early 1970s were part of a broader conversation about the potential for information technology to empower and integrate isolated and marginalized populations into a more inclusive American society.

The communications platform now known as cable began as a quick fix to problems of patchy access to broadcast media: first in rough terrain where mountains and ravines disrupted the reception of television antennae, and then in urban spaces where concrete canyons presented similar difficulties. In 1949, on opposite sides of the continent, enterprising television users in hilly areas of rural Pennsylvania and Oregon improved their on-and-off TV reception by placing collective antennas high on mountaintops to catch the faint signals coming out of distant cities, and then running cables from the antenna receivers down to the reception-poor valleys below. By the late 1950s and early 1960s, similar configurations took shape in urban environments: a single antenna on top of a tall building became the nexus for a web of wires that connected sets on lower floors and in smaller buildings below. As community antenna television (CATV) networks developed in more regions, enterprising television producers began first to disseminate their own content through these networks, and then to explore the possibility of more interactive communications between CATV users. Hook a camera up to your TV set and the cable could become a space for social interaction and community formation as well.[17]

At first, broadcast television networks like NBC, ABC, and CBS, as well as the regulatory FCC saw little harm in a cable television market that served only 150,000 subscribers at the margins of mainstream media. But as CATV subscriptions jumped by an order of magnitude over the next decade, and cable providers made inroads into larger markets, broadcasters perceived a threat to their market dominance. The Big Three television networks pressured the FCC to issue a nationwide moratorium on new cable television systems, ostensibly to study the social, economic, and political benefits of this new communications platform.[18] The FCC complied. During the freeze, between 1966 and 1972, at least five major commissions evaluated the societal benefits of cable television, including President Johnson's Task Force on Communications Policy, New York mayor John Lindsay's Advisory Task Force on CATV and Telecommunications in New York City, the National Academy of Engineering's Committee on Telecommunications, the Sloan Commission on Cable Communications, and a MITRE Corporation/Urban Institute Symposium on Urban Cable Television.[19] In retrospect, the moratorium allowed existing networks to hit the pause button and figure out how to reconsolidate their own control over television markets—even though its public face was a series of elaborate speculations on the democratizing potential of cable as a common good.

Cable offered new access to information and services to two very different demographics at the same time. It had the potential to combat social isolation in rural areas that were rapidly depopulating, and to bridge the marginalization of urban neighborhoods marked by cycles of racial demarcation and municipal disinvestment in the peak era of suburban white flight. In the years of the FCC moratorium, terms like "ghetto" and "inner city" became increasingly racialized, and increasingly used as a shorthand for public understandings of an unfolding urban crisis. Minority-majority neighborhoods had suffered visibly in the aftermath of civil unrest that flared in the Watts neighborhood of Los Angeles in 1965; in Baltimore, Detroit, Newark, and other cities during the "long, hot summer" of 1967; and again after the assassination of the Reverend Martin Luther King, Jr., in 1968. When the Kerner Commission presented its powerful report that year, it considered isolation—including media isolation—to be a key feature in the social structures, economic disparities, and political tensions that continued to inscribe racial difference into the fabric of American cities. The disconnection of the inner city was not a natural feature of urban geography, but a product of broader societal forces that demarcated and segregated the terrain. "White society is deeply implicated in the ghetto," the Kerner Commission concluded. "White institutions created it, white institutions maintain it, and white society condones it."[20]

In the wake of the Kerner Commission report, several cable enthusiasts hoped interactive television might offer a new tool for civic engagement that would increase civic participation and community engagement among urban minority populations, and weave the "Black ghetto" back into the fabric of city life. "With cable television the Blacks (and many other minorities) continue to dream of the new day dawning," FCC commissioner Nicolas Johnson pronounced in 1972, "a new day when the monopoly of the entrenched broadcasters will be dissolved in CATV's economy of abundance." In this new era of participatory television, he continued, "any man with an idea will have access to the communications system; any man will be able to talk back to his television set."[21] Advocates of participatory TV promised that community control of this "electronic superhighway" would bring equity in education, housing, transportation, emergency services, and crime prevention. Healthcare, especially, was singled out as a particularly important area for intervention.[22]

In retrospect, it seems extremely naïve to suggest that an information technology could reverse the accelerating vector of postwar urban segregation, a process whose origins can be traced in racially restrictive covenants, federal redlining maps, and brutal policies of urban renewal that systematically circumscribed and denied services to the Black communities they pathologized.[23] All the same, at the time, Kenneth Bird's telemedical experiments in Boston and Reba Benschoter's telepsychiatry link in Omaha were put forward as models for a new kind of healthcare system that would better integrate previously segregated communities of patients. By 1973, more than 2,500 medical "transactions" had been successfully carried out through Bird's expanding telemedical network in the Boston area.[24] In Omaha, Reba Benschoter and Cecil Wittson expanded their initial two-hospital link between the Nebraska Psychiatric Institute and the Norfolk State Mental Hospital to a broader set of connections involving all Veterans Affairs (VA) hospitals in Nebraska. In 1968, the National Library of Medicine's new Lister Hill Center for Biomedical Communications reached out to Dean Siebert at Dartmouth College to provide connection between the Dartmouth-Hitchcock Medical Center and the Claremont General Hospital. This two-hospital link soon expanded into an interstate network connecting several institutions across Vermont and New Hampshire.[25]

Federal funding for community telemedicine was not limited to HEW: it quickly spread to defense and aerospace agencies as well. Defense contractors had already landed lucrative contracts translating counterinsurgency techniques developed in Vietnam into domestic policies for anticipating and preventing civil unrest in urban America. Aerospace contractors and the National Aeronautics and Space Administration (NASA), under pressure from the Nixon administration to show how everyday Americans benefited from public investment in the space agency, now made the equally bold promise that high-tech devices designed for outer space could address the health disparities of the "inner city."[26]

The Lockheed Missile and Space Company worked with NASA in the late 1960s to develop a system to monitor the health status of astronauts on long-haul missions to Mars or beyond. In the early 1970s they issued a series of reports for the Nixon administration that highlighted their telemedical technologies as a new basis for connecting urban, rural, and suburban health systems. NASA and Lockheed theorized that any health system could be conceived of in abstract spatial terms. The smallest unit of primary care (which they called an Area

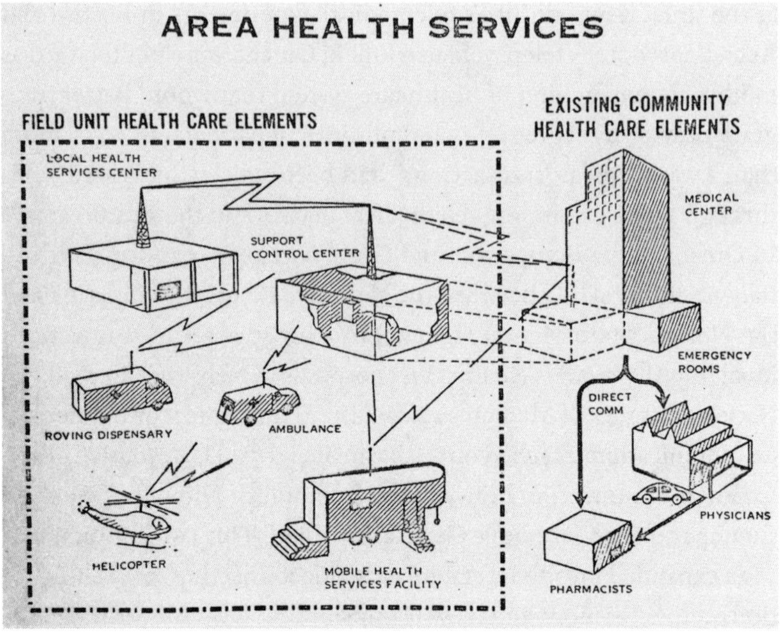

FIGURE 5.1. NASA's Area Health Services Field Unit (AHSFU) concept, using telecommunications to connect community primary care with secondary and tertiary care facilities, as developed with Lockheed Missile & Space Company. Courtesy STARPAHC Collection, Arizona Health Sciences Library, University of Arizona, Tucson.

Health Services Field Unit, or AHSFU) could be linked to institutions of secondary care (local hospitals, with basic laboratories, operating rooms, and radiology facilities), which in turn could be linked to still larger tertiary care facilities (with subspecialty practices and intensive care units for the care of complex patients). NASA-developed telecommunications systems and systems engineers could help ensure that any individual seeking care at any point would have access to the full benefits of an integrated healthcare system (fig. 5.1).

In a large rural area, a single AHSFU (fig. 5.2) might consist of a dozen Local Health Service Centers wired to one another by cable television and connected to a series of Mobile Health Service Facilities by microwave transmission or satellite links. In a "medically deprived urban area" (fig. 5.3), a Field Unit might radiate outward from a central medical center through a series of cable-ready community

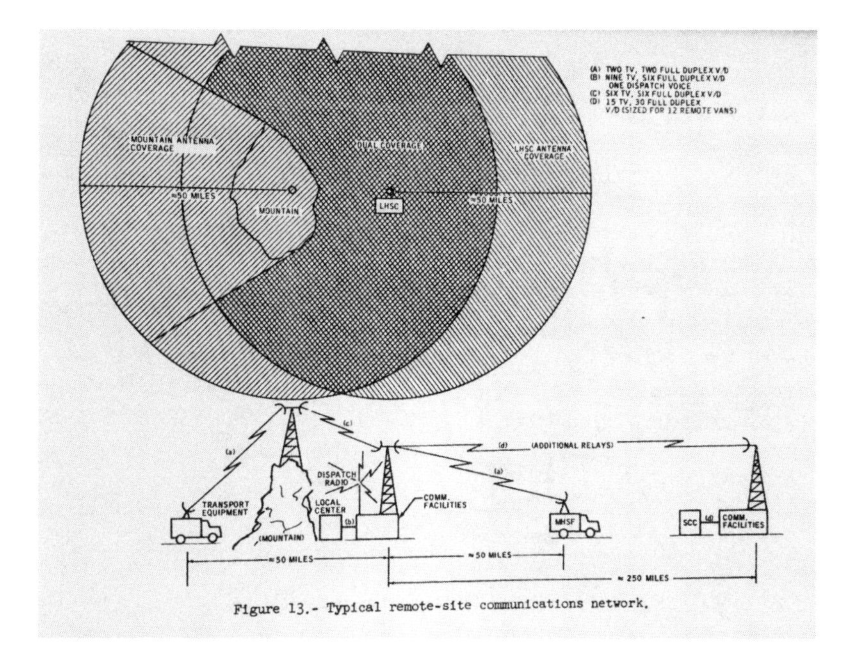

FIGURE 5.2. "Typical remote-site communications network," from NASA's 1972 report on the Integrated Medical & Behavioral Laboratory Measurement System (IMBLMS) Area Health Services Field Unit project. Courtesy of History of Medicine Division, National Library of Medicine, Bethesda, MD.

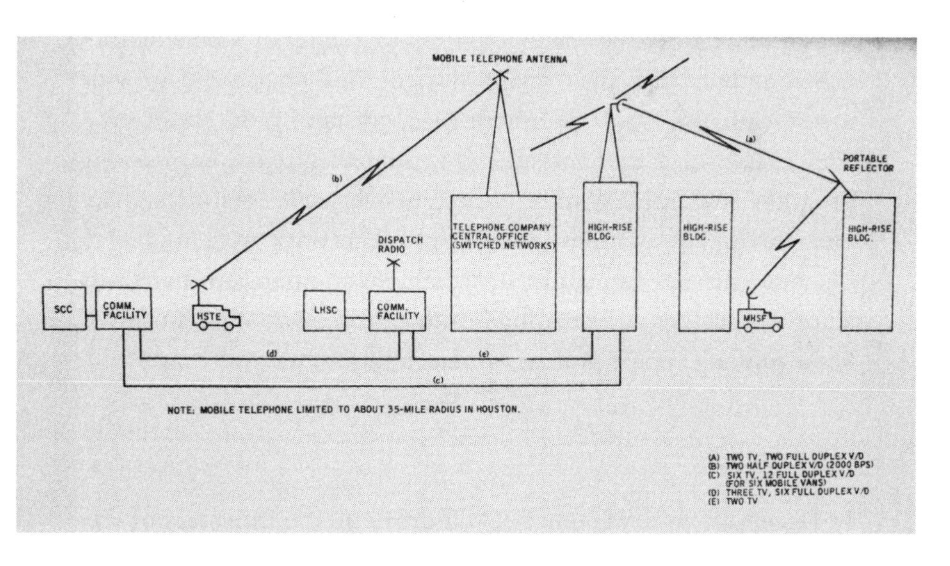

FIGURE 5.3. "Typical inner-city site communications network," from NASA's 1972 report on the Integrated Medical & Behavioral Laboratory Measurement System (IMBLMS) Area Health Services Field Unit project. Courtesy of History of Medicine Division, National Library of Medicine, Bethesda, MD.

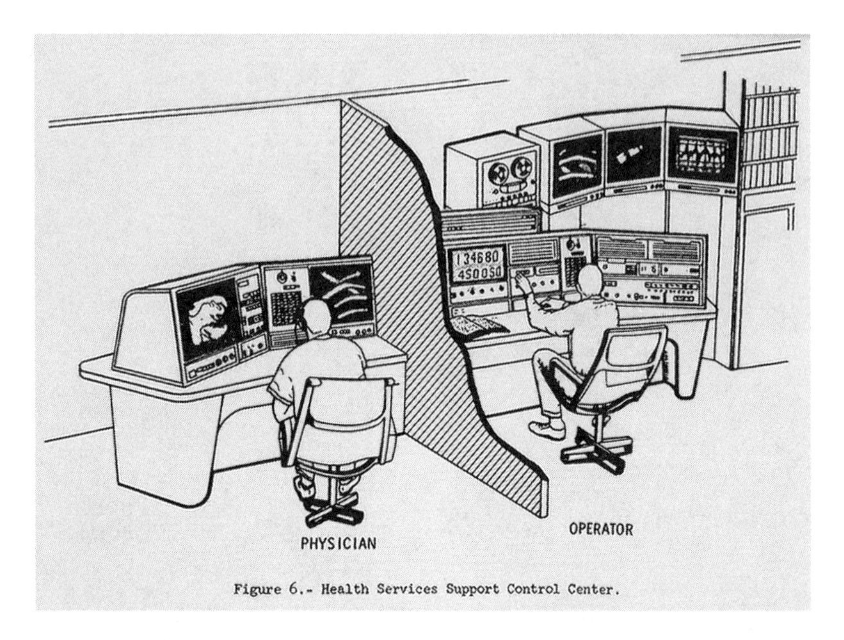

FIGURE 5.4. Physicians working alongside operators in prototype telemedical booths, from NASA's 1972 report on the Integrated Medical & Behavioral Laboratory Measurement System (IMBLMS) Area Health Services Field Unit project. Courtesy of History of Medicine Division, National Library of Medicine, Bethesda, MD.

health centers that also had radio links to a series of Mobile Health Units, ambulances, and "roving dispensaries" that could provide medications (fig. 5.4).[27] Kenneth Bird, consulting on NASA projects, called for a new "massive government-initiated engineering program" within HEW to be created to oversee the testing and implementation of a nationwide telemedical network, with funding to "facilitate the emergence of telemedicine as an extension of virtually all of our current and pending health care delivery mechanisms."[28] Some funding appeared almost immediately.

* * *

In December 1971, Maxine Rockoff drove up the Interstate 95 corridor to Parsippany, New Jersey, to visit AT&T and Bell Labs with her supervisor, Bruce Waxman. Rockoff was committed to studying the role of information technology in expanding access to healthcare. She held a doctorate in mathematics and was one of the few women involved in designing early experiments in medical computing at the University of Pennsylvania in the 1950s. By 1971 she had moved to the Washington, DC, area to direct a small program on healthcare technology assessment in the Department of Health, Education, and Welfare. There she worked with Waxman to assess the value of new information technology in medical practice.[29] Along with four technical consultants, the pair were going to witness a demonstration of Bell's much-vaunted Picturephone, a two-way television device that communicated image along with a telephone call using standard telephone wires.

This was the first in a series of hundreds of trips Rockoff would make over the course of the 1970s as chief of the Logistics Unit of the Health Care Technology Division (HCTD) of the National Center for Health Services Research. The Logistics Unit was looking for "clinical impressions": concrete demonstrations that could document the value of communications technologies to improve access to healthcare. Funds were limited, but they could support promising pilot projects. By the end of the decade, Rockoff had directly administered roughly a million dollars of HCTD funding for seven central sites split between urban (Chicago, New York, Cambridge, and Cleveland) and rural locations (New Hampshire, Vermont, Minnesota). Her extended travels took her farther afield to assess other federally funded projects, such as those supported by the Regional Medical Program in rural Maine, by the National Science Foundation in Miami–Dade County, by NASA in Alaska and Arizona, and others. Over the course of the decade she helped convene a series of telemedical conferences in Boston, Ann Arbor, Miami, Chicago, Washington, DC, and Tucson. Rockoff's work helped frame expectations of what a national plan to develop telemedicine systems for community health might look like.[30]

The seven pilot projects initially funded through Rockoff's Logistics Unit roughly addressed two kinds of barriers to access: the social geography of the "inner city" (attributed to racial and ethnic concentration, isolation, lack of dependable public transportation, and economic barriers), and the physical geography of rural scarcity (attributed to lack of sufficient density of specialists and necessary equipment across vast distances and challenging physical terrain). In both cases, the prospect of accessing medical care through interactive television promised new forms of social engagement to communities. Telehealth, they hoped, could cure the problem of "medical ghettos."

While the polymath sociologist W. E. B. Du Bois had appropriated the term "ghetto" as early as the 1930s to describe the concentration and segregation of Black neighborhoods in American cities, in the first half of the twentieth century the term was more commonly used to describe first Jewish and then other well-demarcated ethnic enclaves in urban life. By the end of the 1960s, however, "ghetto" had become synonymous with largely Black and Latino neighborhoods whose infrastructures and tax bases had eroded in the decades of white flight. As the term became specifically associated with Blackness, its meaning also shifted from a purportedly value-free sociological category to a pejorative one. This slippage is part of what the writer James Baldwin was referring to in 1965, when he claimed that depicting Harlem as a ghetto was "one of the most damaging misuses of a concept that has ever come about in the United States." So, too, with the term "inner city."[31]

White liberal healthcare reformers (and many African American and Latino colleagues as well) understood the poor health outcomes in these "inner city" neighborhoods to be the result of a long series of structural forces and embedded racism. Most of them had read the 1968 Kerner Commission report alongside the 1969 report of the Harvard Medical School Commission on Relations with the Black Community; several would also have been aware of an influential book called *Medicine in the Ghetto* recently published by the Harvard cardiothoracic surgeon John C. Norman. The Harvard study revealed that the average physician practicing in Boston's "predominantly Black ghetto" of Roxbury was sixty-six years old and "in one out of two cases, a graduate of a medical school which no longer exists." This was no accident. A coalition, led by elite medical schools, the American Medical Association, and the Carnegie and Rockefeller Foundations had shut down all but two medical schools that trained African American physicians in the first half of the twentieth century.[32]

Data from Chicago now showed a 200 percent difference in preterm births and rates of tuberculosis in poor Black neighborhoods compared to affluent white ones, with even more striking differences in the rates of venereal disease and cervical cancer. As Lloyd Ferguson, the first African American physician to be appointed an assistant dean at the University of Chicago Medical School, noted, "A dual system of health services delivery exists in Chicago—one for the poor or near-poor black ghetto dweller, another for the more affluent white suburbanite."[33] Insufficient access to medical care compounded these disparities. "Because of factors such as poor or non-existent transportation, inability to afford care, health care provider hostility, etc.," the physician and information technology scholar Joel Reich concluded, the average inner city resident "is unlikely to leave the ghetto to obtain health care."[34]

Scholarly analyses of the "medical ghetto" as a problem for medical sociology and health policy circulated in the late 1960s, but often at arm's length from direct experience. It is worth pausing to note the confidence with which Reich, then based at the University of Washington, could make these causal claims between poor health and poor access to healthcare in Chicago based on his reading of data from these committee reports. Reich correlated "the lower level of health quality in ghetto and rural areas" to lack of access to primary healthcare.[35] Many physicians working in urban academic medical centers would have been operating from a similar background in the early 1970s. Many physicians who identified with principles of race liberalism were shaken by the violence of the 1968 uprisings and the recognition that many communities of color living in the shadow of their teaching hospitals regarded their medical neighbors with suspicion and animus.

As John Norman told the group of physicians and civil rights activists gathered in the idyllic New Hampshire resort of Wentworth-by-the Sea to discuss the problems of the "Medical Ghetto," any conversation about ghettos in 1969 needed to address the problem of systemic racism. "Reduced to essentials, the prototype ghetto today, urban or rural, is a Negro community with all of the attendant social illnesses of economic, political, moral, and medical exploitation."[36] Emphasizing the fundamental Kerner Commission conclusion that racism—rather than poverty per se—was the basic cause of the health disparities of the American ghetto, Norman repeated a quote from one commission member: "we've got to see it as a system and attack it as a system."[37]

Norman had convened this meeting. He was one of a small number of Black physicians who had graduated from Harvard Medical School by 1969, and despite his own professional success, he understood himself to be directly bound up in systems of institutional racism in ways that were vexing in multiple ways. "The 'system' has permitted perhaps one but never more than three American Negroes to graduate from Harvard Medical School in any given year," he continued, "and has allowed virtually none to train for specialization in any of its 18 affiliated teaching hospitals. None has remained to practice in Boston where the ghetto population has now reached 85,000."[38] Norman also chafed at the limits of the Harvard Commission on Relations with the Black Community, on which he nonetheless agreed to serve.[39] While "society's usual reaction when faced with problems of great magnitude is to form commissions or committees to attempt definition, study, analysis, and consultation," he concluded, "in the preoccupation with theories, guidelines, solutions, and programs, regardless of how well-intentioned these efforts may be, it should be remembered that there are *people* in the ghetto to whom dignity and pride are as important as they are to those who grandly assume the responsibility for ordering the lives and habits of others."[40]

Norman was addressing a motley collection of attendees that included deans of elite medical schools; heads of civil society foundations, medical associations, and pharmaceutical firms; prominent journalists inside and outside of the African American community, including the editors of the *Boston Globe*, *Newsweek*, and *Ebony*; representatives of the Urban League; and noted community organizer Saul Alinsky. Over a few days, a heady mix of Frantz Fanon and Ivan Illich shared the stage, sometimes more and sometimes less amicably, with indexed and tabulated health services data and quotations from the Moynihan Report on the pathologization of the "Negro Family."[41] Several African American conference participants described the additional burden of reassuring their colleagues that talking about racism in medicine need not be a direct threat to their own racial or professional privilege. As Charles L. Sanders, managing editor of *Ebony Magazine*, asked the largely white physician audience:

> Should I, as a black man, now recite once again the familiar litany of condemnation? Should I call you all "honkies," disrupt your speeches with a demand that Harvard Medical School and the Boston Globe and the National Center for Health Services Research and Development—and, oh yes, the AMA—give tomorrow, or at least next week, say, $100 million toward an effort to stop black babies from dying? Should I perhaps set afire this magnificent conference hall? I can assure you that I shall do none of these things. But I think that you ought to know that there are some blacks, some far less disciplined ones, who are not so comfortable as I, who would just love to be with us here at beautiful Wentworth-by-the Sea and who would do at least one, or two, of the things that I would not do. I wonder if any of them have sneaked in?[42]

Were American physicians really ready to talk about racism, Sanders wondered, as something endemic not only in the lives of their patients but in the structure of the medical profession itself? If not, then this conference was just another conference producing another report and little more.

James Baldwin's beloved Harlem became an iconic site both for demonstrating the health disparities of the "medical ghetto" and for imagining technological solutions to resolve them.[43] When New York mayor John Lindsay apologized in 1969 that "no service of the City Government was allowed to become so inadequate and obsolete in the last 20 years as municipal health services," he added that telecommunications offered a new means to right this wrong.[44] Harlem Hospital, the center of a vibrant community of Black physicians, had suffered from severe Medicaid cutbacks in the early Nixon years, generating problems with continuity of care, long waits for appointments, and overworked staff.[45] Later that year Harlem became a central focus for the citywide $12 million New York State Ghetto Medicine Program, which contracted private voluntary hospitals to provide care for urban residents affected by Medicaid cuts.[46] At Mount Sinai, a private hospital in East Harlem that had just opened a new medical school, Kurt W. Deuschle used this moment to revisit what it meant to think of Mount Sinai as a "ghetto hospital."

The child of German immigrants, Deuschle began his medical career in 1954 working for the US Public Health Service at the Division of Indian Health hospital at Fort Defiance, Arizona. Working with the Cornell physician Walsh McDermott to address steep health disparities among the Navajo, he came to think of the Navajo Reservation as a "Third World within the continental United States." Indian lands were as underdeveloped in access to medical care as they were in economic investment and social integration. He saw a similar situation in the health status of Harlem, in particular, and urban America, in general. In 1969, inaugurating Mount Sinai's new Department of Community Medicine, he invited his mentor to speak about technology and community health.

McDermott defined community medicine as a new field dedicated to "the idea that it should be possible for professionally expert and knowledgeable people to see how medical science and technology can be applied for their benefit."[47] The relationship between experts and the communities they served was mediated through the selection of appropriate technologies.[48] The trained eye of the community medicine specialist could "maintain a continuous scan, like a radar beam, across the whole field of biomedical science and technology," then "stand off and look at a community, as determined by neighborhood or geographical lines, and try to see it as it is." With this combination of expertise and awareness, a specialist could choose which diagnostic tests, antibiotics, or vaccines had the most capacity to improve the health of a given community.[49] The community medicine movement was not a rejection of technology but an embrace of the *right* technology to fit the needs of the community.

The role of the community medicine specialist was to serve as a sort of medical Prometheus delivering the spark of new technology to benefit the common man. One of Deuschle's most promising Prometheans was an African American physician and East Harlem resident named Carter L. Marshall. Marshall, a second-generation physician, emphasized the interplay of biological and social forces in shaping community health. He was aware that many East Harlem residents often resisted setting foot in the hospital until they were so ill as to be beyond the reach of medical help. He hoped that the right kind of communications platform could help bring them into care at an earlier stage of illness—and at an earlier stage of life.[50] Marshall envisioned a pediatric outreach program that would stretch directly from the Mount Sinai Hospital into the structure of Harlem's tenements and public housing projects themselves, using medical television.

"There are two ways you can look at problems that involve the delivery of health services," Marshall told the *New York Times*, shortly after establishing the telemedical link between Mount Sinai and a community health center in the Wagner Homes projects at 121st Street and Fifth Avenue. He explained that one would be to fix the structure of the healthcare system itself, the other to use technology to circumvent these fundamental problems. "Our interest here," he continued, "is how we can adapt technology to the delivery of health services, regardless of the organizational framework."[51] Pointing to the Kerner Commission Report on Civil Disorder and the Sloan Commission Report on Cable Communications, Marshall suggested that Harlem was an ideal site to study the potential for cable television to reduce disparities in access to healthcare. "We at Mt. Sinai," he stated, "are in a unique position vis-à-vis CATV in that we are located in the franchise area of the TelePrompTer Corporation" and could form a "unique partnership between a voluntary medical setting and the private sector of CATV."[52]

The timing could not have been better. Mayor John Lindsay had just declared New York to be a model city for exploring the social benefits of cable connectivity. TelePrompTer had just announced a new storefront studio on 125th Street. The company also pointed to recent programming on sickle cell disease as evidence of its commitment to use cable television to advance the health of the Black community. Reverend C. T. Vivian, head of the Harlem-based Black Center for Strategy and Community Development, called TelePrompTer's move "the first time to my knowledge that an independent Black group has been able to develop a relationship with a CATV company that would allow for creative programming out of the Black condition on an independent basis."[53] When TelePrompTer came under fire for not being a Black-owned business, it noted that more than half its work force was made up of African American and Puerto Rican New Yorkers.[54]

TelePrompTer's dedicated coaxial lines linked Mount Sinai's high-tech hospital with a community clinic in the Wagner Homes projects funded by New York's Ghetto Medicine Program. The Wagner Homes clinic served 1,300 children by 1972, but lacked the budget to support a full-time physician. With continuous cable links, Marshall hoped full-time medical services could become available to more residents of Wagner Homes and surrounding public-assistance housing projects, encouraging Harlem residents to seek out formal healthcare.[55] Each station had a Telemation camera with 9-inch direct video monitor and a 19-inch receiver, with wide-angle and zoom lenses and microphones (fig. 5.5). The clinic was staffed by nurse practitioners (who spent most of their time at the Wagner Homes clinic) and pediatricians (who spent most of their time at the Mount Sinai Hospital).

"The question frequently put to us," the clinic team recalled "is how can a physician or nurse at one end of the cable see into the child's ear or determine color of a skin lesion via a black and white system?" The answer was complicated. The camera, the television, the

FIGURE 5.5. Demonstration of the telemedical link between (a) Mount Sinai Hospital and (b) the Wagner Homes community health center in East Harlem.

cable links—all of the gear itself was not enough to make telemedicine work. It was the paramedical staff of the clinic, specifically the nurse practitioners who were physically present with the patients in the Wagner Homes, that formed the crucial link in the circuit. "This is so," they concluded, "because the personnel at the Clinic can serve as an extension of the provider at the Medical Center by describing the condition of the middle ear, the morphology of a skin lesion and the result of palpation of a patient's abdomen."[56] The extension of the physician to underserved communities via telehealth turned out to be reliant on the attending nurse or physician assistant. For telehealth to work, clinics needed more than technology. They needed qualified staff.

They also needed to build trust. Marshall was surprised to find that most children and parents were immediately receptive to getting direct medical care over the television. In several cases, patients were *more* communicative with their healthcare providers over television than in person. He speculated that either "the child is probably very familiar with television," and views any occurrence on the screen as entertainment, or "the nurse is less threatening on television . . . she cannot, for instance, give an injection over the television!"[57] Harlem's *New York Amsterdam News* echoed this assessment: "The children love it . . . and are more willing to come to the clinic now that they can be 'on television.'"[58] In an early report to the National Cable Television Association, the Mount Sinai team explained their plans to expand the cable link to other health stations, schools, daycare centers, and—ultimately—into every home in Harlem.[59]

Yet many Black and Latino parents in East Harlem were as wary of Mount Sinai's intentions as their children were of its injections. Well before the overt medical racism of the US Public Health Service's Tuskegee Syphilis Study was exposed in 1972, activists in the Black Panther Party and the Young Lords had criticized mainstream, private academic medical centers like Mount Sinai as indifferent at best and "genocidal" at worst toward the minority communities that lived alongside their campuses.[60] Many Harlem residents understood that recent shifts in federal and state healthcare funding, including the

New York State Ghetto Medicine Program that funded private hospital outreaches to community health centers like Wagner Homes, also threatened funding for public institutions like the municipal Harlem Hospital.[61] Mount Sinai was singled out by East Harlem Health Council chair Robert Palese as a private entity receiving "huge sums" under the Ghetto Medicine Program, "without answering the health needs of the poor people in the community."[62] TelePrompTer's Harlem storefronts and Black-oriented programming were criticized by *New York Amsterdam News* columnist Cinnamon Dee as a veneer of community concern plastered over an underlying ambition to gain monopoly control of the Harlem cable market. The ribbon-cutting of the TelePrompTer storefront on 125th Street, Dee argued, was just "another grand rip-off of the community."[63] If the cable company did not return revenue to the community, Dee asked, how could it be understood as anything but a new experiment in economic extraction?

The Mount Sinai demonstration received extensive publicity. It was a success story that showed how the "sociocultural gulf that separates inner-city residents from health care resources" could be bridged by cable television.[64] But the Mount Sinai team and their federal funders were less clear whether it was the communications technology or the personnel involved that actually served to bridge that gulf. If the real goal was to provide good care to as many urban residents as possible, the technical role of coaxial cable itself might be less important than the structural role it played in enabling nurse practitioners to operate more autonomously in neighborhood community health centers. Imagining a network of five Harlem clinics linked to a central consultant at the Mount Sinai Hospital, economist Charlotte Muller calculated that each link might take roughly 5 percent of a supervising physician's time. For the cost of one physician, twenty nurse practitioners in twenty clinics could then be allowed to see more patients unaccompanied.[65] By the time Muller's analysis was published in 1977, however, the television link between Mount Sinai and Wagner had already been cut.

One of the final visits conducted via TelePrompTer involved Dr. Nicholas Cunningham, a pediatrician at Mount Sinai, and Betsy Voss, a nurse practitioner at Wagner Homes. Voss held a two-year old child in front of the camera so that Cunningham could evaluate the seriousness of her diarrheal illness. "For more than two years," the *New York Times* reported, "Dr. Cunningham spent his mornings before the console without pay, available to Miss Voss or the two other nurse practitioners at the Wagner clinic for consultation on any but the most routine pediatrics cases." But by 1975 the research contract was complete. Having proved the efficacy of telecommunications with nurse practitioners, the federally funded contract was terminated. As a spokesperson for the National Center for Health Services Research drily noted, the demonstration was successful at its stated goals and now "the project has gone about as far as it can go."[66] Federal funding for telemedicine was about demonstration, not implementation.

Much of this could have been predicted in the 1969 conference on Medicine in the Ghetto where Deuschle first described the role of the community medicine physician as an expert who stood apart from a community and selected the proper technologies to improve their health. "When a demonstration model designed and operated by whites proves to be unsuitable, no one makes a fuss about the inefficiency or inexperience," Dr. John Holloman, a physician and civil rights activist associated with Harlem Hospital, had warned. "Someone simply writes it up as a part of the study, presents it at a conference, and then a new model is designed and tried until the best way is found to make things work properly. We [the Black community] must have the equal right to make mistakes, to fail, and to try again."[67] Yet in these evaluations of communications technologies in community health, the purse strings were rarely held by those in the community. As a result, even when demonstrations succeeded, they failed all the same.

* * *

As telecommunications technologies were tested as remedies to the isolation of the "medical ghetto," the meaning of that term expanded to encompass rural as well as urban spaces. By 1969, the accelerating decline of economic and demographic indicators in rural counties, coupled with underdeveloped transportation and communication networks, aggravated preexisting barriers to accessing healthcare. Speaking for the Department of Health, Education, and Welfare, Dr. Joseph T. English described Lowndes County, Alabama, as a "rural ghetto . . . strewn over 716 square miles; there is not a single hospital; roads are so poor that you need a four-wheel drive jeep to traverse the terrain after a heavy downpour; medical resources consist of three physicians, one of whom has not been practicing medicine for many years and another who is in his late 70s."[68] Black sharecroppers made up 80 percent of the population in Lowndes County, and the work of civil rights organizers, including Stokely Carmichael, had made Lowndes a key origin site of the Black Panther Party. Rural or urban, the "medical ghetto" was still a marker of racial segregation.[69]

If urban health disparities came from the concentration of poverty in densely populated areas, rural health disparities were understood to be the result of a reverse process through which the American countryside was rapidly losing its population to suburbs and cities. One out of every four rural residents in the United States now lived in poverty linked to the accelerating decreases in population. One eight-county area of West Virginia, which had contained 80 physicians per 100,000 persons in 1938, could count only 53 physicians per 100,000 by 1960, and fewer than 30 in 1974.[70] By the late 1960s, responses to the scarcity of physicians in rural areas called for the deployment of a new kind of health provider: the medical extender.[71]

The medical extender concept emerged out of three different allied health professional movements, each fighting for greater autonomy on state and national levels. Nurse practitioners (NPs) were registered nurses whose additional training qualified them for more independent roles, including the ability to prescribe drugs. Physician assistants (PAs) and community health medics (CHMs) began as military medics with additional training to take on augmented roles in civilian life. By 1970, the American Medical Association officially recognized the physician assistant as a qualified healthcare professional with enough academic and practical training to be able to provide care to patients under close supervision and direction of a physician.[72] By June 1973, thirty-three states had established legislation permitting the practice of physician assistants and nurse practitioners.[73]

In the early 1970s, however, all NPs, PAs, and CHMs in medical practice required direct supervision from a physician. A nurse practitioner could perform some tasks of doctoring independently (for example, conducting a history and physical, prescribing certain drugs), but was not considered an independent medical practitioner under the laws of any state. In a high-volume urban community health center like the Wagner Homes clinic, where a single physician might supervise several NPs or PAs at once in a single building, the use of medical extenders could easily help increase access to care for urban residents. For widely dispersed rural populations, telemedicine promised a new means for physicians to directly supervise their medical extenders across vast distances without needing to spend hours driving from one clinic to another.

Early proof of concept came from the Dartmouth-based INTERACT program, which trained medical extenders to link health facilities across the Vermont–New Hampshire border in the upper Connecticut River Valley. By circulating physician assistants through different rural sites, and supporting them with telemedical links, the system helped to redefine what physician supervision meant in practice, and to multiply the clinics covered by a limited number of physician personnel. The results suggested that these systems worked best when remote physicians could supervise seasoned, more experienced medical extenders with longer training and demonstrated capacity for autonomous work.[74]

The Dartmouth data raised a follow-up question: if interactive television empowered seasoned medical extenders to act autonomously as part of a broader primary healthcare system, could it also enable the use of even less skilled healthcare workers?[75] Maxine Rockoff was interested in Harvard physician Anthony Komaroff's new clinical practice algorithms, developed to enable rapidly trained, new healthcare workers—whom he termed "5-week wonders"—to act as medical extenders without any formal NP or PA or CHM training.

These minimally trained physician extenders, after just five weeks of training, could use formally scripted protocols (with physicians available via cable TV backup) as an "expanded manpower/technology experiment to see how effective and cost-effective certain manpower/technology tradeoffs might be in delivering health services in geographically dispersed facilities."[76] Komaroff's language extended the industrial logic of systems analysis further into the processes of diagnosis and treatment. Automating these processes, he promised, would help generate a larger pool of autonomous health professionals at lower levels of training.

Other rural telehealth projects focused on how telecommunications technologies might reverse the flight of young professionals from rural to urban areas. Among the rural experiments Rockoff visited and supported in the early 1970s, the New Rural Society Project of Windham, Connecticut, stands out as one of the more radical visions for cable television in revitalizing rural community health. Its founder, Peter Goldmark, had recently retired as president of the Columbia Broadcasting Service (CBS) television network, during which time he served as the chair of the National Academy of Engineers 1971 commission on Communications Technology for Urban Improvement. Shortly afterward he left New York City for rural Connecticut. With federal grant support from the Department of Housing and Urban Development (HUD), Goldmark embarked on a "remarkable and visionary experiment," as Maxine Rockoff described, "to make a rural region (Windham, Connecticut) into an attractive and desirable place to live by using the most advanced communications technology available."[77]

Goldmark positioned the hamlet of Windham as the testbed for the New Rural Society Project: a bold plan to use interactive television to reverse the flight from rural areas that was occurring in parallel to white flight from urban centers. The key platform of the New Rural Society, he claimed, "is the redirection of modern technology, particularly in the telecommunications field, toward society's critical needs." New technologies could make rural areas desirable sites for professionals to live in once more. Especially in the field of health and medicine, the New Rural Society Project contended that existing communications devices like interactive television, broadband, and digital computing could improve access to healthcare resources in sparsely populated sections of the country.[78]

The failure of the Land Policy and Planning Assistance Act of 1973 dashed Goldmark's hopes to bring the New Rural Society into existence on a national level, but Maxine Rockoff was impressed by his vision all the same. She was even more impressed by Bonnie Kraig, the head of health projects at the New Rural Society, who had managed to garner support from every single medical practitioner in the Windham region to advocate for full cable linkage of health services in the area.[79] "If we take the initiative for 'building' several model health systems," Rockoff wrote to her supervisions at HEW, "then at least a part of this activity should be devoted to the establishment of a means that would let any other group learn from our experiences in order to 'build' their own system . . . At the very least we should provide the means by which the Ms. Kraigs can become aware, efficiently, of the whole 'innovations smorgasbord.'"[80]

Rockoff asked colleagues at HEW whether any of the 144 National Health Services Corps sites would be good pilot locations for similar models of rural health via cable television.[81] Soon the Lakeview Clinic in rural Waconia, Minnesota, emerged as a key site for federal support for research into the value of rural telemedicine—not as a racially segregated "rural ghetto," but as a racially unmarked site for demonstrating the feasibility of dispersed care. Waconia was seen as a "typical" rural site; it was self-contained, well defined, and easily studied; and it represented "the largest system we think we can understand and develop a paradigm for initially."[82] Led by physician Jon Wempner and constructed by a local cable company called Community Information Systems, the Waconia project served a large rural area with a mostly white population, which was in turn served by a dwindling population of white general practitioners. Mobile video carts in each clinic allowed for real-time cable transmission of close-up and wide-angle views of the patient, along with X-rays, ECGs, chart materials, and electronic stethoscopes and microscopes. The principal goal of the demonstration—and the source of the "Face of God" problem discussed in the last chapter—was to evaluate how doctors, nurses, and patients did or did not accept the value of medical visits via cable television.

Like the urban Mount Sinai system, the rural Waconia system both tested the viability of telemedical care in an underserved community and enhanced the autonomy of nurse practitioners through telemedical links. Within a few years, however, it was clear that, although the system seemed to work well, and doctors and patients were willing to accept it in practice, neither party clearly preferred telemedical visits over driving a longer distance for an in-person visit. Even the increasing gas prices of the oil crisis did not clearly change this calculus.[83] Though telehealth was technically viable, at the conclusion of the initial study in 1975, there was not enough enthusiasm in Lakeview for maintaining the system, and it was removed. Once again, federal demonstration projects failed even as they succeeded. As the technical evidence supporting telemedicine accumulated, the economic and social rationale to support the widespread expansion of telehealth networks fizzled.

It is telling that the most substantial and sustained federal support for rural telemedicine in the 1970s came not from the diminished funding of HEW but from the larger budgets allowed by NASA and its aerospace contractors. A pair of research satellites, ATS-1 and ATS-6, were used to deploy first satellite radio and then satellite television to enable Alaska Native community health medics to work under remote physician supervision across the vast expanse of the central Tanana Region in the rural Alaskan interior.[84] By the early 1970s, NASA's new Integrated Medical and Behavioral Life Monitoring System (IMBLMS)—a telemedical suite initially designed for long-haul, manned missions to Mars—was ready for field trials on earthbound populations.[85] NASA and Lockheed found willing partners among the leadership of the Papago Tribe of southern Arizona (now the Tohono O'odham Nation) and the Indian Health Service, which had recently established a new Office of Research and Development on the Papago Reservation.

More funding existed to create a system for a few astronauts who *might* be heading to Mars than could be cobbled together to study the role of telemedicine in urban and rural health on its own merits. With a budget more than three times as large as all of Rockoff's projects combined, the resulting Space Technology Applied to Rural Papago Advanced Healthcare (STARPAHC) demonstration was by far the largest federal investment in telemedicine and the fullest test of NASA's Area Health Services Field Unit concept.[86] It basically took the telemedical system being designed for the medical bay of the space shuttle, stuffed it inside an off-brand Winnebago, and made slow orbits around the Sonoran Desert, picking up sick Papago and treating them as if they were sick astronauts. When a local journalist published an article portraying STARPAHC as an exploitive project that extracted experimental value from American Indian research subjects, tribal leaders objected. "The Papagos were rather miffed with one newspaper headline," read an article in the *Sarasota Herald Tribune*,

> which said the Indians were guinea pigs for astronauts. And a Washington bureaucrat once visited the reservation and asked the people how they could allow themselves to be "ripped off" by the project. That brought a laugh from Ralph Antone, a former tool and die maker who returned to the reservation a few years ago and now is a member of the tribe's Executive Health Staff. "Those bleeding hearts," he chuckled. "If there is anyone getting ripped off it sure isn't my people, it's the government. We are getting the best health care available for free. How is that a ripoff?"[87]

Long-term commitment to providing access to medical care on the Papago Reservation was never the primary motivation for NASA or Lockheed. But Papago electronic engineers like Peter Ruiz and community health medics like Rosemary Lopez, who took active roles in the project, saw STARPAHC as an opportunity to build infrastructure that could improve health outcomes for the Papago Tribe all the same. Ruiz, who grew up off the reservation and became an electronic engineer in the Air Force, traveled to Lockheed's headquarters to learn how to fabricate, implement, and ultimately operate the STARPAHC system and later became director of natural resources for the tribe.[88] Lopez, the first formally trained CHM on the reservation, helped to design and name STARPAHC, staffed the Mobile Health Unit as a healthcare provider, and would later go on to become secretary of Health & Human Services for the Tohono O'odham Nation.[89] When the STARPAHC Mobile Health Unit was unveiled at the 1974 Sells Rodeo, it was identified as "Rosemary's Baby"—a play on the title of the 1968 horror film (fig. 5.6). STARPAHC may be a monster of sorts, the name suggested, but it was *their* monster.

STARPAHC worked surprisingly well, but that was not enough. The Mobile Health Unit fit Walsh McDermott's concept of an appropriate technology for community health: the right technology for the right population. During the life of the project, an indigenous population living in a remote rural region in which transportation and geographical dispersal were significant barriers to healthcare found increased access to medical care. Data collected by IHS project director Dr. James Justice showed broad acceptance by providers and patients on and off the reservation.[90] Yet after the successful 1976 simulation test of the IMBLMS system, its funding was not renewed. While the Indian Health Service had been hopeful that HEW would continue to support the project after NASA departed, the Mobile Health Unit was mothballed in 1978 and quietly sold to Arizona Northern University, where it lived out a second career with a campus media logo on its side, televising college football games.[91] The STARPAHC telecommunications tower stayed on the reservation and provided lasting benefit to the communications infrastructure of the reservation, but use of that, too, was eventually ceded, in this case to the US Border Patrol.

From NASA's perspective, STARPAHC had accomplished both of its core missions. The IMBLMS system was cleared for use as the future medical bay of the space shuttle *Columbia*, and the viability of rural telemedicine was established. And NASA, along with Lockheed, controlled the budget. For the Indian Health Service, its legacy

FIGURE 5.6. STARPAHC Mobile Health Unit, (a) interior and (b) exterior, in the 1974 Sells Rodeo Parade, shortly before beginning operation. Peter Ruiz Collection, Archives, Himdag Ki, Tohono O'odham Nation Cultural Center and Museum, Sells, AZ.

was more mixed. While the project briefly positioned the struggling agency as an incubator for technological solutions to rural healthcare delivery, the applicability of these technologies at other IHS sites and international health projects through the US Agency for International Development (USAID) and the Peace Corps never materialized. STARPAHC did provide Tohono O'odham living in remote villages more convenient access to healthcare services for a while, but it could not contribute a lasting solution to the persistent problem of access to healthcare on the reservation, nor could it undo the social determinants driving broader health disparities on American Indian reservations. Without sustained political and economic support for implementation, a demonstration was just a demonstration.

* * *

In the early 1970s, interactive television promised to rejuvenate isolated communities and link urban and rural ghettos, to restore social services and forge new forms of community solidarity. Soon, however, community activists like Francille Rusan noticed that ownership patterns for CATV were shifting from "community antenna television to a new and unprecedented pattern of corporate antenna television." Once the FCC moratorium lifted, and cable television became profitable, it became a big business like any other. As the promised veneer of benefits to minority-majority communities fell to the wayside, Rusan complained, "The Black, the brown, and the poor have yet to enter the ring."[92] Instead, cable markets consolidated into monopolies of increasing scale. TelePrompTer was sold to Warner Cable, which then became part of the vast Time-Warner media empire. Home Box Office (HBO) began to broadcast films through national cable networks in 1975, and Turner Broadcasting Systems followed suit in 1976. Well before Ted Turner's Cable News Network (CNN) launched as the first twenty-four-hour cable news channel in June 1980, cable had shifted from a field of limitless local interaction into another medium of monopoly and mass consumption.

All the same, the urban and rural telemedical demonstration projects coordinated by Maxine Rockoff had unequivocally shown how telemedicine could substantially improve access to care. The delivery of high-quality medical care through television was both technically feasible and socially acceptable to patients and providers. But no long-term public or private financing was forthcoming at the end of this first generation of telemedical demonstrations.[93] Yes, a handful of industry actors benefited from telemedicine in the short term. Lockheed landed a nearly $4 million contract from NASA. TelePrompTer gained national publicity for its work in Wagner Homes. But in the end, the populations each project purported to serve were left without care. In retrospect, it is tempting to see the social goods promised by cable firms and defense contractors as a diversionary tactic, a way to distract the public gaze as private interests consolidated their grip over yet another new market for goods and services.

These disappointments did not surprise members of the marginalized communities for whose benefit such cable solutions had been so publicly imagined. If my analysis of these failures mirrors the skepticism voiced by Dr. John Holloman at the 1969 Medicine in the Ghetto conference, it is not because failure was inevitable, or because Holloman and his colleagues were ignorant of the challenges they faced. It was just business as usual. "Is poverty good business?" Holloman had asked the crowd of well-meaning, liberal-minded physicians struggling to formulate responses to the health disparities of American social marginalization. "Many ghetto residents," he continued, "view the establishment of satellite clinics, neighborhood health centers, and the like with some suspicion." The problem wasn't that the services weren't needed or desired, but that the "ghetto resident" was all too often conceived of as a passive user rather than a stakeholder in the system's long-term implementation. "Often when we look about us," Holloman concluded, "it is very easy to discern the correlation between interest (where none existed before) in our community and the large sums of monies available for study, demonstration, renewal."[94] Those monies, controlled by outside interests, were never promised for the long term.

To take Holloman seriously requires a more nuanced view of the appeal of the technological fix. It is not enough to look back with derision at those who would seek in telemedicine a viable answer to the problems of segregation, discrimination, dispossession, and economic erosion that shaped health disparities across urban and rural geographies in the United States. It is more important to ask: why do technological fixes continue to fail to serve the underserved, despite promises to the contrary? Wary as he was of demonstration projects, Holloman nonetheless held out hope that *this* demonstration might yet produce meaningful community responses to enduring social problems. He repeatedly pointed to the emerging technology of telemedicine as a possible solution, insisting that "two-way closed-circuit TV can be established to link the ghetto physician to the medical center, services that could be put into operation almost at once."[95]

Here Holloman and his contemporaries should be regarded not as naïve, but just as pragmatic as any twenty-first-century actor trying to advocate for health equity using the e-Health, m-Health, and wearable technologies of the present moment. As a new vehicle for health equity, telemedicine offered possibilities for community action that were not pie-in-the-sky but directly at hand, not "top down" but community led and community owned. Holloman understood racism as an omnipresent structure of daily life—he could easily chart the social determinants of health and disease that had been layered into the racialized geography of American cities by 1969. Yet if the longer-term solutions to these problems required far larger budgets and time scales, why not begin with smaller projects that could be operationalized on a shorter time scale, through laying a few miles of coaxial cable and hooking a camera to a television set?

Although they failed to erase disparities in access to medical care, the telemedicine demonstration projects of the 1970s accomplished several of their stated tasks, along with a few goals not originally intended. By the end of the decade, these projects built new evidence for the feasibility, quality, and acceptability of remote primary healthcare. Television medicine was acceptable to many physicians and to many patients, and it could be conducted with privacy, confidentiality, reliability, and a high degree of quality for diagnosis in general care, with special evidence of application in psychiatry, cardiology, dermatology, and radiology. A few of the demonstration sites—most notably Mount Sinai's Wagner Homes project—collected economic data attesting to the relative high value and low cost of telemedical systems built with "off the shelf" cable technologies. These data were mobilized by community activists in their efforts to keep these demonstration projects going after the research funding ended.[96]

These projects used underserved communities in urban and rural areas to validate the technology of telemedicine. They also simultaneously used the technology of telemedicine to validate autonomous roles for nurse practitioners and physician assistants in primary care. One unintended consequence of the telemedicine studies of the 1970s was a new body of evidence that demonstrated the competence of these practitioners working largely on their own. The timing of these demonstrations—during the period when the licensing and regulation of NPs and PAs was being reconsidered by state and federal bodies, and grudgingly supported by the American Medical Association—could not have been more crucial.[97] In Kenneth Bird's hopeful reading of Marshall McLuhan's *Understanding Media: The Extensions of Man*, telemedical systems were meant to extend the reach of the individual physician. And yet to the degree that his ideas found purchase in the spread of telemedical demonstration projects in the 1970s, these "extensions" found their most robust legacies not in new devices but in new roles: medical extenders, whose increasing role in healthcare would prove in the short run to be more robust than the medium of television medicine itself.

Cable television did not end health disparities, but for a little while it did enable federal funders and a coalition of stakeholders to come together around a shared goal of engineering equity in access to healthcare. Although two-way television failed to flatten the widening health disparities evident over the course of the 1970s, this was not merely due to foolish idealism on the part of those who hoped it might. A new technology offered reason for optimism, a call for political and economic support, and a lever to use on the broken mess that was the US healthcare system, to try to move it just a bit toward a more equitable and functional status. As Maxine Rockoff summed up her experience with supporting telemedicine demonstration projects in the mid-1970s, each new telecommunications technology offered a new chance to imagine a more functional health system. "The challenge is clear and immediate," she added, "to design the communication systems and redesign the health care system so both are as we want them to be, not as serendipity would have them."[98]

This challenge still stands unmet today because Rockoff was not fighting just against "serendipity" but against entrenched commercial and professional interests. The fact that promising technologies of the past have been insufficient to resolve the inequalities of access in American healthcare does not mean that actors today can resist the allure to use the technologies of our own present to imagine a better future. But we should not for a moment believe that any given communication platform will by itself produce the kind of social change required to resolve health disparities.

Perhaps the most important lesson we can learn today from revisiting the telemedical demonstrations of the past is that demonstrations alone are not enough. Any given technology can hold the power to narrow the gaps in health outcomes between haves and have-nots—or to widen them anew. The key difference between these two poles has less to do with the technological platform itself than with the presence or absence of sustained commitment to realize more equitable access to healthcare. Without political will, without this longer-term attention to implementation, without the drive to follow through, we are left with a series of demonstrations of what might have been.

6

THE PUSH-BUTTON PHYSICIAN

As the medical profession has found itself yanked into electronic modes of practice—a screen in every examination room, an electronic health record for every patient—doctors have discovered a new nostalgia for a paper world they once took for granted. Many older physicians who now resent their computer screens miss, on some level, the manila chart folder with its garish alphabetical stickers, the bound volumes of the patient's hospital record, the physical freedoms afforded by one's own pocket prescription pad. Young physicians, too, complain of screen fatigue, and sometimes wish they could just look at a patient and write on a simple piece of paper.

Yet for those doctors and engineers who first imagined a new medium of electronic medicine in the 1950s, the situation was reversed. Paper was the problem. Journal articles were accumulating at a rate too fast for any human indexer to track. Paper charts got lost between the clinic and the hospital. Potentially lifesaving information did not reach the bedside in time to be put to use in patient care. The computer offered a lifeline, elevating the flow of medical information out of these torpid backwaters and into the ethereal realm of digital data. In a world of digital health, knowledge would travel at the speed of electricity, and medicine would become a science of pure information. The right knowledge, at the right time, and the right place, would save lives.

This is what Vladimir Zworykin hoped. His Center for Medical

Electronics cast aside earlier work on radio pills in favor of the brighter horizons of computer medicine. The former vice president of R&D for RCA Laboratories laid out his case to policymakers and patrons with some simple calculations. Let's assume, he suggested, that most individual parts of a medical record could be summed up in roughly 500 words each. At a rate of 30 bits per word, this would come to 15,000 bits per patient per hospital admission. If an average person had 20 hospital admissions over the course of their life, that meant 3×10^5 bits of electronic memory could store a lifetime's worth of electronic medical data. Multiply that by a world population of 3×10^7 and a storage capacity of approximately 10 trillion bits would suffice for every physician in the world to have nearly instantaneous access to any patient's medical records. As if the fundamental problem of medical information were simply scale. Big numbers for the human brain to retain in active memory, but not too big for the electronic brain of the digital computer.[1]

Existing magnetic storage devices could already carry this much information and more. Counting the words per entry in the compact *Merck Manual of Diagnosis and Therapy*, Zworykin calculated that the average disease required 10,000 words per entry, or 3×10^5 bits, while the more exhaustive *French's Differential Diagnosis* averaged 1 million bits per disease. Multiplying this figure by the number of diseases listed in the 1952 *Standard Nomenclature of Diseases and Operations*—10 million maladies in total—he arrived at the same magic number of 10 trillion bits to contain up-to-date diagnostic information for all diseases known to humanity. This number, Zworykin proudly declared was already within range of existing RCA Bizmatic computers.[2] When he took the proposal to Washington to speak with leadership in the US Public Health Service, he brought beautiful, hand-drawn blueprints for a giant magnetic cartridge system of spinning disks and thousands of magnetic reading heads, fifty horizontal trays, with twenty cartridges in a tray. The prototype Zworykin proposed could easily reach the desired number of 10^{13} bits of digital memory. As Zworykin and others would find, though, storing data would not be the central problem facing a computerized vision of medicine.

Digital media were poised to revolutionize medicine. RCA's new magnetic drives were ultra-vast, ultra-fast, and able to link medical records and diagnostic criteria in ways that transcended the limits of paper media and human comprehension. A network of computers linked first across North America, and then around the world, could make paper journals obsolete, and form "a central electronic *Index Medicus*" to store the growing body of medical knowledge in condensed form.[3] Physicians could send lists of symptoms and physical findings to these computing centers, and with the push of a button receive a list of likely diagnoses.[4] Eventually, the electronic brain of the computer would move beyond diagnosis and recommend medications, procedures, and further testing as well. "This is not a fantasy," he concluded, "but a definite goal toward which various groups throughout the world, generally restricting themselves to particular disease areas, are now working."[5]

Electronic medical journals, electronic diagnostic machines, electronic medical records. Zworykin was not the only one who saw the digital computer as a tool to transform medicine into a science of pure information. Robert Ledley, a physicist and dentist who chaired a National Academy of Sciences commission on the Role of Computers in Biology and Medicine in the 1950s and 1960s, also described the uptake of the digital computer as an epoch-defining shift for a profession formerly bound in paper. "Computers make extensive intellectual demands on their users," he observed; they could help physicians think, but they would also change *the way* physicians think in the process.[6]

Physicians, however, were not immediately convinced that incorporating mainframes into medicine would be in their own interest.[7] By the end of the 1960s, the promised revolution in medical information had faltered. A joint effort by the American Hospital Association and Boston's Massachusetts General Hospital (MGH) to enact "doctor-machine symbiosis" was a technical success but a practical failure. When it launched in 1962, the project had the directive to link hospitals across the country via telephone wires to remote "on line" computers and to integrate medical informatics, electronic medical records, and diagnostic analytics into a total information system. When the project shut down in 1968, MGH project lead Guy Octo Barnett described the effort as a form of technological hubris, which posited the computer as "the solution for almost all problems confronting the health professions." As the revolutionary promise of mainframe medicine was oversold, an "awesome gap between the hopes or claims and the realities of the situation" had emerged instead.[8] In the same year, a speaker at one of the final IBM Medical Symposia recalled the naïve belief, a decade earlier, that computers could simply remake medicine into an information science. "The planners and the dreamers were there, and they talked and talked," he recalled. "I must say here that their talks were in an unmistakable tense—the future." A decade later the future present had not matched up to the future past.[9]

If the reality of the first decade of medical computing did not live up to its initial promises, it was still a heady and productive moment. Much of this was set in motion in the late 1950s, as federal funding changed the trajectory of computers in medicine. Even as these bulky, room-sized and then refrigerator-sized devices were gradually infiltrating public and private workplaces with their punch cards, paper tapes, and magnetic reels, the use of computers had been lagging in the medical field.[10] Yet after Ledley's initial National Academy of Sciences report, Congress provided $50 million in emergency funding to help disseminate computers into biomedical research and practice.[11] In the many efforts that followed, computers did not replace paper—not yet—but they did substantially change the way medical information was produced and circulated and put to use.

* * *

Computers first entered medicine as a solution to the problem of data. By the late 1950s, fueled by increases in federal funding through the expanding National Institutes of Health, the weekly output of medical research well exceeded the ability of any given medical professional to digest. The volume of published articles in the biomedical sciences accelerated exponentially in the postwar decades. Yet as Ledley pointed out, "the great advances in medical knowledge of the past few years have not been matched with a parallel advance in making this knowledge available to the practicing physician."[12] As long as medical knowledge was bound in paper media, doctors needed to travel to libraries or maintain regular subscriptions to a growing number of medical periodicals merely to keep up with rapidly changing standards of practice. Even the librarians at the National Library of Medicine, who maintained the quarterly *Index Medicus* and the *Current List* of biomedical publications, were overwhelmed by the number of new journals emerging on a weekly basis.[13]

In the late nineteenth century, similar concerns over growing medical knowledge had led to the creation by John Shaw Billings of the Surgeon General's Library, the institution that ultimately became the National Library of Medicine (NLM). Billings was a polymath: physician, architect, and librarian. The nineteenth-century crisis of information he faced in building the *Index Medicus* was repeatedly invoked in the twentieth century. "What will the libraries and catalogues and bibliographies of a thousand or even a hundred years hence be like," Billings had asked in 1881, "if we are thus to go on in the rate of geometric progression, which has governed the press for the last few decades?" The math was clear: if the present rate of increase went unchecked, there would soon come a time when "libraries will become large cities, and when it will require the services of everyone in the world not engaged in writing, to catalogue and care for the annual product."[14] To manage this future, Billings favored the information technology of the index card, connected to the index catalog. These paper tools functioned as a form of preelectronic computer, collating card by card all of the powerful components now associated with databases: storage, classification, analysis, and retrieval.[15]

Yet paper tools took the medical information specialist only so far. Where Billings in the 1880s had found it challenging to index 864 medical journals, by 1962 the National Library of Medicine now regularly processed 5,500 medical journals. As the library moved to its new building attached to the Bethesda campus of the National Institutes of Health, director Frank Rogers envisioned a system of digital computers to store, analyze, and retrieve all relevant indexes and metadata from the world's medical literature. He called it the Medical Literature Analysis and Retrieval System, or MEDLARS. All relevant information from international medical journals for the year 1963 would be transferred to punched paper tape for computer input. By January 1964, the Honeywell 800 mainframe at the heart of MEDLARS could support nearly 40,000 demand searches per year.[16] MEDLARS would double the processing speed of human indexers to meet the rising pace of medical publishing. It was the first large-scale information retrieval project to provide both bibliographic access and copies of documents by computer.[17]

MEDLARS became a symbol of the transformational role computers would play in the circulation of medical knowledge. "Giant electronic brains which never sleep are digesting several hundred thousand medical documents each year," the Louisville *Courier Journal* reported in 1964 (fig. 6.1). The MEDLARS digital outputs were "pouring out this vast knowledge to a medical world thirsting for compact, yet thorough, new data on the diseases of mankind."[18] Yet the

PUSH-BUTTON BRAINS PRODUCE . . .

FASTER FACTS FOR DOCTORS

The National Library of Medicine collects and dispenses knowledge electronically

HUMAN INSTRUCTIONS are sent from this keyboard to the Medlars computer system in the background.

FIGURE 6.1. Early reporting on MEDLARS for the general public emphasized that "Push-Button Brains Produce Faster Facts for Doctors."

FIGURE 6.2. Photograph demonstrating MEDLARS human computers as well as digital computers. "At right, the girl feeds paper taped information into a 'reading' device. Center girl uses keyboard to 'tell' computer (in rear) how to index and store on magnetic tape the data from the reader."

workings of MEDLARS depended on human–machine interface—including a staff, almost always depicted as women (fig. 6.2) to read, key, and manage the equipment. Giant reels of magnetic tape with full copies of MEDLARS data soon circulated to other medical libraries throughout the country to decentralize holdings.[19] Within a few years, plans were afoot to expand access to every medical library, medical school, hospital, and ultimately every doctor's office in the country.[20]

As MEDLARS branched out to form a nationwide network for biomedical information, it also became an instrument of soft power in the Cold War.[21] By 1967, as the State Department took interest in the project, MEDLARS tapes were shared with the World Health Organization in Geneva, Switzerland; with the Karolinska Institute in Stockholm, Sweden; and with sites in the United Kingdom, and negotiations were beginning for another site in Sydney, Australia.[22] MEDLARS access was also dangled to Czechoslovakia, Thailand, India, and other nonaligned nations through the US Agency for International Development in partnership with the Rockefeller Foundation.[23]

MEDLARS was celebrated as a model form of medical computing, but new concerns emerged. There was a Catch-22 in verifying the quality of MEDLARS performance. If the problem that MEDLARS was designed to solve was that there was too much information for the human brain to retrieve, how could a human brain know whether or not the MEDLARS computer was retrieving information properly? "It is extremely difficult to estimate the recall ratio for a 'real-life' search in a file of over half a million citations," information scientist F. W. "Wilf" Lancaster complained, when asked by MEDLARS staff to peer into the workings of the system and see how well it performed. This was harder than it looked. The only way to do so completely would be to have a human examine and assess each document retrieved—and that would take several decades! Instead, Lancaster formed a list of twenty representative test users to request MEDLARS searches; he then followed up with these users to compare their search results to those of a local librarian. In a move evocative of a sort of bibliomanic John Henry, the test pitted human against machine.[24]

Lancaster's study was the first major evaluation of a computer-based library information system. In just under 300 test searches, MEDLARS searched the world's medical literature with 57.7 percent recall and 50.4 percent precision, with roughly 175 citations per search.[25] Yet 238 of the searches documented at least one form of recall failure. While these failures were often located in the indexer or searcher, on a broader level they exposed a paradox in electronic information science. The more exhaustive the search, the less precise it would be. Conversely, any increase in precision came at the cost of claims of completeness.[26] As the MEDLARS chief described to NLM staff, "the thing that has made the Lancaster Study a monumental accomplishment was the thoroughness with which each failure was analyzed."[27] On the whole, however, MEDLARS worked at least as well as a human librarian.

The success of MEDLARS gave rise to new forms of failure. The continuing exponential growth of research output in the 1960s soon outpaced the capacity of the original mainframe computer. Shortly before MEDLARS went live, the head of NLM's Data Processing Section warned that the growth curve of the new system would be outdated within five years. If the process of automating medical information was not continuously updated, it would fail altogether.[28] By 1966 more than fifty MEDLARS searches had been backlogged by more than three months.[29] Requests that initially took no more than two days to conduct were now taking two weeks or more to fulfill. What good were vast stores of information for clinical application if the material wasn't accessible? Proposals for a larger, faster MEDLARS II helped to patch this problem through the purchase of larger, faster IBM mainframes and implementation of a form of direct online access via telephone wires to enable searches by individual end-users.[30] But the process revealed a hidden cost of the shift from paper to computer indexing: where paper cards could last for centuries, the shift to digital media brought with it new cycles of dysfunction, upgrade, and obsolescence.

The uptake of computers in the 1960s did not displace the paper medical journal. At a conference in 1964 the head of the National Library of Medicine assured the congregating librarians that "MEDLARS is a new instrument of the research library, not a replacement."[31] Librarians noted with some ambivalence that if all information did make the leap from paper to digital, the ensuing automation of medical information might ultimately displace the medical library from its central location in academic medical institutions. Several years later, recalling his assessment of MEDLARS, Lancaster told an interviewer that he personally had always had a very skeptical view of technology. He refused to use ATM machines or telephone answering machines and word processors, or even typewriters. As proud as he was of his work evaluating computers as tools for information management, in the years that followed, medical libraries had become increasingly depersonalized—and depopulated—with the increasing uptake of these tools.[32] Even John Henry did not outlast the machine that threatened his job.

For many computer enthusiasts, the prospect of a fully digital medical library seemed close at hand. Among those who picked up on the potential of digitizing not just the index but the full texts of all future articles was Vladimir Zworykin. "While a conventional library has the required property of enormous memory capacity," he complained, "the knowledge stored in it is entirely inaccessible when the life of a patient hangs in the balance." Now that the world's medical knowledge was being converted into digital form, it should be linked directly to computers in clinics and hospitals, and made available to the clinician to help with diagnosis and therapy in real time.[33]

* * *

Zworykin's early efforts to promote computers in clinics and hospitals elicited strong resistance from doctors who feared the automation of clinical work, especially diagnosis. In 1960, the scientist and illustrator Athelstan Spilhaus depicted a graphic version of this future for his regular Sunday comic strip "Our New Age" (fig. 6.3). Soon, Spilhaus claimed, patients would be able to hook themselves up to computers, push buttons to answer questions about their symptoms, have their key vital functions measured by sensors, and send that information to a central processing unit. In the blink of an eye, "a diagnostic computer will receive the telemetered information, and after checking by physicians . . . will automatically prescribe a course of action while you wait!"[34]

If this vision of mainframe medicine was meant to provide a utopian future for patients as consumers of healthcare in a digitally mediated world, it evoked a dystopian future for the profession of medicine. On one level, these computers served as metaphors for the encroaching mechanisms of healthcare bureaucracy, embodying long-standing fears of "socialized medicine" in which doctors would themselves become cogs in a larger machine instead of independent private practitioners. But on a more visceral level, they also threatened to displace doctors from the locus of their greatest area of expertise—the rational practice of diagnosis and therapy—and relegate them to the role of technicians. Computers threatened the autonomy of the individual practitioner and the profession in general.

Nor was this just the stuff of Sunday comics. At the New York Hospital, just across the street from Zworykin's Center for Medical

FIGURE 6.3. Telemetry, telemedicine, and the diagnostic computer: the future of computer medicine as imagined in 1960 in the popular comic "Our New Age" by Athelstan Spilhaus. Courtesy of Matt Novak.

Electronics at the Rockefeller Institute, psychiatrist Keeve Brodman had already begun to use mainframe computers to assist in diagnosis.[35] Brodman believed that medicine would only become a more objective science if the process of obtaining information from patients was standardized and automated. His earliest efforts involved a paper tool called the Cornell Medical Index, a roughly 200-question form that provided a comprehensive intake of symptoms and signs of disease. By the mid-1950s, Brodman and colleagues adapted the Cornell Medical Index into a digital Medical Data Screen, using the hospital's new IBM Model 704 computer to correlate data collected on hospital admission with diagnoses patients received when they were discharged from the hospital.[36]

Digesting information from nearly 9,000 patients admitted between 1948 and 1956, the mainframe could use a self-adjusting program to devise its own diagnostic criteria for the sixty most common diseases in the study group. As Brodman described, "the machine system compares the symptom of a patient to the syndromes in its memory," and could compute in less than a second the likelihood that a given patient matched a given disease.[37] Initial results were promising. As the computer adjusted its diagnostic criteria based on the "raw data" it tabulated, its accuracy came to resemble that of a credentialed physician. Overall, 5 percent of diagnoses made by the computer were false, but so were 2 percent of the diagnoses made by human physicians—and in some categories the computer performed better as a diagnostician than the doctors did.[38]

Brodman did not suggest that his computer could or should replace physicians, but he believed it could effectively simulate diagnostic thinking. He had created a new kind of Turing test, which other doctors read as a provocation. Physicians, especially during the so-called Golden Age of American medicine in the mid-twentieth century, liked to see themselves as the embodiment of reason, rationality, expertise, and judgment—especially in the art of diagnosis.[39] Yet physicians had, by the late 1950s, also been shown to be fallible, biased, swayed by pharmaceutical advertisements and sales representatives to prescribe based on persuasion rather than evidence.[40] If a self-adjusting set of probability tables could produce diagnostic outputs similar to those of human physicians, could a commercial, high-speed data-processing system teach itself to make reliable diagnostic predictions that would be free of human bias? Perhaps, Brodman suggested, the computer could serve first as a model for understanding how physicians arrived at a diagnosis, and then as a tool to help them do a better job of it.

After all, who really knew what happened in a doctor's mind when faced with a patient? The British physician and inventor Firmin Nash suggested in 1954 that diagnosis typically takes one of three paths. Most often, diagnosis follows a simple pattern recognition, as subconscious processes trigger the immediate recognition of a disease. If I see that you have a scarletina rash, then I conclude you have scarlet fever. At other times, diagnosis requires conscious recall of descriptions of diseases from medical school lectures, textbooks, journal articles, or past cases. If I see that you have a rash I don't exactly recognize, but it looks just like a picture of scarletina rash that I find while thumbing through a dermatology atlas I have in my office, then I determine you have scarlet fever. According to Nash's breakdown, only a small subset of cases require the third option: a fully conscious and comprehensive analysis. In those cases, a cluster of key symptoms forms a starting point to generate a broader list of possible diagnoses, which are then winnowed down, step by step, using deductive logic.

Most symptoms or signs of illness, like fever or cough, are not unique. They can be found in many diseases, just as the attributes of red hair or attached earlobes could each be found in many people. But just as fewer people have red hair *and* attached earlobes, fewer diseases have both fever *and* cough. Each symptom adds more possible diagnoses to the list—but unique constellations of multiple symptoms help draw the diagnostician closer to the right answer.[41] Nash adapted this mathematical process of "conjugating the manifestations" from a textbook, *Differential Diagnosis*, by Johns Hopkins chief of medicine A. McGehee Harvey. The ideal physician as described by Harvey would first gather signs and symptoms through history and physical examination, then select two or three central features as hallmarks of the case, list all the diseases in which these central features occur, and then "reach final diagnosis by selecting from the listed diseases either (1) the single disease which best explains all the facts, or, if this is not possible, (2) the several diseases, each of which best explains some of the facts."[42] Harvey's model appealed to Nash as a means of mathematically expressing the rational deductive process of diagnosis as a set of formal rules—even if the complexity and context of the clinical encounter was reduced in the process.

Harvey's *Differential Diagnosis* was not the sort of book you could carry around in your pocket. So Nash worked to build a diagnostic slide rule that would convey the relevant data in a more portable form. Using cardboard and celluloid, Nash built a mechanical computer he called the Logoscope.[43] Nash's diagnostic device could differentiate among roughly 300 diseases, using interchangeable cardboard strips each labeled with a different symptom or sign. Each strip contained horizontal hash marks (which he called "spectral lines") that corresponded to a vertical index of diseases. When a set of symptom strips, representing the profile of a given patient, were inserted into the Logoscope, the differential diagnosis would be reduced to the small number of diseases in which each of these hash marks matched up into an unbroken line (fig. 6.4).

Nash admitted the Logoscope was a crude device. He described it as a diagnostic bus or streetcar, not a taxicab: "the user must realize that it will often take him somewhere near his diagnostic destination, and so save him much effort; but he will not usually be taken from door to door." At least one physician seems to have taken the claims of the Logoscope more literally than Nash intended, reporting a series of fifty cases in which the device was relevant to fewer than twelve patients, "and even then it did not always indicate the entire clinical diagnosis."[44] For Nash, however, the limits of the Logoscope were

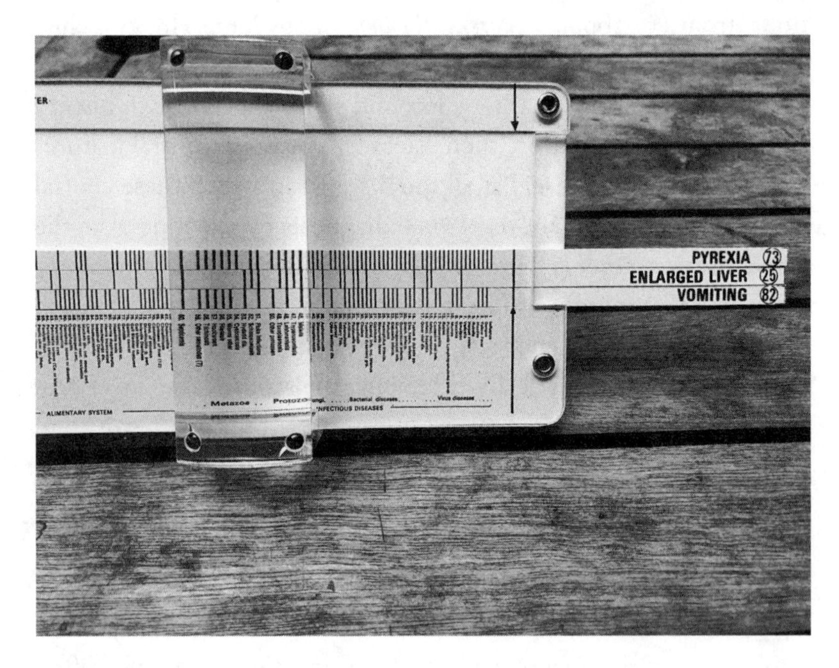

FIGURE 6.4. The medical slide rule: Dr. Firmin Nash's diagnostic Logoscope. Image from author's collection.

part of the point. The device illustrated problems inherent to clinical medicine: in real time, diagnostic decisions were often made from incomplete descriptions, posing hazards for both mechanical and human computation. Nash warned users that even the later Mark III version, which could process up to 700 million different combinations, could not be considered as "a sort of medical Aladdin's lamp, which, when rubbed, gives the user the diagnosis."[45] But then again, the same could be said about the practicing physician.

Both Brodman and Nash saw a future for diagnostic computers as electronic "thought-saving devices" that would assist clinical practice.[46] Both began with paper technologies to standardize and simulate the logical processes of diagnosis. Both hoped that the analytic capacity of electronic computers would help physicians make sense of the increasing amount of data they faced, as an expanding array of screening, imaging, and laboratory tests increased the amount of information collected around any given patient. Much as the industrial age had brought several labor-saving devices into the home and workplace that putatively freed up time for industrial and domestic workers, the diagnostic computer would handle the drudgery and free up the clinician to think about what really mattered to his patients. As we have already seen in the case of the telephone, these "labor saving" devices work to generate new forms of labor as well.[47]

Nash and Brodman offered two very different approaches to *how* a thinking machine might help to make diagnosis into a decision science. Nash's slide rule inscribed the logical processes of master clinicians like A. McGehee Harvey into the concrete form of a mechanical computer. Brodman, in contrast, used automated forms to input raw data into diagnostic tables that computers could adjust through iterative associative processes. This divergence anticipates a powerful and important split in the history of artificial intelligence (AI), between what would later become known as "expert systems" approaches and "machine learning" approaches. In the early years of AI, most computing departments favored approaches based on approximating human intelligence, in which an algorithmic approach to a problem attempted to replicate the knowledge base and logical process of a given expert—in this case, a physician—using the mathematical notations of formal logic. But in subsequent years, approaches to AI turned toward machine learning, in which dynamic algorithms altered their own processes based on associations found along the way. This split in diagnostic computing delineated what would become an enduring divide in the history of computing, between formalists and instrumentalists.[48]

In practice, though, many early attempts to build diagnostic computers alternated between the two approaches. For example, in the late 1950s, one group of hematologists specializing in the diagnosis of blood disorders began with Nash's expert-system methods but then later drifted into the domain of machine learning. The field of blood disorders, which depended on the quantified results of common laboratory tests, seemed an especially promising arena to test out automated diagnosis. At New York Hospital, hematologist Martin Lipkin adapted a series of edge-punch cards into a mechanical computer that replaced Nash's slide rule with a card-sorting device. Each card represented a different disease; the full stack represented all possible diseases. Their edges were notched to represent a different set of signs and symptoms from the clinical history, physical examination, basic blood tests, and bone marrow biopsy. A set of steel rods could be passed through the punched holes of the stack to correspond to the findings of a given patient; those cards skewered by all the rods represented all possible diagnoses for the patient at hand.[49]

In collaboration with Zworykin's Center for Medical Electronics, Lipkin replaced these punch cards with magnetic tape and the processing power of an RCA Bizmatic mainframe computer.[50] As the Cornell–RCA team reported, "we attempted to simulate the thinking of the doctor."[51] Though their initial efforts echoed Nash's model, Lipkin's team shared Brodman's hope that diagnostic computers might improve on existing diagnostic practices by developing their own, more accurate approaches based on raw data. A year later they devised another model that began with a set of twenty diseases and a pool of just under fifty patients and extracted and encoded data from their case histories, physical examination, and laboratory evaluation.

Jordan Baruch, an engineer from the firm Bolt, Beranek & Newman who coined the term "doctor-machine symbiosis," suggested that this model of diagnostic computing could develop its own feedback loop to organize and modify its own probability tables.[52]

"If we are going to seriously consider machines as diagnostic aids," IBM's Richard Taylor added, after learning of this self-modifying diagnostic computer, "we ought to take the point of view that probably the machine will not go through its 'thinking process,' if you will, in the same manner that a human does."[53] It did not matter whether the method a computer used to arrive at a diagnosis reflected the ways that physicians understood their own diagnostic practice, as long as it was accurate. "There is no reason to make the machine act in the same way the human brain does," Taylor concluded, "any more than to construct a car with legs to move from place to place."[54] Implicit in comments by engineers like Baruch and Taylor was the idea the computer might do the job *better*. While the "car with legs" metaphor resonated with the worldview of computer scientists, it did not go over well with physicians. People were not cars. Doctors did not want to cede their role in a knowledge economy to machines—especially not to machines whose process for arriving at answers could not be explained according to the deductive processes through which physicians understood themselves to be experts and exemplars.

Another New York hematologist, B. J. Davis, partnered with IBM to create a different self-adjusting algorithm with potential to diagnose blood diseases better than human hematologists. Davis's partner, IBM engineer Taffee Tanimoto, had been working on computer programs to make genus/species distinctions to serve as a definitive field guide to the plants of the New York Botanical Gardens. With only a few modifications, it could be used to produce a field guide to hematological disorders instead. "In medicine, there are clinical patients and their clinical case histories," Tanimoto noted, while "in botany, there are plant specimens and their morphological properties."[55] In either case, the computer could help to make genus/species distinctions by using machine learning to weight the most relevant distinguishing elements between two otherwise similar entities. Silver maple vs.

sugar maple? Count the lobes on the leaves. Microcytic anemia vs. macrocytic anemia? Count the mean corpuscular volume.[56]

Instead of programming a diagnostic process to replicate the categories used by expert authors of hematology textbooks, Tanimoto and Davis fed "crude medical data" into their IBM 704, and watched how the computer lumped and split the symptoms and signs into clusters on its own.[57] The algorithm in their model, published as a flowchart in *Archives of Internal Medicine* (fig. 6.5), was designed to adapt its own tables so that each confirmed diagnosis/symptom pairing would add to the relative weighting of given symptoms in future diagnoses. The machine diagnosed blood disorders as well as a physician using a textbook, but used its own logic to do so.

Physicians found Davis and Tanimoto's algorithm to be both rational and unreasonable. It was rational in that the steps could be followed mathematically to produce results that were demonstrably accurate and efficient. It was unreasonable in that clinicians could not see their own principles of diagnostic reasoning in the process. In the postwar decades, physicians had fashioned their considerable professional self-image as symbols of deductive thinking and cognitive power. This algorithm suggested a future in which AI might displace physicians from their perceived role as model experts.

It was, in a word, a threat. A month later the same journal published a lengthy editorial on the dangers of automation by William S. Middleton, the medical director of Veterans Affairs (VA). Middleton described a near future in which medical students were unfamiliar with auscultation, reading, and even writing their own chart notes, because the acceptance of diagnostic computing had devalued critical thinking and deskilled the profession. Like Sherlock Holmes, Middleton argued, the physician should be an exemplar of deductive problem-solving. Medicine still had a chance to "escape automation," he warned, if physicians resisted the computer. If they failed, the profession of medicine would "perish from the face of the earth and in its stead a new breed of coldly impersonal, accurate technicians will take its place in the sun."[58] Medicine was a symbolic nexus between art and science that relied on the critical reasoning skills of individual

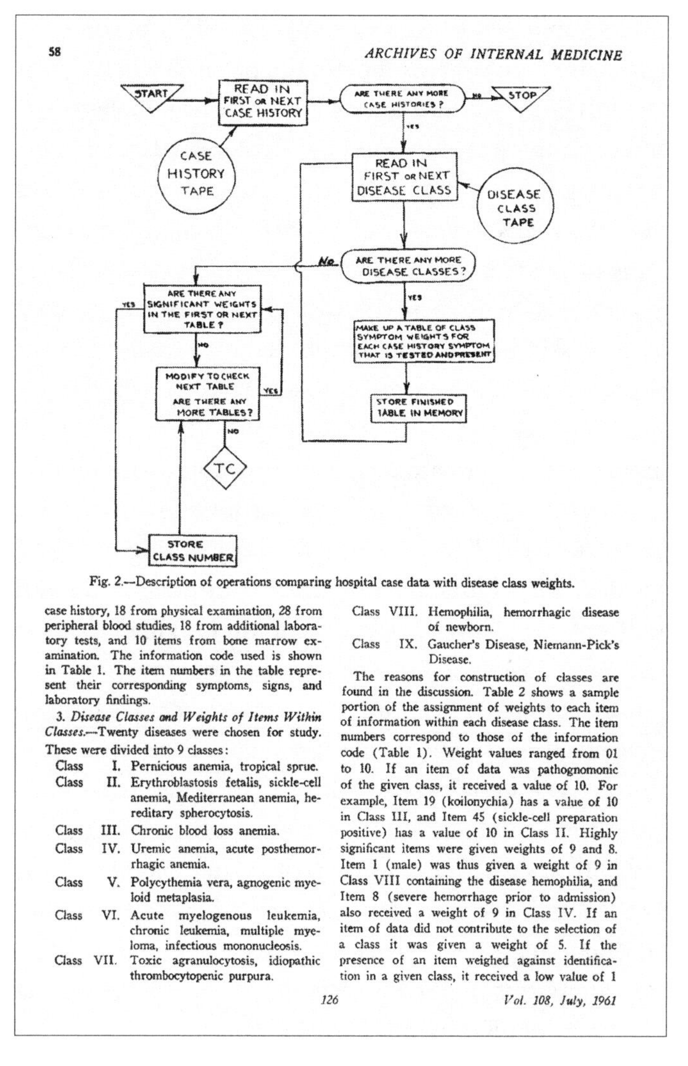

58 ARCHIVES OF INTERNAL MEDICINE

Fig. 2.—Description of operations comparing hospital case data with disease class weights.

case history, 18 from physical examination, 28 from peripheral blood studies, 18 from additional laboratory tests, and 10 items from bone marrow examination. The information code used is shown in Table 1. The item numbers in the table represent their corresponding symptoms, signs, and laboratory findings.

3. *Disease Classes and Weights of Items Within Classes.*—Twenty diseases were chosen for study. These were divided into 9 classes:

Class I. Pernicious anemia, tropical sprue.
Class II. Erythroblastosis fetalis, sickle-cell anemia, Mediterranean anemia, hereditary spherocytosis.
Class III. Chronic blood loss anemia.
Class IV. Uremic anemia, acute posthemorrhagic anemia.
Class V. Polycythemia vera, agnogenic myeloid metaplasia.
Class VI. Acute myelogenous leukemia, chronic leukemia, multiple myeloma, infectious mononucleosis.
Class VII. Toxic agranulocytosis, idiopathic thrombocytopenic purpura.
Class VIII. Hemophilia, hemorrhagic disease of newborn.
Class IX. Gaucher's Disease, Niemann-Pick's Disease.

The reasons for construction of classes are found in the discussion. Table 2 shows a sample portion of the assignment of weights to each item of information within each disease class. The item numbers correspond to those of the information code (Table 1). Weight values ranged from 01 to 10. If an item of data was pathognomonic of the given class, it received a value of 10. For example, Item 19 (koilonychia) has a value of 10 in Class III, and Item 45 (sickle-cell preparation positive) has a value of 10 in Class II. Highly significant items were given weights of 9 and 8. Item 1 (male) was thus given a weight of 9 in Class VIII containing the disease hemophilia, and Item 8 (severe hemorrhage prior to admission) also received a weight of 9 in Class IV. If an item of data did not contribute to the selection of a class it was given a weight of 5. If the presence of an item weighed against identification in a given class, it received a low value of 1

126 *Vol. 108, July, 1961*

FIGURE 6.5. Machine learning in hematology, c. 1961. Courtesy of American Medical Association.

physicians: a fundamentally human task that would be degraded by automating its processes for the sake of mere efficiency. The diagnostic computer was a poison pill, promising short-term benefits at the cost of generations of future physicians.

For the Cornell hematologist Ralph Engle, however, both the hopes and fears found in Sunday-morning cartoon images of medical computing were overblown. The computer might become an aid to diagnosis, but it could never replace the physician as diagnostician, because most of medicine did not fit into the neat tables of data and decision trees. There could be no single, automated assembly-line approach to diagnostic computing. Rather, there were many pathways by which physicians arrived at diagnoses, and computers could help with some of them. Engle found the limits of diagnostic computers to be as instructive as their successes. The hematology computer worked precisely because the input could be easily standardized, and the categories of illness could be separated clearly. But adapting the hematology computer to other parts of medicine, like psychiatry, rheumatology, and dermatology, would involve forms of input that were much harder to standardize, and categories of disease that were far more protean and harder to automatically distinguish.

Engle described a fundamental irony embedded in the early history of diagnostic computing. Different kinds of diagnoses entailed different amounts of certainty and uncertainty. On one end lay a relatively automatic set of decision trees—like the diagnosis of streptococcal pharyngitis—in which clear rules gave clear answers. On the other end lay a tangle of protean disease categories like lupus, vasculitis, and syphilis, which could take on vastly different appearances in different individual patients. In areas of greater diagnostic certainty, computers could easily produce correct diagnoses, but they were rarely needed. In areas of substantial diagnostic uncertainty, where aid from computers might be welcomed by perplexed physicians, they could not yet offer useful guidance.[59]

Physicians were not really being asked to cede professional or economic control to these machines, not any time soon. Diagnostic computers had a more profound impact, Engle argued, as a metaphor to help physicians gain more insight into their own thought practices. The limits of the diagnostic computer served as a model for understanding limitations in human cognition, and in understanding that the physician's self-image as a rational expert who consciously deduced the diagnosis of any given patient was true only in a minority of instances. Much as Nash had used his slide rule to ask challenging questions as to how doctors think, Engle used his experience building a hematology computer to point out the very different processes involved in what physicians called diagnosis.

The medical mind continued to serve as a model for further generations of artificial intelligence. Engle's expanded HEME computer at Cornell used Bayesian algorithms, but over the course of the 1970s Jack Myers's INTERNIST-I computer at the University of Pittsburgh and Edward Shortliffe's MYCIN computer at Stanford became index examples of expert-system approaches to artificial intelligence. Yet on the whole, by the end of the first decade of medical computing, the mismatch between where computers were accurate and where they were needed struck a sobering note in the prospect of diagnostic computing. The vast majority of patients could be diagnosed by immediate pattern recognition, and diagnoses for most of the rest could be made with help from a colleague or a textbook. IBM's Richard Taylor estimated that only 1 percent of all cases could really benefit from computer diagnosis, and even in these rare cases most physicians seemed unwilling to admit a computer into the clinic for these purposes. Without more evident proof of benefit, the average physician would continue to resist the computer. "To date," Taylor concluded, the physician "has not, we understand, tolerated the thought of having a keyboard in front of him."[60]

* * *

Vladimir Zworykin believed that physicians would only truly begin to appreciate mainframe computing once they could see the benefits of the electronic medium in day-to-day clinical work. Once computers automated all parts of clinical interaction that could be automated, he argued, then physicians would find they had more time to focus on the important human tasks that computers could not do. More of the patient's time with the doctor could be engaged in talking and listening, receiving the advice and reassurance only possible in person-to-person connections. As the use of computers made medical informatics into more of a science, medical practice would become more of an art. This benefit would be most visible, he suggested, as digital computers made the tedium of medical documentation more efficient.

Zworykin insisted that the fundamental relationship between doctor and patient would only be enhanced, never degraded, by the presence of the computer in the clinic. In 1962, he asked an audience of physicians to imagine a typical medical clinic fifty years in the future, in which physicians were freed of asking the same questions, conducting the same physical exam over and over again, since a standard history and physical would be automatically performed by a human–machine interface. In the year 2012, he predicted, any "Mr. Jones" stopping by Middletown Clinic for his annual checkup would see the computer before he would see his doctor:

> As he enters he is ushered into an examination booth, inserting his coded social security card into an appropriate slot for identification of the examination record. A series of standardized questions concerning his physical condition are then flashed on a screen in front of him and he records his answers by means of yes-no push buttons. Weight, temperature, respiration rate, electrocardiogram, reflexes, and other data are registered directly. Blood, breath, and urine specimens are inserted into analytical machines, which add corpuscle counts and chemical data to the record. While Mr. Jones goes on his way, the examination record is transmitted in coded form to the regional electronic health record storage center, where it comes to form part of Mr. Jones' permanent health record.[61]

Middleton's dystopia was Middletown's utopia. While Zworykin admitted his predictions might sound like the stuff of fantasy, several pieces of this future were already coming together. "The use of electronics for making both health records and the sum total of medical knowledge more immediately accessible to the practicing physician," he concluded, "appears so necessary that it can be regarded as inevitable."[62]

Zworykin had been impressed by a prototype electronic medical record developed at Tulane University using an IBM 650 computer bought with federal funds in 1958.[63] By 1960, the Tulane group could already claim three successes. First, they determined that it was possible to place an entire clinical medical record on magnetic tape. Second, hundreds of thousands of these records were available for use by different users at the same time. And third, freed from the manila folder, patient charts became part of an open-ended digital database for a clinical research protocol they called the Medical PROBE. As a research tool, Medical PROBE turned "man-months" of labor coding case histories into "machine-minutes," and allowed new modes of research to be conducted on medical records.[64]

If every clinical encounter could become part of a vast integrated database, the medical record would no longer be bound to individual patients or individual clinics or hospitals, but instead would become a vast, interconnected research organ making clinical practice into a form of networked data science.[65] "The computer in its true form," an editorial in *JAMA* concluded, "is nothing more than an extension of the person's mind."[66] As they altered their prototype with faster processors and larger data banks, the Tulane team urged physicians to consider the computer as a tool for reevaluating the nature of the medical record itself. Accessible, generalizable, precise, and portable, the electronic medical record was ready for large-scale demonstration in a major hospital.

In 1962, the American Hospital Association and the National Institutes of Health announced the launch of the Hospital Computer Project of the Massachusetts General Hospital (MGH), Boston's oldest and best-known general hospital. The online computer system, designed by Jordan Baruch of Bolt, Beranek & Newman, had 48K of core memory. It could support sixty-four simultaneous users, each using a wood-paneled teletype unit connecting by dial-up terminals to an external mainframe.[67] It was a technical success but a practical failure. The physician and computer scientist G. Octo Barnett, who became the MGH project lead for the Hospital Computer Project in 1964, soon found himself at odds with BB&N colleagues who had little interest in how doctors and nurses could make use of computers in day-to-day life. Under the strain of regular patient care, the BB&N systems regularly broke down, and the electronic medical record database was often offline for hours at a time. "Demonstration programs," Barnett recalled years later, "however impressive, and promises for the future, however grandiose, could not substitute for the reality that all our computer programs could only function in the hothouse atmosphere of parallel operation for several days to several weeks and operated by our own research staff."[68] Showing that an electronic medical record *could* work was not enough to convince physicians and nurses to trust it with the vital aspects of patient care. Paper charts remained crucial to the functioning of hospital wards throughout the project.

Baruch and Barnett's attempt to create a total hospital computer system in MGH attracted substantial media attention. "The computers are coming to the health professions," *Medical World News* announced in a feature article on the Hospital Computer Project, "beset by a flooding tide of paper work, growing personnel shortages, and demands for patient care that threaten to overwhelm them, computers offer unique and revolutionary solutions."[69] Yet when the same journal revisited MGH a year later, the project had been called off, and was already considered to be a fiasco. Barnett detailed the frustrations he faced trying to build a total electronic information system within a hospital. He blamed not just the limitations of existing computer technology, but also the mutual incomprehension between physicians and engineers, hospitals and computer firms.[70]

The MGH Hospital Computer Program was a highly visible top-down failure for electronic medical records in major academic hospitals. But other efforts in smaller community hospitals were somewhat more successful. In 1966, Lockheed Missiles and Spacecraft systems engineers submitted a proposal to the US Public Health Service to develop a "total systems approach" to build their own proprietary Medical Information System.[71] Lockheed's first electronic medical record was tested in the El Camino Hospital of Mountain View, California, and the project continued after Lockheed spun off its medical informatics division to Technicon Medical Information Systems.

Technicon was able to monetize the electronic medical record as a product and a service. Unlike the initial MGH Hospital Computer Program, the Technicon system was operational more than 99 percent of the time. Two-thirds of all clinical staff initially opposed the total medical information system, but as the project came to an end, they found they had learned to use the systems. In 1974, when El Camino's physicians and nurses were asked if they would be willing to pay full fees to continue using the Technicon system, the electronic medical record had become part of the everyday life of the institution. Clinical staff voted overwhelmingly to pay a subscription fee to keep the system. Other hospitals, including NYU, Temple, Loyola University of Chicago, and University of California–Irvine, followed suit. In the public sector, Technicon was selected to design a similar system for the Clinical Center of the National Institutes of Health, which soon generated more than 110,000 patient records. Many federal hospital systems, including the Indian Health Service and the Veterans Affairs system, adopted some of these early electronic medical record systems as well.[72]

Yet Technicon's success was the exception, not the rule. Most attempts to develop a total hospital information system were dead in the water by the end of the 1960s. As hospitals retreated from these all-encompassing visions of digital healthcare, computer systems became integrated into hospitals and clinics in a more piecemeal fashion. Computers offered some benefits to individual hospital departments, but they did not suddenly displace the paper chart or transform all clinical interactions into electronic data. Octo Barnett, who built a successful laboratory at MGH and MIT for computers from the wreckage of the Hospital Computer Project, pursued flexible approaches to computerize the hospital as a series of parts rather than as a whole. This modular approach tended to start with an area that was particularly well suited to automation—such as hospital billing or the reporting of laboratory results. Additional modules, like searching an automated analysis of drug interactions for the hospital pharmacy, could then be rolled out later as the system expanded. Computers could enter hospital practice bit by bit, rather than all at once.

Along with his colleague Neal Pappalardo, Barnett developed a specialized language for these modular systems called MUMPS (Massachusetts General Hospital Utility Multi-Programming System), which laid much of the groundwork for later electronic medical records systems. Pappalardo left MGH to found Meditech, one of the first commercially successful electronic medical records firms, and MUMPS became the basis for its systems. A MUMPS Users Group, founded in 1972, eventually grew to include a variety of academic medical centers and the Veterans Affairs system, as well as Meditech, Epic Systems, and GE Healthcare. The lasting impact of this modular approach can still be seen in its use by the three biggest commercial suppliers of electronic medical records in North America.[73]

Physicians soon complained that these modular systems were developed for billing offices and laboratories, and not for patients and their physicians. Modular systems contained many redundancies, and often increased, rather than decreased, the amount of text produced for each patient. Like the telephone and the radio-pager, the electronic record made more work for doctors, not less. As the electronic health record has become an inevitable feature of clinical care in the twenty-first century, studies also indicate that it has increased—rather than decreased—the amount of time clinicians spend documenting their efforts.

Donald Lindberg saw this coming. Lindberg, who would become a key figure in supporting the role of computers in medical informatics at the National Library of Medicine, was already wary by 1968 that computerization of records might increase, rather than decrease, the volume of the medical chart. "There is a hope growing among physicians," he noted, "that computer systems can be used to cut through the bewildering mass of handwritten pages which constitute the traditional patient record or chart and eliminate the 95 per cent of the record which is irrelevant." But for the electronic record to help, rather than hinder the direct care of patients, healthcare practitioners needed to ensure the records would be designed with patient care in mind. Lindberg would devote decades of his career at the National Library of Medicine to achieving this goal.

This idea was echoed by Lawrence Weed. A young internist at Case Western Reserve University, Weed called for the design of computerized records that could help teach physicians to become better doctors. For Weed, the paper chart was a morass. It interfered with the ability of doctors in training to learn how to think scientifically. Important information got lost or misplaced. Paper, a slow medium, hampered doctors from applying the deductive logic necessary to the diagnostic and therapeutic decisions they faced. Paper charts led to inconsistencies in the recording of clinical information, inconsistencies in diagnostic thinking, and inconsistencies in access to new medical knowledge. Electronic systems offered "a more organized approach to the medical record" that could help make medical training and medical practice more scientific.[74]

Weed proposed the "automation of a problem-oriented medical record," as a solution to the vagaries of the paper chart.[75] Paper charts simply allowed too much variation for how a given doctor entered clinical information on any given patient.[76] By making the medical record into a more functional database, computers could help make medical practice into a more efficient data science. If the form of the medical record could be standardized around the patient's past, present, and future problems, it could become a far more powerful tool for helping physicians think broadly about their patients' well-being over longer periods of time. Like Zworykin, Weed insisted that computers would enhance, rather than detract from, the human elements of medicine.[77]

Weed's vision of a more scientific practice of medicine appealed to many physicians, but his use of computers did not. Instead, his effort to transform medical records through medical computing had its greatest impact, at least in the short term, in producing new forms of paper charts. Weed's synthesis of medical informatics, medical diagnoses, and medical records transformed the original components of the patient note (chief complaint, history of present illness, past medical history, and so forth) into a new set of computer-inspired categories (database, problem list, initial plan, progress notes).[78] For Weed, this involved a careful distinction of the kinds of data gathered during the history and physical. Subjective data (S) reflected the qualitative and quantitative description of symptoms appropriate to the problem, while objective data (O) reflected the clinical findings on examination, imaging, and laboratory analysis. The calculating mind of the physician would then make a general assessment (A), which allowed a reflexive chance to question the accuracy of the original problem, before formulating a plan (P), which would be revisited in feedback loops in future visits. Thus the SOAP note—an iconic feature of the paper chart in late twentieth-century medical practice—was born out of a series of speculations of the role of computers in medicine.

The SOAP note is still widely used today in American medicine, and it is an enduring reflection of Weed's philosophy. The computer, he believed, should be understood as a "coupling agent" between patients, physicians, and information. By collecting a database of problems, the electronic medical record could help highlight what was needed to solve or treat them, and bring to bear resources the ordinary physician might not otherwise have at hand. "Looked at in this way," Weed elaborated, "the physician's memory becomes just one of the available information sources." Instead of being limited by the predilections or limitations of a given physician's knowledge base, "the very tool that the physician uses to do his work (e.g., the computer system) should have built into them the parameters of guidance and the currency of information that he needs to do his work well."[79] To Weed, as to Zworykin, Engle, and others, the computer offered a means to acknowledge and transcend the limitations of the individual physician for the benefit of their patients.

Weed continued to advocate for total hospital information systems, pitting his efforts against Octo Barnett and the modular systems of MUMPS and Meditech, and others who would build the electronic medical record in a piecemeal fashion. Left unchecked, this "proliferation of automated systems within parts of a hospital complex" would lead to further sprawl.[80] Medically speaking, the sum of the parts would be much costlier than a system rationally designed as a coherent whole. The computerization of single components of the hospital complex, like pharmacies and laboratories, would just lead to more bloat in the system, more unnecessary prescriptions, and more unnecessary daily labs. Only with a cohesive, problem-based approach guiding the logic of the electronic medical record would all these parts come together to produce a more rational form of medicine with the patient at the center.

* * *

Weed's work spread quickly, but the medical profession adopted only those parts that meshed well with the ideal of the independent physician-expert. His 1968 paper on "medical records that guide and teach" was an instant classic, and his subsequent book, *Medical Records, Medical Education, and Patient Care*, sold out several editions. The problem-oriented medical record became standard at Case Western Reserve; when he moved to the University of Vermont in 1969, it was rapidly adopted by interns and residents and private physicians associated with the hospital. At Vermont he developed a fully computerized Problem-Oriented Medical Information System (PROMIS) that spread from general medical wards to the gynecology service to other parts of the hospital.[81] When Emory University School of Medicine hosted a conference on "Teaching Methods and Patient Care with Emphasis on the Weed System" in 1971, the 300-plus people in the audience represented three-quarters of all medical schools in the United States and Canada, as well as the American College of Physicians, the Veterans Affairs hospital system, several US armed services medical branches, and NASA.[82]

Weed's vision for PROMIS recalled Zworykin's three-part vision for integrated medical computing. First, PROMIS was intended to coordinate a total information system for each individual patient by combining the perspectives of all members of the care team in one platform. Second, the use of computer databases could restore the art of medicine by enabling "the conversion of the physician from a memory machine and oracle to one of the principal components of a guidance system in medicine." Third, the computer interface would provide a dynamic set of feedback loops to improve the efficiency and accuracy of diagnosis and treatment.

Paper was the problem. The profession of medicine had been mired for centuries in a sluggish medium that sequestered knowledge from practice. Like other early supporters of medical computing, Weed planned an electronic medical record that could integrate with newly digitized medical literature at the point of care delivery. This, he imagined, would lead to better diagnosis, to a more current reading of the medical literature, through a series of virtuous loops of continually improving medical knowledge and clinical practice.[83] "It is now possible," he added, "to develop a library of displays which can be logically linked, by a touch of the physician's finger," so that the doctor at the bedside could be in constant contact with the latest medical research and the advice of colleagues.[84] Like Zworykin a decade earlier, Weed hoped to couple the electronic possibilities of current medical literature with diagnostic guidance as a given patient's problems unfolded on an electronic medical record in real time.

Yet the full promise of PROMIS did not come to pass.[85] Only a limited number of physicians adopted Weed's software. Most preferred to write their notes by hand. Instead, the problem-oriented medical record would find its broadest uptake not in the form of electronic medical records but in the millions upon millions of SOAP notes still being written by hand in patient charts even at the turn of the twenty-first century. The SOAP note owes its existence to Weed's experiments with medical computing, but it became most popular as a paper tool. This form of mainframe medicine would profoundly change the way many doctors thought and shared information about their patients—but not the medium in which those thoughts were recorded.[86]

By the early 1970s, most efforts to build a single computer system for all of the informational needs of a hospital or health system had broken into smaller pieces. As Jordan Baruch observed, computer systems in medicine now had to serve three very different sets of users, with very different needs, in the fields of clinical care, research, and administration. Clinical care systems required lots of different modules for different aspects of medical care, with heavy emphasis on the interface that healthcare providers would use. Researcher systems needed very large, long-term memory and sophisticated programming configuration. Hospital administrators, in turn, wanted scalar business systems like those found in any other cost-conscious industry.[87] When these goals came into conflict, the vision of the computer as a tool for administration and finance trumped both patient care and research. If you couldn't serve three masters, Baruch suggested, you served the one who paid the bills.

Speaking at the tenth and final IBM Medical Symposium in 1970, keynote speaker Warner V. Slack struck a somber note. In the early 1960s, Slack had led efforts to use computer terminals to enable patients to write their own clinical information into their medical records. But in spite of successes like MEDLARS and failures like the MGH Hospital Computer Project, his overall takeaway from the first decade of medical computing was one of chastened hopes and more pragmatic engagements. Even if the computer had brought a few real benefits into medical science and medical practice, he concluded, "the concept that the mere presence of a machine will result in good things—that if you plunk a big computer down in the middle of a hospital, automatically good things will accrue—this is wrong and has done harm."[88] After a decade of effort, he concluded, "there aren't too many sick people in the world who have really benefitted from computers as yet—unless perhaps to get their bill a little earlier—which, from the patient's viewpoint, is of moot value."[89]

The problem wasn't that computers weren't useful in medicine. Their usefulness had been oversold and misplaced. The computer had been presented as a revolutionary technology that would suddenly transform the experience of being a patient, the daily work of being a doctor, and the efficiency and accuracy of healthcare institutions in general. The real change that mainframe computers brought was more evolutionary—not a rupture with paper technologies, but continuous and convergent with them.

Paper remained the dominant medium of medical practice through the end of the twentieth century. While the successes of MEDLARS in the 1960s eventually led to the all-electronic medical library of the twenty-first century, for much of the twentieth century MEDLARS, MEDLARS II, MEDLINE, and PubMed represented digital wayfinders for medical professionals who still worked in print. Despite the fears that computers would usher in a post-physician era, diagnostic machines did not displace the doctor or the prescription pad in the late twentieth century. Instead, with the failure of early total information systems, medical records found piecemeal uptake in different parts of the hospital and clinic—especially in places like the billing office, the admissions office, the pharmacy, and the laboratory. For the next half-century these electronic media overlapped and converged with thick tomes of paper charts that remained, until very recently, the most durable means of archiving clinical experience, research findings, and diagnostic thinking.

Computers offered new tools to produce, circulate, and use medical knowledge. They offered new metaphors to understand the process by which physicians diagnosed their patients and approached complex clinical decisions. They offered new ways to rethink how relevant information could be collected from a given patient, organized rationally, and shared with others. But these changes found most traction where they could be integrated within existing paper practices in the library, clinic, and hospital. These forms of success were less newsworthy, harder to see, more challenging to connect directly to the computer. They were also easier for subsequent generations of digital medicine enthusiasts to forget.

7
THE AUTOMATED CHECKUP

More data leads to better health. This equation is the central promise of precision medicine, a term introduced to the American public in the 2015 State of the Union address, as President Barack Obama announced a national initiative to "use new data systems to tailor therapeutics" and personalize medical decision-making based on "individual differences in people's genes, environments, and lifestyles."[1] It is the promise of medical informatics: "the science of how to use data, information and knowledge to improve human health and the delivery of health care services."[2] It is the promise of a growing market of smart devices that continuously draw data from our persons and into cloud computing to predict our health futures. Even if it is not always clear whether these predictive analytics are more useful to our own physical well-being or to the financial well-being of businesses tapping into a new resource, it appears to many that a new assumption in healthcare is at work. More data, better health.[3]

Yet this idea is not new. A half-century earlier, RCA's Vladimir Zworykin had predicted that in 2012—that is, by the end of Obama's first term—an automated medical history, physical examination, and laboratory tests would be uploaded to a central computer well before a patient saw their doctor. Zworykin's vision of personalized medicine through an automated "examination booth" with "yes-no push buttons" and specialized sensors to collect data on individual human bodies may have sounded like science fiction in 1962.[4] But it was built in brick and mortar, and put into practice just two years later, in a clinic in Oakland, California.

The first Kaiser Permanente Automated Multiphasic Health Testing Center opened its doors in the summer of 1964 with a mission to automate the generation, storage, and analysis of personal health data for preventive medicine. Automation was built into the structure of the clinic. Like the popular H&H Automat restaurant chain, which reconfigured its dining room so customers could select the meal they wanted without having to wait for an individual waiter, Kaiser's new health center was designed to speed the extraction of relevant medical data from patients without having to wait for an individual physician. A patient entering Kaiser's automated health testing center would receive a stack of IBM computer cards from the receptionist (fig. 7.1), which served as tokens to enter the clinic's twenty stations. Each station encoded an encounter with a different human or mechanical sensor that related to a different part of the medical history, physical examination, or laboratory testing typically found in an annual doctor's visit. The doctor did not need to be there at all.

It took two to three hours to complete the circuit. After depositing a punch card with demographic information at Station 1, you would undress and put on a paper gown for Stations 2 through 8, which included an electrocardiogram (ECG), glucose tolerance test, skinfold somatotyping, chest X-ray (and mammography if indicated), and a tilt-table test. Dressed once again in your street clothes, you would face new sensors in Stations 9 through 15 that recorded your visual acuity, pupillary escape, tonometry, spirometry, retinal photography, and audiometry onto other punch cards. Station 16, the "History Chamber," contained a substack of 207 punched cards, each bearing a single question that could be answered either yes or no, with which a computer would process your health risks. Do you suffer from headaches? Has anyone in your family had a history of diabetes? Do you need to get out of bed to urinate in the middle of the night? Stations 17 through 20 extracted blood and urine for automated multiplex laboratory analysis that would be fed directly into the central computer, which might tailor a few additional tests before you'd leave. Perhaps

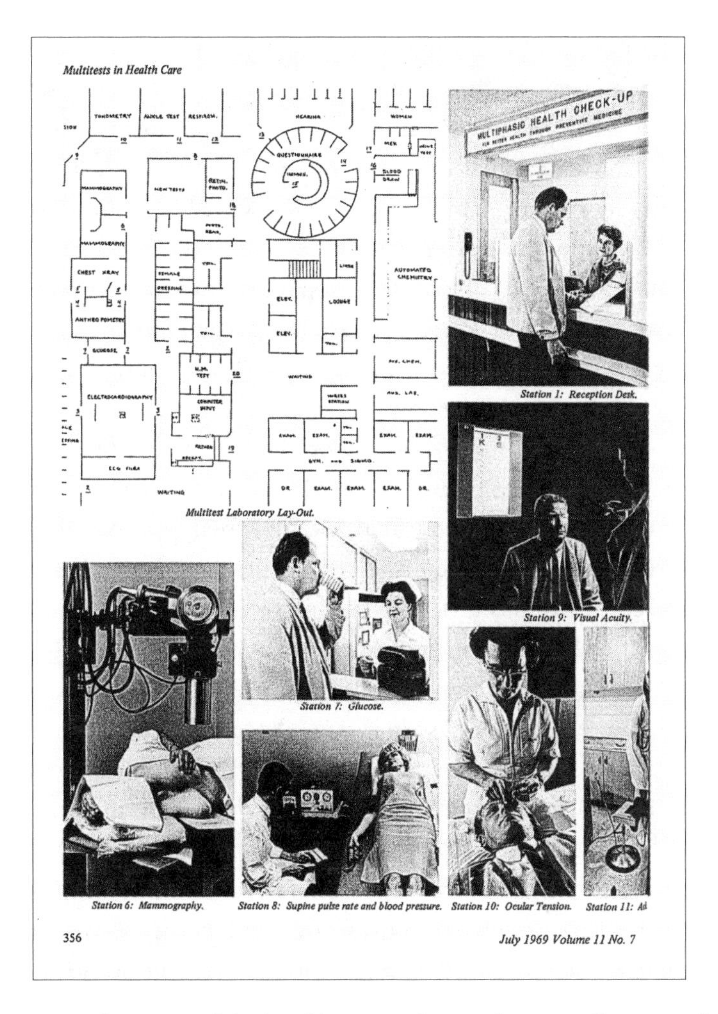

FIGURE 7.1. The automated checkup: blueprint and station-by-station illustration of Kaiser Permanente's first, twenty-station automated clinic in Oakland, CA. Courtesy of Springer BMC.

you would meet criteria for a full gynecological examination and Papanicolau smear, or a sigmoidoscopic evaluation of the colon, administered in a separate annex just before the exit. As you walked out the door, the data would be analyzed, printed out, and sent to your physician for a discussion about what these data meant for *you*.[5]

Computers promised new precision in prevention. "The combination of automation and computers," one Kaiser physician claimed, yielded "speed and precision unmatched by even the most highly trained personnel . . . in a manner not humanly practical, nor even possible."[6] To other physicians the prospect of automated health checkups immediately raised fears of a new form of "assembly-line medicine" in which the role of the doctor would be reduced to that of a mere technician. Patients, however, seemed less concerned by this. "I have to admit," one of the first patients processed in the Oakland clinic recalled for *Reader's Digest*, "that when I first heard about the center, I didn't know much about it, and some of the terms that go with it—'automated,' 'computerized,' 'multitest,' and so forth—stirred some rather interesting images." But the computer, he added "is something that the typical patient in the center never sees, and I don't think he is even aware that the data is being processed as it is taken from him."[7] Like the diner at the automat, the patient at the Automated Multiphasic Health Testing Center saw only the direct interface with its cubicles, not the system as a whole.

The automated checkup blurred the public health function of screening for chronic disease in a general population with the medical function of diagnosing and treating the hidden ailments of a specific individual. It reframed the goal of preventive care as a process of data collection that focused on individual bodies and individual risks and largely elided the social world. Kaiser Permanente's philosophy of better health through better data—or "Total Health"—was formed in explicit opposition to the expansion of nationalized healthcare, and prioritized a vision of health that located risks in the individual body rather than the broader community. The original Oakland site, and the many other such sites that followed, were designed to be a streamlined interface between patients, the data that could be extracted from them, and the algorithmic manipulation of personal health information.

As Kaiser's model spread, the automated checkup was promoted by public health officials and policymakers as a device to deliver access to preventive healthcare to all Americans. In the decade after the first automated clinic opened its doors in 1964, more than half a million patients had their health status automatically extracted, stored, and digitally analyzed in the Bay Area multiphasic testing centers of Kaiser Permanente alone. The automated clinic promised to provide preventive care, at a lower cost, to urban and rural communities that had been unable to access preventive medicine of any kind before. In the US Senate and the House of Representatives, lawmakers proposed a new "Preventicare" federal health entitlement based on the Oakland model to complement the newly implemented Medicare and Medicaid. The US Public Health Service announced a pilot program of federally funded automated multiphasic health testing centers. By the end of the decade, many physicians and pharmaceutical firms saw automated prevention as an inevitable future for American healthcare. The American Medical Association issued a statement of support, and encouraged individual physicians to buy new devices to perform automated screenings in their own clinics. Medical electronics and device manufacturers, anticipating a boom market, lined up to produce prototypes.

Mainframe medicine was a new form of media. And all new media deal in futures.

* * *

The computerization of preventive care started with a health fair in a union hall. In 1951, the state of California launched a new project with the Oakland-based Permanente Health Plan to rethink the early detection and prevention of chronic disease in outwardly healthy-appearing people. Permanente, one of the first prepaid health plans in the United States, was founded by the shipbuilder Henry Kaiser and the surgeon and health system designer Sidney Garfield to help maintain the health of East Bay and Portland shipyard workers during the production boom of World War II. After the war, the health plan saw a rapid influx from civil servants' and workers' unions motivated by affordable options for preventive care. When the head of the International Longshoremen & Warehousemen's Union brought all 6,000 union members and nearly 9,000 dependents into the Permanente Health Plan in 1950, one condition was that the plan find a way to rapidly screen these newly insured patients for chronic health risks using the best measures available.[8] Garfield reached out to Lester Breslow, a rising star in the California Department of Public Health.

Breslow was in the process of restructuring the state's approach to preventive health, as the chief causes of death shifted from acute infectious diseases toward chronic, indolent conditions like heart disease, cancer, and stroke.[9] New screening tools like portable chest X-rays, Papanicolau smears, urinalysis, and blood tests offered the means to detect hidden tuberculosis, precancerous lesions, asymptomatic diabetes, and high cholesterol in apparently healthy individuals well before any symptoms appeared.[10] But these tools were being used in a piecemeal fashion in individual clinics by private physicians, and Breslow wanted to find a way to engage all of them, all at once, to screen large masses of people.

The system he called "multiphasic screening" first took shape in a warehouse in San Jose, combining several individual screening tests into a single protocol administered to 1,000 workers from four local corporations.[11] Any abnormalities detected would be referred to physicians for final confirmation. This division demarcated the public health function of screening (conducted en masse to detect signals of possible pathology in a population) from the medical function of diagnosis (conducted one-on-one to definitively identify and treat pathology in an individual).[12] As the AMA took increasing interest in his work, Breslow repeatedly stressed that "the multiphasic screening procedure is not a substitute for a visit to a physician," who should still be visited annually whether symptoms were present or not.[13]

Garfield asked his colleague Morris Collen, chief of medicine and head of research at the Permanente Health Plan, to help Breslow develop a larger multiphasic screening event on the docks of Oakland.[14] Between June and November 1951, nearly 4,000 union members passed through an expanded multiphasic health screening at the International Longshoremen's & Warehousemen's Hall in San Francisco. Twelve stations measured height, weight, vision, and hearing; performed chest X-rays and electrocardiograms; measured blood pressure; performed blood tests for syphilis, hemoglobin, urine sugar, and urine albumin; and filled out standardized medical history forms. Nearly one out of five apparently healthy workers were found to have a previously unknown diagnosis, many of which were amenable to treatment with preventive measures.[15]

How many other hidden diseases, Collen wondered, could be detected by adding more screening tests to the lineup?[16] The addition of any new screening test added potential health benefits for patients but also caused a geometrical increase in the computational challenges of screening and storing, retrieving, and analyzing the data. Collen, who had trained as an electrical engineer before entering medical school, looked to the promise of digital computers like the IBM 1440 Data Processing System recently installed in Kaiser Permanente headquarters. After Garfield sent him to attend Zworykin's International Conference and Exhibition on Medical Electronics in New York in 1961, Collen made important contacts at IBM, returned to Oakland, and drafted a grant application to the US Public Health Service to "automate and improve our existing manual multiphasic screening program."[17] Collen and Garfield asked the staff at IBM/Oakland to help them build a new multiphasic testing facility from the ground up that foregrounded a computational, data-centered approach to individual preventive care.[18]

The design of Kaiser Permanente facilities had consistently emphasized functional architecture, with a special interest in creating spaces that used communications technology and systems theory to optimize the flow of people and information. *Time* magazine described Garfield's original Kaiser Foundation Hospital in Oakland as the first "Push-Button Hospital." The new Automated Multiphasic Health Testing Center could be described as the first punch-card clinic: engineered to efficiently extract individually relevant data from masses of patients who would flow through it.[19]

Collen and Garfield envisioned the spaces and surfaces of the Automated Multiphasic Health Testing Center as a man–machine interface. The automated checkup would serve as a gateway into the healthcare system for all newly insured patients. Though the clinic required human workers to guide the patient through the three-hour circuit and strap them into the diagnostic machinery, these were largely paramedical rather than medical roles.[20] The clockwise circuit of numbered cubicles scripted a consistent flow of information from patients as they moved from station to station. Most of this information could be collected mechanically, like obtaining a sandwich at an automat. Even the complex details of a patient's medical history could be broken down into a mechanical yes/no decision tree, filled out on punch cards, and represented within the binary logic and magnetic memory of the central computer. A better technological interface, it was hoped, would lead to a better flow of medical information—and better care.

As a private health plan that employed its own physicians, Kaiser Permanente could use automated multiphasic testing to blur the line between screening and diagnosis in a way that the California Department of Public Health simply could not. "Since *diagnosis* is defined as the identification of a specific disease," Collen explained, "then as screening becomes more comprehensive, precise, and quantitative, disease detection approximates disease diagnosis and automated multiphasic screening approaches *automated diagnosis*."[21] At times Collen referred to this as "provisional diagnosis," at other times as "machine diagnosis." Either way, to many private practitioners the blurring of screening and diagnosis threatened a crucial firewall between public health and private practice. It also threatened their own role in the art of diagnosis—long protected as something only an individual clinician could do when confronted with an individual patient—into something that could be automated and mechanized on an assembly-line basis.

The automation of preventive screening also changed the role of individuals and populations in creating new definitions of normal and pathological. Pop-up health prevention fairs had previously compared the test values they measured in screening tests against established tables of normal values in other populations. But with a computer built into a permanent testing facility, Collen and his statistician colleagues could use their own measurements to create "likelihood statistics" that made increasingly accurate inferences about the probability of disease in a given community as more data were collected on more patients.[22] Given a bit of time, the clinic's computers could calculate their own definitions of normal lab values from the roughly 4,000 patients the clinic processed each month, building their own relative weights for decision-making, which would become more accurate as more information was collected.[23]

In other words, the IBM 1440 at the heart of the Oakland clinic did not merely store and tabulate data, nor did it merely suggest possible diagnoses through expert-driven flowcharts. It was a machine that could adapt to develop improved decision models to offer better preventive care to each individual according to their medical history, physical examination, and laboratory tests. The machine could learn. The vast troves of medical data it held could produce new norms, new pathways for differentiating pathology, normality, and everything in between.[24]

Yet as the machine learned, person by person, its input was limited to the measurements and yes/no answers of a collection of atomized patients. In turn, as it processed medical histories one by one, the IBM computer built a model of preventive medicine that could only be understood as the sum of the risk profiles of thousands of individuals. Not surprisingly, the form of prevention supported by such data overwhelmingly equated prevention of disease with the use of new medications, surgical procedures, or behavioral change. This model of prevention prioritized the individual, rather than the community or broader social world, as the site of intervention.

* * *

For a moment, it seemed that the automation of prevention would become the basis for a new American healthcare system built around mainframe computers. It certainly seemed that way in September 1966, as Sen. Harrison A. Williams of New Jersey emerged from a trailer parked outside the US Capitol, blinking his eyes. The trailer contained a mobile version of Collen's automated multiphasic health clinic, and Williams reported back to fellow members of the Senate

Special Committee on Aging, "I will say right now that I am a little bleary from the drops that are necessary for the proper testing." Nonetheless he could clearly see that automated multiphasic health testing should play a role in moving American health policy from a reactive approach to "sick care" toward a more proactive form of "health care" that stressed prevention.

Williams convened these hearings to propose a new Adult Health Protection Act—immediately nicknamed Preventicare—that would build a new Medicare program around the Kaiser model of automated multiphasic health testing.[25] A nationwide program of computerized preventive medicine, he claimed, would allow America to "enter an age of health maintenance, rather than relying almost solely on health repair." The bill's co-sponsor, Rep. John Fogarty of Rhode Island, visited the Oakland clinic and declared that scaling up the Kaiser model would "make medical testing services more widely available, to heighten their effectiveness, and I hope, to bring down costs."[26] The automated clinic was a catalyst, allowing policymakers to envision new forms of public health programs that were previously unthinkable.

Yet as the Senate committee heard testimony from deans of medical schools, heads of hospitals, and commissioners of public health in favor of automating preventive medicine, it also heard cautionary notes from physicians who warned that the efficiency and cost-effectiveness of these systems would produce heartless assembly-line medicine. Many of these objections repeated the more generic fears physicians had expressed in relation to other information technologies: the deskilling of the medical profession, the reduction of the physician from an independent professional to dependent technician. Other objections, however, included the more specific concern that expanding access to more and more tests for more and more patients would inundate an already beleaguered medical profession in a sea of meaningless data. "We have a great many tests, we have thousands of tests that would be useful," Dr. Arthur E. Rappaport of the American College of Pathologists testified. "How shall we perform it? Do we have the people to do it? Well, automation promises to give us some relief but what we gain in the relief of the performance of tests we lose by the flood of information." As Norman "Jeff" Holter had found in the early days of biomedical telemetry, having more data was not necessarily better—not if the data overwhelmed the capacity of the clinician to process. Rappaport worried not only that physicians would be unable to handle this deluge of data but also that a phrase that had already become a cliché in computing—garbage in, garbage out (GIGO)—would soon apply to medicine as well.[27]

Collen understood that some physicians were skeptical about the value of automated health testing, but he described this as a part of medical culture: physicians had long dragged their feet when faced with important new technologies. Other diagnostic and therapeutic advances, such as the microscope, the stethoscope, and surgical anesthesia and antisepsis, were first resisted by physicians until they perceived the benefits they actually provided in clinical practice.[28] In his experience, once any physician saw firsthand the benefits of multiphasic screening for their patients, they demanded it. "Just as every doctor now takes it for granted that the laboratory will use the microscope to extend examination capabilities to test specimens," he stated, "so will he soon expect a good laboratory to also add and use automated analyzers and computers." In addition to improving the efficiency of preventive medicine, multiphasic testing also improved the quality of life for physicians by removing drudge work from the clinic. "The computer has been programmed by the physician," he quipped, "to do what the physician used to do but can't stand any more."[29] Collen and others were confident that the automated checkup would soon become just as routine in medical practice as a routine complete blood count or an annual physical exam.[30]

The Adult Health Protection Act never made it out of committee. In retrospect, the fall of 1966 was overshadowed by the midterm elections, which saw substantial erosion of Democratic control of the House and Senate—not the easiest time to lobby for further expansions of federal healthcare funding. But the publicity of the Preventicare hearings drew national and international attention to the Kaiser model of automated prevention.[31] Visits to the Oakland clinic soon became so numerous that the US Public Health Service provided funds for a half-time physician to serve as a "tour guide" to the center. Collen launched a Training Program in Automated Multiphasic Screening to help disseminate the model. Critics warned that prepaid health plans like Kaiser worked because most of the population was healthy and employed, but might not work the same way among sicker and more impoverished Medicare and Medicaid patients.[32] In response, the Public Health Service announced four demonstration projects to test the potential for automated multiphasic health testing of Medicare and Medicaid recipients. The goal of public health funders was to see how well the Kaiser model worked outside of Oakland, and outside of the patient population of the Kaiser Health Plan.

Like the early telemedicine experiments, these federally funded demonstration projects positioned automated testing as a new means to expand the reach of primary healthcare in underserved rural and urban areas. When the federally funded health system of the Tennessee Valley Authority (TVA) started to expand periodic health examinations for its widely dispersed employees in the 1940s, it developed a set of mobile health clinics based on trailers. In the late 1960s the TVA adapted the Kaiser model of automated multiphasic health testing to suit the needs of rural Appalachia in place of urban California. Mobile clinics capable of conducting automated health screenings promised expanded access to preventive care at low cost. By 1968 the TVA's mobile units were equipped with computer punch cards to collect screening data, finding abnormalities and sending full printouts of data to the relevant area physicians. In communities steeped in rural poverty, one TVA spokesperson claimed, mobile automated multiphasic testing centers served "as a point of entry for the uninitiated into the health care system," expanding access to care in spite of a perceived shortage of physicians in rural areas.[33]

Four dedicated US Public Health Service pilot programs tested the role of automated multiphasic testing in increasing equity in access to healthcare in urban centers.[34] In Brooklyn, the Brookdale Hospital Center adapted Kaiser's automated clinic to extend preventive healthcare for Medicare and Medicaid patients in the minority-majority neighborhoods it served. As project head Dr. Leo Gitman described, the Brookdale clinic population differed substantially from the Kaiser Permanente population in age, ethnicity, race, and income. But Gitman believed that all could benefit from this new pathway to preventive care.[35] "Until several years ago," he disclosed, "I was extremely pessimistic over the solution of the problem of preventive medicine, especially in our high-density, low income, multi-ethnic population." There were so many newly insured Medicaid and Medicare patients, with limited health literacy, and few health professionals staffing the clinics in underserved areas. Kaiser's model, along with the support of the US Public Health Service, helped make preventive health seem possible, attainable, for Brookdale's patients. "The publications of the Kaiser-Permanente Group caused tremendous excitement for those of us struggling with this problem," Gitman concluded. "We are no longer pessimistic."[36]

As a profile in *Look* magazine summarized, the adoption of these "computers that will keep you healthy" enabled Brookdale to add 25,000 new patients a year. What started with screening often became a long-term primary care commitment.[37] In addition to the Brookdale project in New York City, the Public Health Service also supported pilots of automated multiphasic health testing in Providence, Milwaukee, and New Orleans. Each of these urban centers stood to benefit from the promise that automation would help expand access to care. Each was also under pressure to show cost-effectiveness as well. Kaiser's costs at the Oakland clinic started at $25 per person, but it was estimated that this unit cost would decrease to $15 per person as more clinics opened.[38] Yet the multiphasic health testing centers built outside of the Kaiser system could not reproduce the same demand or cost-effectiveness. By 1969, each of these clinics still handled fewer than 1,000 patients per month on average, and the cost per patient was correspondingly higher—roughly $60 per patient.[39]

Public health officials were still hopeful that if the concept of community-based automated testing centers caught on as a nationwide effort, the higher volume would make them cost-effective.[40] Yet by then public health agencies were only one of several stakeholders interested in the automated medical future. Several private industries now saw their own futures tied to the expansion of multiphasic screening as well.

* * *

Automated prevention was a mass market from the beginning. From the moment Morris Collen first contacted IBM/Oakland to help Kaiser build its first punch-card clinic, the firm was explicit that "to IBM, this project initiates a first step into a large potential market for automated mass screening."[41] A conference on automated preventive medicine sponsored by the National Academy of Engineering drew nearly 200 attendees who later recalled how they "felt that they were participating in the birth of an explosive new health-care industry, geared to the accomplishments of the space age and aimed at revolutionizing the health-care delivery process of the United States." Market forecasts predicted that sales of automated multiphasic health testing equipment would reach nearly $1 trillion by 1980.[42]

Correspondence files preserved by Morris Collen overflow with letters and pamphlets from new companies whose business plans hinged on expanded government, insurance, and corporate spending on devices to automate preventive care. As the 1960s drew to a close, these entrepreneurs were confident that automated multiphasic health testing would soon become an expected feature for any major medical center.[43] Some of them sought to replicate the equipment already in place in Oakland, while others added new bells and whistles. A vice president of the Medequip Corporation of Park Ridge, Illinois, wrote to Collen describing the Interex 1024, "a free standing console with television-like display that interacts with a patient for self-administered medical history taking" and that could replace the paper cards of the Oakland clinic with a computer terminal.[44] This was just one of a broader line of products Medequip was bringing to market in anticipation of a growing demand for automated multiphasic health testing centers.[45] In 1968 Searle Pharmaceuticals developed a new subsidiary to manufacture and install automated testing facilities for hospitals and health systems. At its peak in 1971, Searle Medidata was installing AMHT suites across the country at a steady pace.[46] Smaller firms, such as Health Screening Services of Denver, Colorado, sprang up to offer automated multiphasic health testing services to local physicians, including mobile or fixed installations that, for a flat fee, came equipped with a team of nurses and medical auxiliaries to run them.[47]

Collen collaborated extensively with domestic and international manufacturers, as well as with physician associations and state agencies around the world. He served as a consultant to groups in Sweden, Switzerland, England, France, Germany, Italy, Denmark, Finland, Australia, New Zealand, and the Soviet Union. But his earliest and most durable international corporate collaborations took root in Japan. Many large Japanese employers had started experimenting with multiphasic health screenings of populations of employees in the 1950s, around the same time that Collen and Breslow developed their first prototypes for multiphasic health testing in Oakland. In Japan the practice was called *ningen dokku*, which translates as "human dry dock." *Ningen dokku* practitioners suggested that the human body, like a long-haul ocean-going vessel, should be lifted out of the water at least once a year for a full inspection of all vital parts. Initial versions required a full week of intensive hospital-based testing, though by the 1960s a "fast human dry dock" had been compressed to two full days.[48] Like the Executive Physical programs available at elite American hospitals, *ningen dokku* were initially conceived as private facilities for paying customers. As a result, however, only a tiny fraction of Japanese adults who might benefit from comprehensive health screenings had the means to access them.

In the early 1970s, executives at the Toshiba Corporation, an electronics and computing firm that ran a hospital for its own employees outside Tokyo, became interested in automating a *ningen dokku* for rank-and-file employees.[49] They arranged a visit to Collen's clinic in Oakland to adapt the Kaiser model for their workforce. The flowchart looked a bit different: early in the circuit, Toshiba employees would submit samples of feces for occult parasites, and ingest a chalky flask of barium for a full gastrointestinal (GI) series of X-rays, neither of which was included in Kaiser's clinics.[50] Gastric cancer and other GI malignancies were more common in Japan compared to California, as were intestinal parasites related to the more fish-forward Japanese diet. Toshiba's first automated clinic opened in 1970, and began processing clinical evaluations and follow-up for the firm's 115,000 employees and their families.[51]

The automated *ningen dokku* were readily accepted by Toshiba employees. To meet rising demand, Toshiba developed a mobile version using a train of shipping containers on five large trucks.[52] Similar employer-based automated health testing facilities soon spread to other large Japanese corporations. In 1973, as the Japan Society of Automated Multiphasic Health Testing and Services began to hold annual meetings, Morris Collen looked with admiration and envy toward widespread professional, corporate, and state support of these clinics, formally endorsed by the Japan Medical Association as a means of integrating preventive health and computing. More than 100 Japanese hospitals had already been computerized. By 1985, it was predicted, 250,000 Japanese private homes would contain computer terminals, and all major cities would contain automated health testing centers.[53]

Yet even in the earliest publications of the Toshiba group, a note of trepidation was mixed into the general chorus of enthusiasm for automated health testing. Two out of three apparently healthy Toshiba employees were found, in the extended battery of medical tests, to have at least one abnormality. Toshiba scientists raised concern in 1972 that "although the GI x-ray test finds most of the abnormal cases during screening, approximately half of these cases will be later diagnosed as normal when the detailed examinations using photofluorography and/or gastric camera photography are performed."[54] There might be hidden costs in subjecting a healthy population to a large battery of tests, especially if some of those who screened "positive" for cancer turned out to be "false positives": not actually sick at all. Sometimes more data might just lead to more anxiety. For Collen, this concern recalled a comment that Sen. Maxine Neuberger, the chair of the Senate Special Committee on Aging, had asked during the Preventicare hearings in 1966. "If you depend on the IBM card," she pointed out, "you might come up with cancer you don't have."[55]

* * *

Automated multiphasic health testing might be an efficient and cost-effective way to use information technology to expand access to preventive medicine. But it wasn't clear whether this program of more tests for more people produced better health, or just increased anxiety. Toshiba's findings suggested more testing could also function to increase the ranks of the "worried well," by diagnosing healthy people with cancers they didn't have. At Kaiser, however, Sidney Garfield described automated testing as a crucial technology—for separating the "worried well" from the "asymptomatic sick." Kaiser embraced the metaphor of the healthcare system as a machine, and the role of new information technologies in fine-tuning its performance for all users.

From Garfield's perspective, the benefit of automated multiphasic health testing was obvious. "Although no long-term evidence exists that the course of disease is influenced by multiphasic testing," he wrote in 1970, "this is irrelevant," since the status quo was already so much worse. The then-current system of fee-for-service American medicine served neither the sick nor the healthy. The recent expansion of federal health entitlements under Medicare and Medicaid, whose passage he had opposed, had created an influx of "worried well" patients. To focus on a few false-positive cancer diagnoses was to miss the value of redirecting many more of these healthy people *away* from the medical system. Critics might say that a screening test that cost $2,000 to detect a single true positive case of breast cancer was too expensive. But a system that cost $4 per person to reassure 499 women that they did *not* have cancer, and cost $4 to help one woman benefit from early diagnosis of a life-threatening disease had as much or more value in detecting health as in detecting disease.[56]

For a little while, Garfield's vision of Total Health seemed as though it might attract as much support from American corporations as the *ningen dokku* had among Japanese corporations. A few large technology firms in the United States began, like Toshiba, to build their own, in-house automated multiphasic clinics to separate the worried well from the early sick among their employees, supported by their own health maintenance plans. By 1971 Lockheed Aircraft Company and IBM had both opened up long-term, computerized multiphasic screening programs on their campuses at Sunnyvale, California, and Endicott, New York, respectively.[57] Kaiser's vision was echoed by other prepaid health plans around the country that became the first generation of health maintenance organizations (HMOs), including the Group Health Cooperative of Puget Sound and the Health Insurance Plan of Greater New York. These centers would become key to establishing the value of preventive mammography for breast cancer. They would be key players in the passage of the Health Maintenance Act of 1973, perhaps the centerpiece of healthcare policy for the Nixon administration.

Kaiser's vision of Total Health Care might have looked a lot like socialized medicine to the casual observer. Indeed, Kaiser physicians had for a time been blacklisted by the California Medical Association based on this allegation. But Garfield was adamantly opposed to national health insurance. He saw a federal guarantee to "health care as a right" as one of the gravest threats to the American healthcare system, because it led to a sense of entitlement and overuse rather than a common sense of collective responsibility for health. "The real role of multiphasic testing in medicine," he concluded in 1970, "is inherent in its critical relation with 'medical care as a right.'"[58] Kaiser's automated clinics were explicitly engineered to *prevent* rather than promote socialized medicine, and both Garfield and Collen saw automation as an aid in this process.[59]

Automated health checkups could make the American healthcare system more efficient by separating access to healthcare from access to doctors (fig. 7.2). Every person, sick and well alike, could have an automated exam to test their overall health status. Only those who were actually sick would then need to see their doctor. A report from the federal Department of Health, Education, and Welfare found

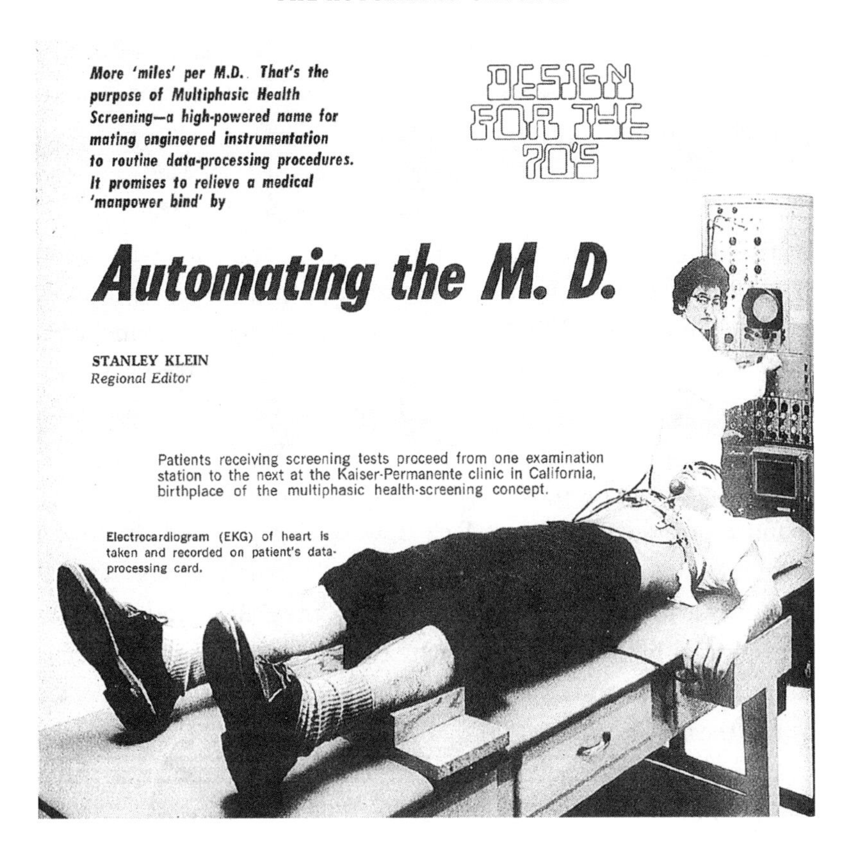
More 'miles' per M.D. That's the purpose of Multiphasic Health Screening—a high-powered name for mating engineered instrumentation to routine data-processing procedures. It promises to relieve a medical 'manpower bind' by

DESIGN FOR THE 70'S

Automating the M. D.

STANLEY KLEIN
Regional Editor

Patients receiving screening tests proceed from one examination station to the next at the Kaiser-Permanente clinic in California, birthplace of the multiphasic health-screening concept.

Electrocardiogram (EKG) of heart is taken and recorded on patient's data-processing card.

FIGURE 7.2. The automation of early diagnosis was frequently reported as a means of automating the functions of the physician, as in this article from 1970. Courtesy Machine Design.

automated multiphasic health testing to be a promising solution to mobilize nonphysician providers, including nurse practitioners and medical assistants, to expand primary healthcare. Yet the report warned that "while some of the public may regard it as a boon to health care, others may look upon some aspects of it as a way to check up on their physician or as a way to avoid seeing a physician at all."[60] Columbia psychiatrist Jerrold S. Maxmen welcomed automated checkups as the first step in an inevitable progression to a more efficient "post-physician era." "In the 21st century," he predicted, "doctors will be rendered obsolete by a collaboration between the computer and a new breed of health care professionals—the medic."[61]

As a physician who welcomed a post-physician era, Maxmen was decidedly in the minority. More physicians shared the concerns of Richard Bates, a contributing editor to *Medical Economics*, who warned that automated testing in conjunction with Medicare and Medicaid expansion would lead to a new era he called Machinicare. In time, "computer-manned centers in every large city, where millions of people will bare their chests, their veins, and their innermost psychic secrets," would fully displace the livelihood of the independent doctor. Federal bureaucrats loved Machinicare because it appealed to the machinery of federal bureaucracies. Pharmaceutical companies and device manufacturers loved Machinicare because it would "require a lot of expensive gadgets, and you can bet that the manufacturers will rush to sell them way before anybody has had a chance to report whether they are worth the cost."[62] Bates was specifically warning his readers of the perils of the Kaiser model and the Preventicare bill, but more generally he was wary of the convergence between public health spending and corporate interests in the space of healthcare—a conjunction that Arnold Relman, editor-in-chief of the *New England Journal of Medicine*, would later popularize as the "medical-industrial complex."[63]

In a rebuttal to Bates, Morris Collen dismissed physician fears of medical automation as part of a longue durée of cultural resistance to change by the medical profession. The annual physical exam, which began as a screening for insurance companies, had initially been opposed by many doctors—until they realized that these practices served to draw patients into their clinics rather than drive them away. As doctors experienced the heightened efficiency of computational analysis, Collen was confident that they would eventually see the value of automated screening in light of the new forms of personalized care it would help bring into being for doctors and patients alike. But for independent physicians like Bates, the assembly line of the automated health checkup was a metaphor for the machinery of an increasingly bureaucratic medical system. These concerns were not merely cultural, but also structural: the protests of independent professionals wary of becoming smaller pieces within a larger, more corporate model of healthcare.

When Searle Medidata, Inc., displayed its products at the annual conference of the American Association of Medical Colleges, several physicians expressed unease with the "startling number of what I would call hawkers of medical equipment."[64] *Medical Economics*, a journal oriented toward the needs of the physician as a private businessman, counseled physicians to anticipate the financial opportunities of automated testing for independent fee-for-service clinicians, since "there will be more ailments to care for and more neurotics to treat because of borderline tests."[65] The resourceful private practitioner should be able to get ahead of the curve and possibly even incorporate the technology of multiphasic testing on their own terms. With luck, they might even generate more business or revenue as a result.[66]

Indeed, by the early 1970s, the business of automated testing was already shifting its sights from federal funding and prepaid plans like Kaiser to the fee-for-service market. Several medical electronics manufacturers shifted their market plans away from the full automated clinics installed within large institutions and toward smaller modules designed to integrate multiphasic health testing into existing private medical practices. The value of the technology shifted from a cost-saving device to a new and potentially lucrative source of revenue to the private physician. For a fixed cost of $5,000, a private medical office could purchase a mini "carrel" automated testing battery, which would fit in a corner of their clinic. Invented by Dr. Frederick I. Gilbert of Honolulu (fig. 7.3) and marketed by Searle Medidata, these devices promised to bring in additional revenue to fee-for-service practices, processing up to twenty-five patients per day.

This was an inversion of Kaiser's original logic of cost-effective preventive care. Uncoupling the technology of automated screening from public health policy or the prepaid health plan turned it into just another set of tests doctors would be reimbursed for ordering more of. There was still no proof that more tests would necessarily

Honolulu internist designs testing unit that physicians can install and run at moderate cost

MULTIPHASIC SCREENING CUT DOWN TO SIZE

A group practice in Hawaii has demonstrated that it doesn't take a lot of elaborate apparatus and huge outlays of money to set up a semiautomated unit for screening apparently well patients. Such a facility, the group's director contends, is within the reach of many physicians.

"Doctors tend to view multiphasic health screening only in terms of expensive equipment requiring large numbers of highly trained personnel," says Dr. Fred I. Gilbert Jr., director of the Straub Medical Research Institute in Honolulu. Such physicians usually think of the assembly-line type of operation used at big diagnostic clinics, where a patient goes through some 20 stations for a head-to-toe examination. "But there is another way."

At the Straub Clinic—a group-practice, fee-for-service clinic of 60 doctors that handles about 1,000 patients a day—the machines are, in effect, brought to the patients rather than the other way around. The six-foot-square room known as a carrel—after the library term meaning a place for individual study—contains a photomotograph for evaluating thyroid function, a Vitalor machine for measuring respiratory function, an ECG, an audiometer, a sphygmomanometer, calipers for measuring fat, and a sight scanner. The cost of setting up one such carrel, which can handle 25 patients a day: $5,000.

Group practices involving even as few as four internists would find the system particularly useful, Dr. Gilbert believes. Larger groups can simply increase the number of carrels used. In the Hawaiian clinic they have four of the

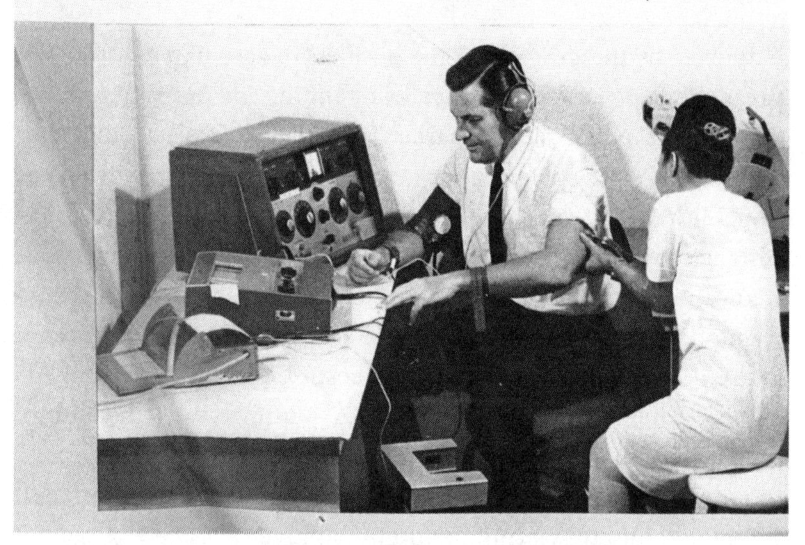

FIGURE 7.3. Dr. Frederick Gilbert (wearing headphones) demonstrating the smaller "carrel" model of automated multiphasic health testing, which was more amenable to use by physicians in private practice. Courtesy Medical Tribune, Inc.

lead to better health, especially when they delivered ambiguous or borderline results to worried well patients. "As the neurotic knows by instinct, no doctor can safely ignore a borderline test," *Medical Economics* reminded its readers, "even if subsequent investigation proves the test result groundless, there must always remain the wispy doubt that something is about to go catastrophically wrong. After all, doctors are merely human and mortal, while computers are shiny and expensive."[67] More tests would just lead to more tests, not to better health.

Morris Collen continued to argue that more data would produce better health—and an approach to healthcare better tailored to the needs of the individual patient. As Kaiser's automated health testing centers gathered routine laboratory and physical data from the thousands of people they processed each month, the contents of their computers could correct and displace the biases of "normal values" that previously came from the hospital laboratories, and had been largely based on otherwise sick bodies in sick places. The automated clinic allowed Kaiser to produce new normal values based on *well people* instead of sick people. And it could be personalized. Once an individual patient had been through the automated testing suite a few times, they could compare their current values against their past values, and become the reference for their own personalized definition of health.[68] The union of computation, automation, and prevention now "permitted application of specific norms to each individual patient" in a way never before possible.[69]

Yet the data accumulating on the overall medical benefits and public health impact of the Oakland clinic did not support the thesis that more data produced better health. In a series of papers, Garfield, Collen, and colleagues at Kaiser compared the health status of populations of patients who did and did not undergo automated screening between 1964 and 1973. While some subgroup difference could be found for indexes like self-reported disability and a small suite of "potentially postponable" conditions, the results did not show meaningful overall differences in disability or mortality between the two groups. Nor did further follow-up analyses a decade later.[70] Other studies in the later 1970s and early 1980s that attempted to document the value of automated multiphasic health testing—or, ultimately, any sort of general periodical health exam—largely foundered as well.[71]

These mounting concerns over the lack of proof of benefit from automated health screening were joined by new evidence of the potential risks of overtesting, overdiagnosis, and overtreatment. In the late 1960s and early 1970s, a fierce debate erupted over the costs of diagnosing individuals with conditions they did not feel, and which might never affect them in their lifetime.[72] "In theory, screening is an admirable method of combating disease," a withering report from the World Health Organization read, but "in practice, there are snags."[73] Many perceived benefits of screening for disease in healthy people were illusory. Detecting a chronic disease earlier could give the impression that a patient lived longer after diagnosis, even if their overall life expectancy remained unchanged. And larger studies of preventive medicine in the 1970s began to spell out the dangers of excessive testing, the harms of unnecessary surgeries, and the false equation between early diagnosis and better outcomes in chronic conditions like breast cancer and prostate cancer.[74]

The boom markets promised for automated health testing devices failed to appear. Outside of Kaiser, insurers were hesitant to reimburse a new set of procedures with no clear evidence of efficacy in preventing disease or extending life. By the end of the 1970s, the World Health Organization (WHO) had moved away from endorsing the comprehensive automation of multiphasic screening. Instead, in the early 1980s, the WHO and the newly founded US Preventive Medicine Task Force promoted a more focused approach to prevention that favored specific proven interventions for discrete conditions: the Pap smear to prevent cervical cancer, the mammogram to reveal breast cancer, the treatment of asymptomatic hypertension, diabetes, and cholesterol.

Popular opinion had shifted away from multiphasic screening as well. The iconoclastic Ivan Illich, whose 1976 critique of the medicalization of society, *Medical Nemesis: The Expropriation of Health*, became a runaway best seller, depicted automated multiphasic health testing as a key example of how modern medicine commodified its subjects. The medical gaze was reductive, extractive, and mechanistic. Taking the body as a set of material items to be weighed and measured, it erased the existential complexity of individuals and the communities they lived in, and ignored the role of political, economic, and social forces in promoting health. Illich had no love of Kaiser's Oakland clinic. *Medical Nemesis* described automated testing as the "poor man's escalator into the world of Mayo and Massachusetts General," an "assembly-line procedure of complex chemical and medical examinations" that "transforms people who feel healthy into patients anxious for their verdict."[75]

* * *

In retrospect, the early 1970s proved to be the high-water mark for visions of automated prevention based on multiphasic testing centers built around mainframe computers. Yet many elements of this vision persist in the present day, and have been resuscitated in a new dream of algorithmic health based on AI, neural nets, mobile sensors, and cloud-based data trails. For the automated health testing centers of the 1960s and 1970s, however, the bust followed close on the heels of the boom—at least in the United States. "[We had] the altruistic thought that through our creativity we could improve the system of health delivery," the aspiring AMHT device manufacturer H. R. Oldfield recalled. Market analysts had claimed in 1968 that multiphasic health testing was "coming like a big freight train—and the medical community won't be able to block it even if they want to."[76]

And then, he sighed, "almost as suddenly, the party was over." Looking back at earlier predictions of trillion-dollar markets for automated multiphasic health testing, Oldfield admitted that "it is possible that the 'train' may have been trying to travel too many tracks at the same time, or may sometimes have been loaded with the wrong grade of coal."[77] In rapid succession, "the USPHS withdrew its financial support from the four large experimental/demonstration centers, the Congress shelved its bill to provide reimbursement for AMHT procedures under Medicare, and funds were withdrawn from other supporting programs."[78] The outlook for the industry declined further, especially as the critical studies mounted. Private practitioners, lukewarm on multiphasic testing from the start, stopped purchasing mini-carrels for their clinics. What had seemed like an inevitable convergence of private industry, public health, and professional interest quickly frayed and unraveled.[79]

Where manufacturers like Oldfield blamed an erosion of public support for the collapse of the promised AMHT market, Collen blamed manufacturers like Oldfield. To Collen, automated prevention was a promising intervention undone by a premature push by device manufacturers to capitalize on the promise of a new market. The marketers had been allowed to hawk it as a product before the evidence had been collected. "You well know that we did whatever we could to discourage publicity in the lay press," he wrote in a memo to the clinical directors at Kaiser Permanente. "However, this is now a fact of life and I am prepared to accept it. As a Chief of Service and a responsible PMG Administrator, I suggest that you consider that this change in medical practice represents a growing and increasingly significant demand from the public—which is the market place. Whether medical practitioners like it or not, automated multitest laboratories are definitely a marketable process (witness the dozen medical-engineering-commercial enterprises being started to sell the system)."[80] By promoting automated multiphasic screening based on the timelines of speculative capital instead of scientific merit, the long-term potential of the field simply got ahead of its evidence base.

Private markets for automated multiphasic testing suites failed to materialize. Preventicare failed to gain support as a public measure. The US Public Health Service demonstration projects for AMHT, like the demonstration projects for telemedicine, were short-lived. By the end of the 1970s, most American AMHT centers were closing their doors, even as similar centers were booming in Japan.[81] For many American device manufacturers, the growing Japanese market for automated multiphasic systems was a vision of what could have been. Addressing an international conference on Automated Multiphasic Health Testing Services in Tokyo in 1980, a former VP of Searle Medidata observed that the rising number of AMHT units in Japan showed that a future for automated health screening was indeed taking shape, just not in America.[82] Twenty different multiphasic programs were open in Japan at that point. By 1993 the number grew to 240, and as of 2016, the Japan Society of Ningen Dock included more than 5,500 physicians and nearly 500 allied health personnel across 1,649 medical institutes throughout Japan.[83]

Collen, who by that time had taken a leading role in the formation of an international field of medical informatics, viewed these developments as a mixed blessing. Enthusiastic reports from Japanese researchers suggested *ningen dokku* might be credited with extending the lifespan of the average Japanese worker over the second half of the twentieth century—though the data presented were not sufficient to distinguish the role of this specific technical intervention compared to the general improvement in economic conditions over the same time.[84] By 1993—too late to impact the prospects of public health policy in the United States—long-term data from the original population screened at the Oakland clinic (compared to a control group of Kaiser Permanente patients who did not engage with the computer screening), suggested that participating in automated health screening was associated with a 30 percent decrease in death from colorectal cancer, and a 20 percent decrease in overall mortality for "potentially postponable conditions" over the twenty-seven years of study.[85] Conceding that "it is unlikely that AMHTS, as we know it, will again become popular,"[86] Collen still hoped that by the end of the 1990s all large medical centers would be connected in a network of medical informatics that linked electronic patient records with algorithms for more effective preventive medicine.[87]

Several elements of this vision can be found in the renewed vision of precision medicine in the twenty-first century, as the increased use of electronic medical records has created a virtual interface for preventive medicine prompts to appear during any given clinic visit. Should you have a Pap smear? A colonoscopy? A mammogram? Are all of your vaccinations up to date? These decision trees are increasingly automated within the electronic health record. Yet the vision Collen had first proposed—a nationwide system of clinics designed to be automated interfaces between patients and the digitization of their health record—did not come to pass in this country.

In the early twenty-first century, as "big data" became a key buzzword of biomedical research and health policy, medical informatics has seen its star rise as well, with Morris Collen as its patron saint. The lifetime achievement award of the American Medical Informatics Association is named in his honor. Later in his career, Collen referred to his experience with automated clinics as a useful lesson in the broader political and economic challenge of instituting social change in the delivery of healthcare via technological systems.[88] Just as Collen's automated clinics largely erased the social context of the patients they processed, Collen himself had not accounted for the extent to which the social context of American healthcare would create political backlash to his attempts to engineer a more efficient system through automation.

The younger Collen thought that rapid, total change in the informational architecture of medicine could be achieved nearly overnight, with the right technology and the ability to convince physicians, policymakers, and industry leaders that it was worth investing in. He saw in the digital computer a tool to rapidly transform the nature of preventive medicine and access to care. Collen's original Oakland clinic was a symbol of that hope: a machine that, in three hours, could rapidly digitize a person's health status, and recommend algorithms to prevent disease and maintain wellness. Yet this progressive vision of preventive medicine could only understand the health of a community one individual at a time. It could not explain disparities in health by race, by ethnicity, by income level, by social geography. Once an individual had been screened and scanned, and once they understood their individual health risks, what happened to them was their responsibility.

Collen's original automated clinic foundered on a set of premises that will be familiar to many twenty-first-century readers: that having more data is better, that increased testing leads to better health, and that nothing would be lost, only gained, in collecting *in silico* vast amounts of health-related information.[89] It foundered on the premise that the automation of data collection, management, and utilization could serve as a fulcrum to expand access to healthcare to all Americans. It foundered on its reliance on a confluence of public health policy and private-sector market speculation. The older Collen knew better than to believe that redesigning a single clinic around a computer could instantly transform preventive medicine through automation. By then, he had seen many promised digital transformations come and go, never delivering exactly what they promised, but nonetheless slowly transforming patients, preventive care, and predictive algorithms into electronic forms in a more piecemeal fashion.

The failure of automated prevention in the United States is a story of technological transformation unfulfilled, of an inevitable and seemingly inexorable development that did not come to pass. More data did not produce better health, but it did lead to extensive speculation by policymakers, practitioners, and marketers alike. This new technology produced a sense of rapid transformation shared by very different stakeholders, with very different goals. Yet even as the majority of physician organizations, public health organizations, and medical device manufacturers saw an imminent restructuring of the healthcare system around the combination of automation, prevention, and informatics, this agreement of a number of influential stakeholders in US health policy could still be wrong. The failures of this vision of automated prevention offer an instructive window into how information technologies do and do not change our lives: a reminder for twenty-first-century audiences regarding our limited ability to perceive which promises of new technologies will and which will not transform the practice of the present.

Conclusion

THE MEDIUM OF CARE

Historians make a brisk side trade pointing out false claims of novelty. "This thing you think is new," we often find ourselves saying, "is not as new as you think it is." The telegraph enabled a form of networked social media before the internet. Television gave rise to telemedicine a half-century before the COVID-19 pandemic made it a household practice. The landline telephone engendered an earlier form of "tele-" medicine nearly a century before that. Wireless sensors and networked computers were already reshaping the healthcare industry by the mid-twentieth century. Electronic medicine is nothing new. That would be one way to read this book.

But that would not be enough. At its best, this kind of argument helps to puncture the inflated claims of today's technologists, who promote health applications of new devices as unprecedented without bothering to check for the many precedents that do exist. At its worst, though, this form of history turns into a museum of quaintness, a world of "fun facts" for cocktail party conversations that bolster our clichéd narratives of technological progress. Isn't it cute that Hugo Gernsback conceived of telesurgery in 1925 using antiquated devices like vacuum tubes and radio-controlled servos? Isn't it quaint that Norman Holter produced wearable wireless heart rate sensors in 1949, bundled into a bulky backpack? Isn't it fascinating that Vladimir Zworykin had blueprints for an electronic medical record in 1960, depicting a state-of-the-art mainframe computer with less memory than a household thermostat contains today? What a cast of visionaries, these men ahead of their time. If only the tech had been better.

Some problems can be solved with better technology, but not the challenges at the heart of this book. The urban and rural telemedical demonstration projects of the 1970s didn't fail because the tech didn't work. The programs failed to live beyond initial demonstrations because there was neither professional endorsement, economic support, nor political will to build on the promise of telemedical systems to reduce disparities in access to care. Electronic medical records in the 2020s are not much closer to providing a universally portable system of medical information than they were in the 1960s, even after a half-century of exponential growth in processing speed and memory arrays.[1] The challenges here are not merely technical problems. They reside in a gridlock of stakeholders at the heart of the American medical system, whose motivations point in several different directions. These challenges highlight the power dynamics already at play in a society that conceives of healthcare primarily as a speculative marketplace. They underscore the emptiness of recurrent promises that new forms of digital medicine will place everyone on a more even footing: doctor and patient, rich and poor, rural and urban, Black and white.

Yes, the tech is better now. Integrated circuits developed in 2020 have roughly 7 million times the processing power of the most sophisticated microprocessor from 1970. Information technologies are more widely diffused and accepted by patients and practitioners than ever before. By 2018, 85 percent of American households had easy access to broadband internet, and as of 2021, 85 percent of the US adult population had a smartphone.[2] We are more used to spending time online, and to sharing broadly that which previous generations would have guarded tightly as private. Well before the COVID-19 pandemic made telehealth a nearly universal experience of accessing healthcare, the vast majority of American hospitals and clinics had already shifted from paper to computer records.[3] Yet the crisis of the pandemic revealed a nation for which access to telemedicine exacerbated differences between haves and have-nots, rather than resolving them. It revealed a digital divide in healthcare access that has been getting worse, rather than better, as we have come to rely more completely on an electronic medium of care.

The medium of care is not neutral. The histories in this book are not being trotted out for the sake of a parlor game. They are not accidents. Just as there were urgent stakes at play for all who lived through these episodes, there are urgent stakes at play now for those of us who would bring our understanding of the past to better understand the present and to imagine a better future.

* * *

The four medical media discussed in this book—telephone medicine, radio medicine, television medicine, and mainframe medicine—occupy roughly a century between the 1870s and the 1970s. The stories they tell are neither revolutionary breaks nor tales of linear progress but rather cyclical episodes: a series of promises made and not kept, and then forgotten. They end before 1980, at a distant remove from the digital now. They offer a different kind of history than the preambles to the present that most technologists use to recount the development of their own fields.

Take, for example, the corporate history of Epic Systems, which chronicles how the world's leading electronic medical records provider rose to prominence between the early 1980s and the 2020s. Like so many other technology firms today, Epic's website boasts of its origins in a basement and charts its progressive growth over four decades to employ more than 9,000 workers in the greater Madison area and to serve more than 250 million patients worldwide.[4] The story of Epic is also the chronicle of its reclusive founder, Judith Faulkner, named by *Forbes* magazine as the most powerful woman in healthcare and one of the most influential female tech firm founders in the world today.[5]

Faulkner's career trajectory intersects with several elements from the era of mainframe medicine. She arrived at the University of Wisconsin in Madison to study computer science in 1965, just as Warner Slack was developing one of the first computer terminal interfaces for the direct entry of patient records. Enrolling in Slack's graduate seminar on computers in medicine, Faulkner was hooked on this vision of computers in the service of medicine, of information technology empowering patients in healthcare. After Slack left Wisconsin to join the faculty of Harvard Medical School in 1970, she visited him in Boston and met Neil Pappalardo, who worked with Octo Barnett on the Massachusetts General Hospital (MGH) Hospital Computer Project.[6] The initial Hospital Computer Project had ended, but the language of the MGH User Multi-Programming System (MUMPS) was thriving in Pappalardo's newly founded firm, Meditech. Faulkner used MUMPS to launch her own firm in Madison. Her first MUMPS product was a simple application to schedule physician call schedules. Her client, John Greist, as chief medical resident at the University of Wisconsin Hospital, had received complaints from physicians who thought existing call schedules showed signs of favoritism. Faulkner's algorithms promised equity with a computerized code to allocate call schedules by computer, and the complaints faded away overnight.[7]

In 1979 Faulkner and Greist founded Human Services Computing, Inc., renamed Epic Systems in 1983. They now tracked patient data instead of provider schedules. Epic ran out of a basement on Madison's West Side, its core of operations the Data General Eclipse minicomputer made famous in Tracy Kidder's *The Soul of a New Machine*.[8] As mainframes gave way to minicomputers and microcomputers—which were then linked by the new TCP/IP protocols of the branching internet—the computer became a more commonplace object within the hospital. As Epic grew over the 1980s and 1990s, it added new layers of functionality: reimbursement software, staffing software, laboratory reporting software, pharmacy software, ER scheduling software. In 2009, when the US Congress allotted $30 billion for hospitals and clinics to switch to electronic medical records under the Health Information Technology for Economic and Clinical Health (HITECH) Act, Epic was one of the few firms that had all the pieces ready to deploy, including Cadence (an updated scheduling application), Stork (to monitor deliveries and prenatal care), and MyChart (a direct interface for patients).[9]

This was no accident. As a member of President Obama's Health Information Technology Policy Committee, Faulkner had helped to shape the contours of the new law. HITECH accelerated the switch to electronic records with both carrot and stick, providing stimulus funding for health systems that adopted them, while withdrawing reimbursement from health systems that did not. Between 2008 and 2014, the percentage of hospitals using electronic medical records rocketed from fewer than 10 percent to more than 75 percent, and Epic grew in turn to become the leader in both domestic and international markets.[10] Billed as an ideal public-private partnership, HITECH was projected to cost up to $100 billion to enact but to save $80 billion a year once put in place. Public health would benefit from an interconnected network of epidemiological data. Healthcare practitioners would enjoy new freedoms to check up on patient labs and follow-up calls from the golf course. Patients would be assured that key information, personalized to their own health needs, would be instantaneously available at any emergency room, and any doctor's visit.

Yet by the second term of the Obama administration, Epic Systems had only delivered on *some* of these promises. Writing in *Mother Jones* in 2015, Patrick Caldwell denounced the public-private partnership as an "EPIC Fail." Yes, Epic electronic health record (EHR) systems in Michigan permitted the epidemiological research by which pediatrician Dr. Mona Hanna-Attisha shed light on the devastating rise in pediatric lead poisoning in the city of Flint, alerting the state and nation to one of the greatest municipal public health crises of this century. But the built-in lack of interconnections between systems—a residual feature of the competitive healthcare marketplace—made many broader public health applications of electronic medical records difficult, if not impossible. EHRs had promised to save money for health systems, but the $80 billion a year savings had not materialized. Instead, several nonprofit health systems faced such economic distress from the price of implementing Epic systems that they went bankrupt or were acquired by for-profit hospital chains. EHRs had promised to make patient health data portable from clinic to clinic. But the digitization of medical records instead reproduced the fragmentation of the American healthcare system within new digital silos. Yes, physicians and nurses and lab technicians could now access clinical data at a distance. But Epic was not a labor-saving device. The average provider's experience with Epic resulted in *more* time spent charting rather than less, even if some of these hours could now be spent at home, working late into the night once their children were asleep.[11]

The corporate history of Epic offers a heroic progressive narrative that insulates the present from the past. But Epic is neither hero nor villain. It is just the latest in a series of healthcare information platforms marketed to the general public on a series of promises that were not fulfilled. The story of Epic resonates with a set of themes visible in the history of telephone medicine, radio medicine, television medicine, and mainframe medicine. Like these earlier episodes, the rise of Epic promised to achieve broad public health goals by concentrating wealth in a new technology market—doing well by doing good. It hinged on believing that *this* time, investing in the right technology would knit together the broken pieces of the US healthcare system and promote equity in access to high-quality healthcare. And ultimately it suffered from a broad lack of accountability when it failed to deliver on these promises. Like these earlier information technologies, Epic transformed the healthcare landscape, but not as advertised. It has been most successful by becoming routinized as infrastructure.

We should not be surprised to learn that Epic failed to deliver on many of the personal, professional, or public health benefits it promised. It has delivered on some, at least. We should be surprised, however, that none of the framers of US national health policy thought to ask what had happened when previous forms of information technology promised to revolutionize the practice of medicine. In the century or so that we have been seeking and delivering healthcare through information technologies, why do we continue to believe the next platform will simply solve the problems of the last? In this light, the story of Epic is a recurrent symptom of a more chronic underlying problem.

To believe that HITECH would simply make patient data more

transparent and useful in medicine and public health is to believe that the medium of care can be neutral to the interests of existing players. But the medium of care has never been neutral. It is not neutral to the interests of those who would seek to commodify it, and establish a dominant market position—whether this reflects the interests of RCA, IBM, Motorola, or Smith, Kline & French at mid-century, or the interests of Epic Systems, Apple Health, Google Health, or Amazon Care in the present day. It is not neutral to the interests of those whose work practices it seeks to rationalize—whether nurse practitioners seeking increasing autonomy or physicians fearing increasing automation.

Perhaps we should not be surprised that medical records have become a lucrative market. But we should also remember that the purposes and uses of the electronic medical record have changed substantially over the past half-century and are still malleable. Like many other providers today, I use an Epic interface in my clinic every week. Recently the dashboard I see when I log in to a given patient's chart was modified to include a new Social Determinants of Health module. This graphical display is meant to encourage providers to better visualize the many ways in which social context can find its way into the health of our individual patients. A color-coded wheel now alerts me to a patient's risk of financial instability, housing status, food insecurity, transportation needs, environment for physical activity, social stress, and broader community relations. It is a new effort to use the technological interface of the electronic health record to turn attention *toward* the social worlds of patients, not away.

It is not unreasonable to hope that this new medium of care might still help to produce greater health equity. It is, however, unreasonable to think it will do so on its own, without continued oversight, public investment, and political will.

* * *

What if the telemedicine and automated prevention experiments of the 1960s and 1970s had not been just demonstration projects, but instead had received continued oversight, public investment, and political support to achieve their stated goal of promoting health equity? It's not a crazy idea. After all, that's what happened with a similar demonstration project framed around the potential value of television in producing equity in education—another domain associated with public welfare, professional responsibility, and federal oversight.

In 1966, the same year that Reba Benschoter presented her first report of the use of television to extend mental healthcare across the state of Nebraska, Joan Ganz Cooney presented a report on *The Potential Uses of Television in Preschool Education* to the Carnegie Corporation. Cooney proposed that the medium of television could help bridge the substantial disparities in access to preschool education by race, ethnicity, class, and social geography. She then partnered with Lloyd Morrisett at Carnegie on a $1 million pilot project to develop a high-quality childhood television program, representing a multicultural urban audience, to close that gap. When the feasibility study resulted in another $8 million in federal funding and foundation grants, the Children's Television Workshop (CTW) was born—and with it, the early childhood intervention known as *Sesame Street*.[12]

Sesame Street launched in September 1969 with a mandate to broadcast across all national television markets so that children from all socioeconomic backgrounds could access this new intervention. It launched with a commitment to measuring outcomes and refining its strategy to produce measurable improvement in the preschool education of children in general, and children from underserved communities in particular. It has been one of the most enduring successes in using a new media form to reduce existing disparities in access to education. Following that first episode, the familiar crew of Bert, Ernie, Big Bird, and Cookie Monster aired new episodes each week on public broadcasting stations for more than a half-century.[13]

Just as telemedicine has the power to either reduce or augment disparities in access to healthcare, educational television also has as much potential to worsen disparities in access to early childhood education as it does to improve them. Studies in the early 1970s by the educator Herbert Sprigle and the psychologist Thomas D. Cook argued that *Sesame Street* might paradoxically serve to widen educational gaps between "inner city" minority children and suburban white preschoolers. They were accompanied by criticisms from other early childhood educators concerned that federal funding of televised preschool education would reduce the support for brick-and-mortar preschools. The director of the federal Head Start program of early childhood education, Edward Zigler, canceled its funding for the Children's Television Workshop in 1971, claiming that television could not substitute for classroom presence. "There are simply no short cuts to the problem of educating poverty children," Zigler declared. "*Sesame Street* and other programs with surface appearances of education may be thrust at the public as easy answers to a complicated problem."[14] Many professional educators, like medical professionals, were skeptical of technological fixes to complex social problems, especially fixes that undermined those institutions—like the school and the hospital—in which they exerted most influence.

Except *Sesame Street did* work to reduce the equity gap in early childhood education. What's more, Children's Television Workshop conducted regular research on utilization and impact to continuously improve its efforts to reach target audiences. A landmark pair of studies by the Educational Testing Service of Princeton, New Jersey, in 1970 and 1971 showed that in its first two years of operation, the program had a significant and lasting positive impact on its viewers. It helped more children enter school performing at grade level for reading, writing, and mathematics. It helped them stay in school at grade level. The effects, the study authors noted, "was particularly pronounced for boys, black non-Hispanic children, and those living in economically disadvantaged areas." *Sesame Street* continued to receive substantial public and private support throughout the 1970s. By the end of the century, more than 1,000 outcomes studies supported its claims to use the medium of television to build equity in access to preschool education.[15]

Unlike the telemedicine demonstration experiments, the pilot projects that gave rise to *Sesame Street* received early, sustained, and robust financial support from public and private sources. The program was not asked to demonstrate profitability or long-term market projections. It gained widespread popularity and broad-based political support. Most notably, it made consistent efforts to adjust its content, message, and delivery in order to ensure its intervention could maximally reach its targeted underserved audiences. After initial studies showed that *Sesame Street* might have more uptake with suburban, middle-class families than with the residents of urban neighborhoods it was originally designed to benefit, the cast went on a national tour to build attention in target cities. The intervention worked to tilt the demographic profile of viewers.

When a group of African American parents (including Matt Robinson, a founding member of Children's Television Workshop who portrayed the character "Gordon" on the show) criticized its portrayal of the inner city as a whitewashed version "more Westchester than Watts," the casting changed substantially. When a group of Chicano and Puerto Rican activists criticized the lack of Latino representation on the block, CTW not only added the new characters Luis and Maria but also took steps to make *Sesame Street* a bilingual educational intervention. CTW routinely invited its critics to join its advisory committees, so that their interventions would not default to a middle-class, white user profile. Professional educators, invited into the process, overwhelmingly came to support the program.

A three-month demonstration study, built from a grant to evaluate the potential role for television in reducing disparities in access to early childhood education, grew into a permanent fixture of public broadcasting that continues to impact millions of children worldwide. *Sesame Street* succeeded where early telemedicine failed, not because education is inherently more viable through television than healthcare, but because it could garner substantial investment, establish a broad basis of political support, and earn the early support of the educators. Yes, on many levels, comparing educational television with telehealth is comparing apples and kumquats. But this alternate history reminds us of the possibility that any of the interventions described in this book *could* have found broader success in reducing disparities in access to healthcare. The fact that this did not happen does not mean that failure was inevitable. The tech wasn't the problem. The problems were more deeply rooted: the lack of robust financial support for health equity, the lack of political will to find sustained solutions for sick adults who live in poverty, the opposition of a profession fearful of losing control over its work practices.

It is also worth noting that Children's Television Workshop did not present the technology of the television *alone* as the answer to bridging gulfs in access to early childhood education across different segments of American society. Rather, television became a catalyst for building other coalitions and conversations about how our society could realize a more equitable future. CTW actively sought out evidence for where it was failing to improve the lives of the underserved and marginalized communities it was intended to reach, and then tried out new ways to reach them. It invited activists and other vocal critics into its own decision-making process. These conversations were sustained, and they were backed by a commitment to rigorously reviewing what was and wasn't working—where the technology was eroding disparities and where it was shoring them up.

It is not too late for telemedicine to take on a similar accountability in measuring and improving its commitment to building health equity. Just as the electronic health record is still a malleable thing, the emerging infrastructure of telehealth can still be redesigned to reduce disparities in access to care rather than augment them. One promising example is Project ECHO (Extension for Community Healthcare Outcomes), founded in 2003 at the University of New Mexico to link academic specialists and community physicians to use telehealth devices to improve access to high-quality specialty care, especially in underserved communities. A key tenet of Project ECHO was to use the bidirectional links that telehealth technologies provide to enable two-way flows of information about what kinds of needs are being met and what kinds of needs remain unfulfilled—and to follow up on these outcomes. First applied for access to hepatitis treatment in rural areas, this reflexive, hub-and-spoke model has demonstrated the potential for telehealth technologies to reduce and even eliminate disparities in access to care by race, ethnicity, income levels, and zip code.[16]

With support from the Agency for Healthcare Research and Quality (AHRQ) and the Robert Wood Johnson Foundation, Project ECHO has since expanded from New Mexico to a nationwide network of related projects. These stretch from infectious disease to chronic disease and mental health programs, from rural health disparities to urban health equity, and from domestic to international applications. Project ECHO has been a source of optimism for those who would see telehealth as a means to produce greater health equity.[17] In contrast to many other telehealth projects, Project ECHO providers report that their engagement in these community efforts has led them to feel more connected, less isolated, and *less* likely to experience burnout.[18] More recently a collaboration between the University of California–San Francisco and the Commonwealth Fund has developed a Telemedicine for Health Equity Toolkit, delineating practices that healthcare systems can put into place to make sure the medium of care is oriented toward access and equity.[19] These efforts to bend telehealth closer to health equity require the active involvement of many concerned parties. They are not simply a result of newer technologies.

The medium of care is not neutral, but it is not fixed, either. It is what we make it to be. It is what we allow it to be. It is what we demand it must be.

* * *

For the most part, though, instead of demanding that new information technologies produce the equity in access to healthcare they originally promise, we tend instead to forget that those promises were ever made. Then we forget that the technologies for which these promises were made—humdrum, everyday things like telephones and radio-pagers and coaxial cable—were ever new things in the first place. "We live," as Marshall McLuhan put it in 1969, "invested in an electric information environment that is quite as imperceptible to us as water is to fish."[20] We learn not to see the media of care that we are suspended within.

When a new medium of clinical care is initially conceived and promoted to the medical profession and the lay public, it is suspended in a froth of enthusiasm between the interests of speculative research, speculative fiction, and speculative capital. It is newsworthy, and grabs headlines in the popular press and professional journals. Each new electronic medium was initially borne aloft on a wave of speculation, a sea of stories predicting how it would revolutionize medical knowledge and clinical practice. Telephones would make modern specialty care available to all Americans, wherever they were, as long as they could tap into the "web of wires." Radio broadcasts from inside the body would make visible previously hidden diseases, enabling the precision treatment of chronic diseases before they became intractable. Alternately: telephones would undo the privacy of the doctor–patient relationship, while radiotelemetry would expose the privacy of the body to unprecedented forms of physiological surveillance by employers or the state. Most of these speculative hopes and fears simply did not come to pass—at least not on the utopian or dystopian timelines anticipated in the mid-twentieth century.

This is not to say that these technologies did not transform the understanding and practice of medical care. But as we have seen repeated in the chapters of this book, in order to take root in the daily practice of American medicine, new communications technologies needed to produce some alignment between professional self-interest, the rhetoric of popular common interest, and the political economy of market viability, cost-efficiencies, and returns on investment. Both the telephone and the radio-pager were sold to practitioners as devices that would extend their coverage, amplify their practice, and allow them to perform work at their leisure, on their own terms. Yet both forms of technology served to make physicians increasingly accountable, to encroach upon their privacy, and to lead to new forms of burnout. As with telephone and pager, so too with email and the electronic health record. Every new labor-saving device really makes more work of a different kind—a cycle repeated in the efficiencies and burnout associated with telemedical visits today.

These technologies were all viewed as having subversive potential—both by those who stood to benefit from disruption of the status quo and those whose livelihoods were threatened by the process. The telephone and radio-pager extended, but also threatened, the autonomy of the physician in private practice. Wireless monitoring of physiological parameters through wearable technologies, and later through automated mainframe clinics, extended the power of preventive medicine. But it also raised concerns about the amount of data collected, the privacy of those data, and the ultimate benefits and costs for individuals being increasingly screened. Physicians believed that telephone, radio, television, and computer all enabled an extension of their professional authority, but worried they could also lead to loss of control of the conditions of their own labor, or, worse yet, might open up the private spaces of the profession for new forms of public critique. Yet as the imagined universe of applications for any new information technology gave way to those actively supported in practice, it mattered who controlled the purse strings, and who governed what a given practitioner or patient could or could not access. As these media found more regular use, folded into and supporting existing power dynamics, their subversive potential was forgotten. In the case of the most successful dissemination of new media, like the telephone or the pager, collective forgetting is a direct consequence of that success.

At the most fervent moments of speculation, in which the promise of these new platforms became most imminent, it seemed impossible that the utopian or dystopian visions they evoked could be so easily forgotten. Yet the most powerful effects of those prior iterations of new information technology that have been absorbed into the American healthcare system lay in their becoming invisible. Many of these technologies we simply have learned not to see. Not only do we not see the telephone, the radio, the television, or the mainframe computer when we look for histories of medical technology, we also do not see the list of prior promises made when these old technologies were new. Promises of equity. Promises of access. Promises of empowerment. Promises of a new form of healthcare made seamless in a world of pure information. Promises unfulfilled.

* * *

This book began with the domestication of a powerful new form of electronic media, the telephone, which linked homes, doctors' offices, and hospitals in new, often exciting, and occasionally uncomfortable ways. It closed with the vision of a fully automated, computerized clinic that promised to both redefine the personal basis of medical care and reframe the fundamental question of access to healthcare—but which did not come to pass. Both have been forgotten as new media of medicine, but for different reasons. The automated clinic has been forgotten because it failed in spite of widespread enthusiasm across healthcare institutions, providers, insurers, lawmakers, and professional bodies. The telephone has been forgotten because it succeeded so thoroughly in rewriting the script of everyday medical practice that it became invisible.

Across all these episodes it is evident that failures of communications technologies—failures to resolve the power dynamics of intimate clinical relations or the labor functions of physicians or nurses or the steep disparities that fracture access to healthcare across different marginalized populations—do not limit the allure that the next technology *just might* hold the key to short-circuit long-established, socially structured forms of health inequity in the United States. As the physician and civil rights advocate John Holloman pointed out in 1969, even as he was otherwise skeptical of the promises of demonstration projects to resolve the racialized health disparities in American cities, he remained optimistic about the possibilities of communications technologies. The computerized clinic, the multiphasic screening clinic, and "the many potential advantages of the health data bank" could potentially be "utilized as a kinetic force for the improvement of the health of people." So too with radio technologies like diagnosis by telemetry, and other forms of telecare by telephone or by television.

The medical ghetto was real, Holloman insisted, and those who wished to improve the health of all who lived within it could not turn away from the hope that present-day and soon-to-come technologies might provide. Although the telephone and the television would not themselves undo systemic racism, he argued, "telephonic and two-way closed-circuit TV can be established to link the ghetto physician to the medical center, services that could be put into operation almost at once."[21]

These technologies were readily at hand. Why not use them? It was not naïve for Holloman to hope, a half-century ago, that a new tool might break the cycles of disappointment of the past. That the right technology could serve as a catalyst to level the steep asymmetries of access that characterized American healthcare. That it might be a nidus of change, like a grain of sand in the shell of an oyster, to help build something luminous and beautiful. Nor is it naïve for us to want to believe, even today, that the new communications technologies through which medicine is now being practiced might bend the arc of our bloated, inefficient, and unjust healthcare system just a bit closer to that of the equitable and just society we hope to live in.

But what Holloman saw so clearly then, we must now strive to remember: technology alone will never accomplish this uniquely human endeavor. The responsibility is ours to see it through.

ACKNOWLEDGMENTS

This book is dedicated to Marilyn Freedman and Gerard Buter, my grandparents, who have served as the final reviewers and line editors on all of my books. Fiercely independent nonagenarians, they helped instill a love of good science writing early in my life (somewhere between the appreciation of a morning well spent reading the Sunday paper and an afternoon well spent bird-watching on a hike). I have been fortunate indeed to have had them play such an active part in my life these many years.

In a decade or so working on this book I have accumulated a lot of debts, and I will not be able to account for all of them here, let alone pay them back. First and foremost, I thank those who took the time to speak with me and share their personal archives, most notably Rashid Bashshur, Reba Benschoter, Rosemary Lopez, Maxine Rockoff, Peter Ruiz, and Bernard Siquieros. Thanks to those who helped me access unique collections of manuscripts, especially David Piper at the Arizona Health Sciences Library; Barbara Nissen at the Icahn School of Medicine; Lincoln Cushing at the Kaiser Permanente Archives; Laura Tretter at the Montana Historical Society; Sue Topp at the Motorola Solutions Heritage Services and Archive; John Schleicher at the McGoogan Health Sciences Library of the University of Nebraska Medical Center; Bernard Siquieros at the Tohono O'odham Himdag Ki; Marc Brodsky at the Special Collections, Virginia Polytechnic Institute and State University; and the extraordinary staffs of the Center for the History of Medicine at the Countway Medical Library, the Hagley Museum and Library, and the History of Medicine Division at the National Library of Medicine.

We like to think of books as sole-authored affairs, but my work has only been possible thanks to a network of colleagues, both in the Institute of the History of Medicine and the Center for Medical Humanities and Social Medicine at Johns Hopkins, and elsewhere. A stray conversation with Allan Brandt a lifetime ago started my research into the medical history of the telephone; several years later a long plane ride with Mary Catherine Beach helped me expand this inkling of an idea into a book proposal. This book germinated as a Faculty Scholars Program fellowship from the Greenwall Foundation, and has had substantial input from my cohort and other scholars in the program. It reached fruition with an NLM G13 Scholarly Works in Biomedicine fellowship and a Mellon Foundation Sawyer Seminar grant on "Precision and Uncertainty in a World of Data."

The rough components of this project were smoothed out and refined at many conferences and colloquia along the way, as generous scholars took time to point out my errors or make more interesting connections than I had seen at first. It benefited from time I spent as a visiting professor in the University of Zurich (thanks to Flurin Condrau), at CERMES3 in Paris (thanks to Jean-Paul Gaudilliere), in the University of Oslo (thanks to Anne Kveim Lie and Helge Jordheim), and at Keio University in Tokyo (thanks to Junko Kitanaka). Along the way it was helped by friends and colleagues in the American Association for the History of Medicine, the History of Science Society, the Society for the History of Technology, and the Society for Social Studies of Science. I cannot properly thank here all those who made contributions, but I would like to single out Robby Aronowitz, Etienne Benson, Adam Biggs, Gabriela Maya Bernadett, Victor Braitberg, Angus Burgin, Nathan D. B. Connolly, Stephanie Dick, François Furstenberg, Michael Healey, Helen Hughes, David Jones, Matt Jones, Bonnie Kaplan, Tess Lanzarotta, Rebecca Lemov, Mara Mills, Heidi Morefield, John Durham Peters, Judith Vick, Alexandre White, and Keith Wailoo for crucial comments on chapter drafts. Special thanks go to Yulia Frumer, Kim Gallon, Andrew Lea, Joe November, Kirsten Ostherr, Scott Podolsky, Natasha Dow Schüll, John Harley Warner, and Hannah Zeavin, as well as the anonymous referees for their generous comments on how to improve the full manuscript.

I could not have finished this project in the turmoil of the COVID-19 pandemic without the help of gifted and patient editors. A decade of conversations with Karen Darling (and more recently with Tristan Bates) at the University of Chicago Press helped the idea of this book become a reality. In the last mile my predawn sessions with Mara Naselli became an essential part of the rhythm of pandemic life—essential to seeing the project through, and to finding space in difficult times to seek purpose and meaning in the craft of writing. I also thank the physicians, nurse practitioners, and medical assistants I have had the honor to work alongside at the East Baltimore Medical Center through the COVID-19 pandemic, and especially my patients, who have taught me more about care at a distance in this past year than I had learned in the previous decade. And there is no way for me to properly acknowledge my family for seeing each other through this experience in whatever ways we could: my parents, siblings, in-laws, cousins, nieces, and nephews, but especially Liz, Phoebe, Levi, and Milo Greene, who spent the last year or so in constant contact with me around the clock, packed in the same house, commuting outward to our different workplaces and schools on computers that chewed through the same bandwidth, sitting elbow to elbow at our dining room table.

It has been a trying time for all of us, working at a distance. Here's to finding ways to be more present in the years to come.

MANUSCRIPT COLLECTIONS

AGP — Al Gross Papers, 1909–2000, MS 2001-011. Special Collections, Virginia Polytechnic Institute and State University, Blacksburg.

BTA — British Telecom Archives, London, UK.

CWF — Cecil Wittson Files. Special Collections and Archives, McGoogan Health Sciences Library, University of Nebraska Medical Center, Omaha.

DSRC — David Sarnoff Research Center records (Accession 2464.09). Hagley Museum and Library, Wilmington, DE.

HRF — Holter Research Foundation records, 1914–85. Montana Historical Society, Helena.

ISM — Archives and Records Management, Icahn School of Medicine, New York. Courtesy Barbara Nissen.

LEP — Leon Eisenberg Papers, H MS c196. Center for the History of Medicine, Countway Medical Library, Boston.

MCP — Martin M. Cummings Papers, MS C 554. National Library of Medicine, Bethesda, MD.

MFCC — Morris F. Collen Collection. Kaiser Permanente Archives, Oakland, CA.

MGHTC — Massachusetts General Hospital Telemedicine Center Records, 1968–96, AC 1. Massachusetts General Hospital Archives, Boston.

MRP — Maxine Rockoff Papers. Private collection, New York.

MSHSA — Motorola Solutions Heritage Services and Archives, Schaumburg, IL. Courtesy Sue Topp.

NPI — Nebraska Psychiatric Institute Files. Special Collections and Archives, McGoogan Health Sciences Library, University of Nebraska Medical Center, Omaha.

PRC	Peter Ruiz Collection. Archives, Himdag Ki, Tohono O'odham Nation Cultural Center and Museum, Sells, AZ.
RABC	Rashid Bashshur Collection, ACC 2009-060, ACC 2009-011. National Library of Medicine, Bethesda, MD.
RAC	Rockefeller Foundation, RG 1.3–RG 1.8 (A76–A82) (FA209), Subgroup 5: Projects (A78), Series 200: United States, Subseries A: United States—Medical Sciences. Rockefeller Archive Center, Tarrytown, NY.
REBC	Reba Benschoter Collection (unprocessed). National Library of Medicine, Bethesda, MD.
ROHO	Regional Oral History Office. Bancroft Library, University of California, Berkeley.
STARPAHC	Space Technology Applied to Papago Advanced Health Care (STARPAHC) Collections, 1970–79 and 1991, AHSL 2001-01, Arizona Telemedicine Archives. Arizona Health Science Library and Special Collections, University of Arizona, Tucson. Courtesy David Piper.
TEP	Thomas Edison Papers (digital edition). Rutgers, The State University of New Jersey, New Brunswick.

NOTES

Introduction

1 Bernie Monegain, "Global Telemedicine Market Pegged to More than Double by 2016," *Healthcare IT News*, March 14, 2012, https://www.healthcareitnews.com; Robert Calandra, "Telehealth Business: Boom Times, But Profits May Wait," *Managed Care Magazine*, April 2017, https://www.managedcaremag.com. Portions of this material appear in Jeremy Greene, "As Telemedicine Surges, Will Community Health Suffer?" *Boston Review*, April 13, 2020, http://bostonreview.net.

2 Saif Khairat, Timothy Haithcoat, Songzi Liu et al., "Advancing Health Equity and Access Using Telemedicine: A Geospatial Assessment," *Journal of the American Medical Informatics Association* 26, nos. 8–9 (2019): 796–805.

3 Jeremy A. Greene, Graham Mooney, and Carolyn Sufrin, "The Walking Classroom and the Community Clinic: Teaching Social Medicine beyond the Medical School," in Helena Hansen and Jonathan Metzl, eds., *Structural Competency in Mental Health and Medicine: A Case-Based Approach to Treating the Social Determinants of Health* (New York: Springer, 2019), 15–25.

4 Ateev Mehrotra, Kristin Ray, Diane M. Brockmeyer et al., "Rapidly Converting to 'Virtual Practices': Outpatient Care in the Era of Covid-19," *NEJM Catalyst*, April 1, 2020; Lauren A. Eberly, Michael J. Kallan, Howard M. Julien et al., "Patient Characteristics Associated with Telemedicine Access for Primary and Specialty Ambulatory Care during the COVID-19 Pandemic," *JAMA* 3, no. 12 (2020): e2031640, doi10.1001/jamanetworkopen.2020.31640; Ji E. Chang, Alden Yuanhong Lai, Avni Gupta et al., "Rapid Transition to Telehealth and the Digital Divide: Implications for Primary Care Access and Equity in a Post-COVID Era," *Milbank Quarterly* (2021), https://doi.org/10.1111/1468-0009.12509.

5 On the overemphasis on innovation in the history of technology, see David Edgerton, *The Shock of the Old: Technology and Global History since 1900* (New York: Oxford University Press, 2006); Lee Vinsel and Andrew L. Russell, *The Innovation Delusion: How Our Obsession with the New Has Disrupted the Work That Matters Most* (New York: Penguin Currency, 2020).

6 Existing scholarship on the social impact of electronic medicine is concerned more with the present and future than with the past. This is true both of enthusiastic accounts, like Eric Topol's sanguine *The Patient Will See You Now: The Future of Medicine Is in Your Hands* (New York: BasicBooks, 2015) and *Deep Medicine: How Artificial Intelligence Can Make Healthcare Human Again* (New York: BasicBooks, 2019), and of more skeptical accounts, like Robert Wachter's *The Digital Doctor: Hope, Hype, and Harm at the Dawn of Medicine's Computer Age* (New York: McGraw-Hill, 2015), as well as the 2012 Institute of Medicine report *The Role of Telehealth in an Evolving Healthcare Environment* and two excellent ethnographic accounts of the daily practices of telehealth in the Netherlands: Nelly Oudshoorn's *Telecare Technologies and the Transformation of Healthcare* (New York: Palgrave-Macmillan, 2011) and Jeannette Pols's *Care at a Distance: On the Closeness of Technology* (Amsterdam: University of Amsterdam Press, 2012). Most historical accounts of telemedicine have been written by telemedical innovators themselves, such as Gary W. Shannon and Rashid Bashshur, *History of Telemedicine: Evolution, Context, and Transformation* (New Rochelle, NY: Mary Ann Liebert, Inc., 2009).

7 In the history of health and medicine, this approach to media history has brought increased attention to the "paper tools" through which medical experience and medical knowledge have been recorded. See Guenter B. Risse and John Harley Warner, "Reconstructing Clinical Activities: Patient Records in Medical History," *Social History of Medicine* 5 (1992); John Harley Warner, "The Uses of Patient Records by Historians—Patterns, Possibilities and Perplexities," *Health & History* 1, nos. 2–3 (1999): 101–11; Theodore M. Porter, *Genetics in the Madhouse: The Unknown History of Human Heredity* (Princeton: Princeton University Press, 2018).

8 On the intersection of media history and history of technology, see Lisa Gitelman, *Paper Knowledge: Toward a Media History of Documents* (Durham, NC: Duke University Press, 2014). For more general overviews of media history in the electronic era, see Tim Wu, *The Master Switch: The Rise and Fall of Information Empires* (New York: Knopf, 2010); Jeffrey Sconce, *Haunted Media: Electronic Presence from Telegraphy to Television* (Durham, NC: Duke University Press, 2000). Lisa Cartwright's lively discussion of the impact of still photography and cinema on medical science in *Screening the Body: Tracing Medicine's Visual Culture* (Minneapolis: University of Minnesota Press, 1995) was extended thoughtfully to television and other screen practices by Kirsten Ostherr, *Medical Visions: Producing the Patient through Film, Television, and Imaging Technologies* (New York: Oxford University Press, 2013). See also Joseph Turow, *Playing Doctor: Television, Storytelling, and Medical Power* (Oxford: Oxford University Press, 1989); Leslie J. Reagan, Nancy Tomes, and Paula A. Treichler, *Medicine's Moving Pictures: Medicine, Health, and Bodies in American Film and Television* (Rochester, NY: University of Rochester Press, 2009). On the speculative futures of older communications technologies, see Carolyn Marvin, *When Old Technologies Were New: Thinking about Electric Communication in the Late Nineteenth Century* (New York: Oxford University Press, 1990); Jonathan Sterne, *The Audible Past: Cultural Origins of Sound Reproduction* (Durham, NC: Duke University Press, 2003); Lisa Gitelman, *Always Already New: Media, History, and the Data of Culture* (Cambridge, MA: MIT Press, 2007).

9 Norman M. Klein, *The History of Forgetting: Los Angeles and the Erasure of Memory*

(London: Verso, 2008); on the role of forgetting and the production of ignorance in the history of science, medicine, and technology, see Robert Proctor and Londa Schiebinger, *Agnotology: The Making and Unmaking of Ignorance* (Stanford: Stanford University Press, 2008).

10 Vladimir Zworykin, "Where Is Medical Electronics Going?" Address before the Institute of Radio Engineers, March 22, 1956, 6–7, box 77, folder 68, David Sarnoff Research Center records, Hagley Museum and Library (hereafter DSRC).

11 "Device to Detect Illness Foreseen," *New York Times*, March 23, 1956, 22; Vladimir K. Zworykin, "Electronics and Health Care," *Proceedings of the Institute of Radio Engineers*, May 1962, proof copy, M&A 78 39, DSRC.

12 See, for example, Topol, *Deep Medicine; Taking Action against Clinician Burnout: A Systems Approach to Professional Well-Being* (Washington, DC: National Academies of Science, Engineering, and Medicine, 2019).

13 Bonnie Kaplan, "The Computer Prescription: Medical Computing, Public Policy, and Views of History," *Science, Technology, and Human Values* 20, no. 1 (1995): 5–38; Nathan Ensmenger, "Resistance Is Futile? Reluctant and Selective Users of the Internet," in William Aspray and Paul Ceruzzi, eds., *The Internet and American Business* (Cambridge, MA: MIT Press, 2010).

14 Notable exceptions include Bonnie Kaplan, "The Computer Prescription: Medical Computing, Public Policy, and Views of History," *Science, Technology, and Human Values* 20, no. 1 (1995): 5–38; Joseph November, *Biomedical Computing: Digitizing Life in the United States* (Baltimore: Johns Hopkins University Press, 2012); Hallam Stevens, *Life Out of Sequence: A Data-Driven History of Bioinformatics* (Chicago: University of Chicago Press, 2013); Andrew Lea, "Computerizing Diagnosis: Keeve Brodman and the Medical Data Screen," *Isis* 110, no. 2 (2019): 228–49; Hannah Zeavin, *The Distance Cure: A History of Teletherapy* (Cambridge, MA: MIT Press 2021). See also Tess Lanzarotta and Jeremy Greene, "Communications Technologies as Community Technologies: Alaska Native Villages and the Satellite Health Trials of the 1970s," *Technology's Stories* 5, no. 2 (2017), doi:10.15763/JOU.TS.2017.5.9.01; Jeremy Greene, Victor Braitberg, and Gabriella Maya Bernadett, "Innovation on the Reservation: Information Technology and Health Systems Research among the Papago Tribe of Arizona, 1965–1980," *Isis* 111, no. 3 (2020): 443–70.

15 Nancy Siraisi, *Communities of Learned Experience: Epistolary Medicine in the Renaissance* (Baltimore: Johns Hopkins University Press, 2012); Michael Stolberg, "Patients' Letters and Pre-modern Medical Lay Culture," *Med Ges Gesch Beih.* 29 (2007): 23–33; Robert Weston, *Medical Consulting by Letter in France, 1665–1789* (London: Routledge, 2016).

16 Oliver Wendell Holmes, "The Stethoscope Song," in *Poems by Oliver Wendell Holmes* (Boston: Ticknor & Fields, 1848), 272–77.

17 Stanley Joel Reiser, *Medicine and the Reign of Technology* (Cambridge: Cambridge University Press, 1978).

18 See Hughes Evans, "Losing Touch: The Introduction of Blood Pressure Instruments into Medicine," *Technology and Culture* 34, no. 4 (1993): 784–807; Keith Wailoo, *Drawing Blood: Technology and Disease Identity in Twentieth Century America* (Baltimore: Johns Hopkins University Press, 1999).

Chapter One

1 Jason Lim, "Health Tech Startup CliniCloud Secures $5M Seed Funding from Tencent," *Forbes*, September 24, 2015, https://www.forbes.com.

2 Fitz Tepper, "CliniCloud Raises $5M for Its Connected Home Medical Kit," TechCrunch, September 23, 2015; Sarah Buhr, "CliniCloud's Smart Stethoscope and Thermometer Let Doctors Check Your Vitals from the Cloud," TechCrunch, February 12, 2015; both at https://techcrunch.com.

3 "The Week," *Cincinnati Lancet and Clinic* 3, no. 42 (1879): 356.

4 "The Week," 356.

5 "The Telephone as a Medium of Consultation and Medical Diagnosis," *BMJ: British Medical Journal* 2, no. 988 (1879): 897; "Practice by Telephone," *The Lancet* (1879): 819–22.

6 "Lay Use of the Clinical Thermometer," *JAMA* 35, no. 20 (1900): 1283.

7 Carolyn Marvin, *When Old Technologies Were New: Thinking about Electric Communication in the Late Nineteenth Century* (Oxford: Oxford University Press, 1988).

8 Lisa Gitelman, *Always Already New: Media, History, and the Data of Culture* (Cambridge, MA: MIT Press, 2006), 7.

9 H. P. Bowditch, "Recent Progress in Physiology," *Boston Medical and Surgical Journal* 98 (1878): 593–98; "Medical Notes," *Boston Medical and Surgical Journal* 98 (1878): 547–49.

10 Bruce Fye, "Why a Physiologist? The Case of Henry P. Bowditch," *Bulletin of the History of Medicine* 56, no. 1 (1982): 19–29; John Harley Warner, *Against the Spirit of System: The French Impulse in Nineteenth-Century American Medicine* (Baltimore: Johns Hopkins University Press, 1998), 306–7.

11 "The Telephone as an Electric Reagent," *British Medical Journal* 1, no. 899 (1878): 423.

12 On the role of the physiology laboratory in the self-consciously scientific approach to medical education and medical practice in the second half of the nineteenth century, see William Coleman and Frederic L. Holmes, *The Investigative Enterprise: Experimental Physiology in Nineteenth-Century Medicine* (Berkeley: University of California Press, 1988); William Bynum, *Science and the Practice of Medicine in the Nineteenth Century* (Cambridge: Cambridge University Press, 1994).

13 H. Thompson, "A Lecture on the Use of the Microphone in Sounding for Stone," *British Medical Journal* 1, no. 910 (1878): 809–10, emphasis mine. See also "The Telephone," *The Lancet* (1878): 221–24. For a centennial review of mentions of the telephone in *The Lancet*, see S. H. Aronson, "*The Lancet* on the Telephone: 1876–1975," *Medical History* 21, no. 1 (1977): 69–87.

14 J. G. McKendrick, "Note on the Microphone and Telephone in Auscultation," *British Medical Journal* 1, no. 911 (1878): 856–57, at 857.

15 "Setting careful investigation of their capabilities aside, this is sufficiently evidenced by the fact that none of these instruments have come into general use." C. J. Blake, "The Telephone and Microphone in Auscultation," *Boston Medical and Surgical Journal* 103, no. 21 (1880): 486–87, at 486. On Blake's role in Boston medical institutions, see Warner, *Against the Spirit of System*, 308–13; Charles Rosenberg, *The Care of Strangers: The Rise of America's Hospital System* (New York: Basic Books, 1987), 166–68. For his role in otology and the history of deafness, see Jaipreet Virdi, *Hearing Happiness: Deafness Cures in History* (Chicago: University of Chicago Press, 2020).

16 Blake, "The Telephone and Microphone in Auscultation," 487.

17 "The stethoscope was placed on the heart of a person in London and the beats were distinctly heard by physicians in the Isle of Wight, a matter of 100 miles or so distant." "A Telephonic Stethoscope," *Scientific American* 102, no. 25 (1910): 496, 509.

18 W. M. Brown, "The Telephone and Phonograph in Practical Medicine," *The Lancet* (1878): 371.

19 Thomas Edison to Byron Bramwell, December 12, 1887, https://edison.rutgers.edu/digital/document/LB026089. Edison later wrote Bramwell to request a copy of his article in the *British Medical Journal*, Edison to Bramwell, January 30, 1888, https://edison.rutgers.edu/digital/document/D8818ACC. Thomas Edison Papers, Rutgers University, New Brunswick, NJ (hereafter TEP).

20 B. Bramwell and R. M. Murray, "A Method of Graphically Recording the Exact Time-relations of Cardiac Sounds and Murmurs," *British Medical Journal* 1, no. 1410 (1888): 10–16, at 12.

21 Mara Mills, "When Mobile Communication Technologies Were New," *Endeavor* 33, no. 4 (2009): 146–47.

22 Mills, "When Mobile Communication Technologies Were New," 146–47; Virdi, *Hearing Happiness*.

23 J. J. Duncanson, "The Electric Telephone as a Means of Testing or Measuring the Hearing Power," *British Medical Journal* 1, no. 897 (1878): 335.

24 A. J. Balmanno Squire, "A New Modification of the Telephone for Medical Purposes," *British Medical Journal* 1, no. 913 (1878): 934.

25 Clarence A. Swoyer, "The First Physician's Telephones and the First Telephone Secretary in Columbus, 1879," *Ohio State Medical Journal* (1949): 50–52; George W. Paulson, *James Fairchild Baldwin, M.D., 1850–1936: An Extraordinary Surgeon* (Columbus: Medical Heritage Center, 2005).

26 "A General Practitioner's Enterprise." *JAMA* 23, no. 14 (1894): 556–57.

27 "The Use of the Telephone from a Medical Point of View," *The Lancet* (1883): 963–69.

28 Theodore W. Schaefer, "The Commercialization of Medicine; Or, the Physician as Tradesman," *Boston Medical and Surgical Journal* 131, no. 21 (1894): 501–2, at 501.

29 Rebecca Solnit, "The Annihilation of Time and Space," *New England Review* 24, no. 1 (2003): 5-19; for an influential analysis of "time-space compression" based on Marx's analysis, see David Harvey, *The Condition of Postmodernity* (Oxford: Blackwell, 1989).

30 John L. Hildreth, "The General Practitioner and the Specialist," *Boston Medical and Surgical Journal* 155, no. 4 (1906): 79–83, at 79.

31 Nancy Tomes, *Remaking the American Patient: How Madison Avenue and Modern Medicine Turned Patients into Consumers* (Chapel Hill: University of North Carolina Press, 2016).

32 Claude S. Fischer, *America Calling: A Social History of the Telephone to 1940* (Berkeley: University of California Press, 1992); Tim Wu, *The Master Switch: The Rise and Fall of Information Empires* (New York: Knopf, 2010).

33 The author had constructed this telephone "quite against the rules of the Bell Company"; nonetheless, "after much argument I was permitted to use it as a sort of laboratory experiment, with the express understanding that it should not establish a precedent for its use by others." A. E. Rockey, "A Bedside Telephone," *JAMA* 78, no. 20 (1922): 1535.

34 "It's Urgent, Doctor," TCB 475/ZB/ZB21, British Telecom Archives, London (hereafter BTA).

35 "The Telephone," *Boston Medical and Surgical Journal* 171, no. 15 (1914): 573–74.

36 Ellen M. Firebaugh, *The Story of a Doctor's Telephone—Told by His Wife* (Boston: Roxburgh Publishing Co., 1912), nonpaginated.

37 Robert B. Dempsey, "Appreciation of the Telephone," *California State Journal of Medicine* 11, no. 1 (1913): 36.

38 Ruth Schwartz Cowan, *More Work for Mother: The Ironies of Household Technology from the Open Hearth to the Microwave* (New York: Basic Books, 1983).

39 M. J. Konikow, "Physicians and the Telephone," *Boston Medical and Surgical Journal* (1921): 290; "The Moment and the Telephone," *Canadian Medical Association Journal* 4, no. 10 (1914): 907–9; "Telephones," *British Medical Journal* 1, no. 3203 (1922): 811–12.

40 A. J. De Long, "A Doctors' Information Exchange," *Journal of the Indiana State Medical Association* 14, no. 10 (1921): 349–51, at 349.

41 De Long, "A Doctors' Information Exchange," 349–51.

42 De Long, "A Doctors' Information Exchange," 351.

43 "A Wheelchair Business Boy," *Boys' Life*, June 1922, 26. A lifelong radio enthusiast, De Long was also proud of his self-built, three-tube radio set. "My radio," he noted, "has made our home and especially my room the most popular place in the community. Hardly a night goes by but that people come in to enjoy the programs with me . . . My radio has been instrumental in my winning many new friends and consequently new business." On the intersection of histories of technology, histories of disability, and histories of health and medicine, see David Serlin, *Replaceable You: Engineering the Body in Postwar America* (Chicago: University of Chicago Press, 2004); Mills, "When Mobile Communication Technologies Were New"; Virdi, *Hearing Happiness*.

44 De Long, "A Doctors' Information Exchange," 350.

45 "A Physicians' Telephone Exchange," *JAMA* 60 (1912): 1162; "A Physicians' Telephone Exchange," *California State Journal of Medicine* 11, no. 6 (1921): 146.

46 Algernon Jackson, "Calling the Doctor," *Baltimore Afro-American*, May 28, 1938, 21. For a raw and unsettling account of the technological promise of telephones read against the racial violence of the United States during the first decades of telephone networks, see Eula Bliss, "Time and Distance Overcome," *Iowa Review* 38, no. 1 (2008): 83–39.

47 "Health Organizations and the Telephone," *American Journal of Public Health* 31 (1941): 733–34, at 733.

48 "Robot Telephone," *British Medical Journal* 2, no. 4520 (1947): 57.

49 "Doctors' Telephone Service," *British Medical Journal* 1, no. 4497 (1947): 32–33; "The Unattended Telephone," *British Medical Journal* 2, no. 4532 (1947): 117; "The Unattended Telephone," *British Medical Journal* 2, no. 4565 (1948): 28.

50 "A Doctor's Telephone," *The Lancet* 10 (1954); J. Lister, "The Doctor and His Telephone," *New England Journal of Medicine* 252, no. 21 (1955): 908–9.

51 Lister, "The Doctor and His Telephone," 908–9.

52 "Culpable Neglect of Telephone," *British Medical Journal* 1, no. 5182 (1960): 1370–71.

53 Paul N. Edwards, "Infrastructure and Modernity: Force, Time, and Social Organization in the History of Sociotechnical Systems," in Thomas J. Misa, Philip Brey, and

Andrew Feenberg, eds., *Modernity and Technology* (Cambridge, MA: MIT Press, 2004), 185–225.

54 Gertrude L. Gunn, "'Don't Write—Telephone' for Better Medical Records," *Modern Hospital* 80–81, no. 77 (1951): 5.

55 G. A. W Currie and J. Arthur Keddy, "From Doctor to Disc via Telephone," *Canadian Hospital* 50–51 (1959): 50.

56 "Pictures by Telephone," *Science* 62, no. 1610 (1925): 10–11; G. Gallerani, "Apparatus for Transmitting Views of Objects Great Distances," *Bollettino della Società Eustachiana* 24 (1926): 95–106; N. A. Podkaminsky, "Orthodiagraphy and Teleroentgenography of Heart," *Die Medizinische Welt* 3 (1929): 1724–26; J. Gerson-Cohen and A. G. Cooley, "Roentgenographic Facsimile: A Rapid Accurate Method for Reproducing Roentgenograms at a Distance via Wire or Radio Transmission," *American Journal of Roentgenology* 61 (1949): 557–59; J. Gerson-Cohen and A. G. Cooley, "Teleognosis," *Radiology* 55 (1950): 582–87; J. Gershon-Cohen, M. B. Hermel, H. S. Read, B. Caplan, and A. G. Cooley, "Teleognosis," *JAMA* 148, no. 9 (1952): 731–32.

57 J. Gershon-Cohen, "Field Tests of Telephone Transmitted X-Ray Facsimiles during the Past Eight Years," *Journal of the Albert Einstein Medical Center* 4, no. 3 (1956): 110–12; J. Gershon-Cohen, "Telephone Facsimile," *Hospital Management* 83, no. 5 (1957): 58–59; M. S. Williams, "X-Ray Reports by Telephone," *Hospitals* 33 (1959): 63.

58 Jonathan Coopersmith, *Faxed: The Rise and Fall of the Fax Machine* (Baltimore: Johns Hopkins University Press, 2015); Jennifer Light, "Facsimile: A Forgotten 'New Medium' from the 20th Century," *New Media & Society* 8, no. 3 (2006): 367–68.

59 W. E. Rahm, J. L. Barmore, and F. L. Dunn, "Electrocardiographic Transmission over Standard Telephone Lines," *Nebraska Medical Journal* 37 (1952): 222–23; E. Grey Dimond and M. Fred Berry, "Transmission of Electrocardiographic Signals over Telephone Circuits," *American Heart Journal* 46, no. 6 (1953): 906–10.

60 Richard W. Booth, "ECG's Transmitted by Long-Distance Telephone Provide Consultation for Midwest Physicians," *JAMA* 188, no. 12 (1964): 29–30; Constantine T. Cerkez, Gordon C. Steward, B. Bacongallo, and Geo W. Manning, "Telephonocardiography: The Transmission of Electrocardiograms by Telephone," *Canadian Medical Association Journal* 91, no. 14 (1964): 727–32.

61 F. A. L. Mathewson and H. Jackh, "Telecardiogram," *American Heart Journal* 49 (1955): 77–82; J. N. Edison, "The Transmission of Electrocardiograms by Telephone," *Transactions of the Association of Life Insurance Medical Directors of America, Annual Meeting* (1964): 83–95.

62 Joseph P. Melvin, "Telephone Telemetry," *Journal of Mississippi State Medical Association* 5, no. 3 (1964): 84–86; "Brainwaves by Telephone," *Canadian Hospital* 43, no. 11 (1966): 49–50.

63 I. M. Levine, P. B. Jossmann, B. Tursky, M. Meister, and V. DeAngelis, "Telephone Telemetry of Bioelectric Information," *JAMA* 188, no. 9 (1964): 794–98, at 798.

64 S. I. Allen and M. Otten, "The Telephone as a Computer Input-Output Terminal for Medical Information," *JAMA* 208, no. 4 (1969): 673–79, at 673.

65 Blake, "The Telephone and Microphone in Auscultation."

66 Alistair Fair, "A Laboratory of Heating and Ventilation: The Johns Hopkins Hospital as Experimental Architecture, 1870–90," *Journal of Architecture* 19, no. 3 (2014): 357–81.

67 "Use of the Telephone," *JAMA* (1954): 1122.

68 Rudolf J. Pendall, "Every Hospital Has Its 'Telephone Personality,'" *Hospital Progress* (1952): 52–54. The Canadian Medical Association published its own "Public Relations Forum" a year later with a list of "Telephone Tactics" for the physician. L W. Holmes, "Telephone Tactics," *Canadian Medical Association Journal* 73 (1955): 844–45, at 844.

69 "Mind Those Telephone Manners!" *Modern Hospital* 91, no. 3 (1958): 96–97, at 96.

70 Stu Chapman, "Your Telephone: A Timesaver or Troublemaker," *Legal Aspects of Medical Practice* 7, no. 6 (1979): 21–24, at 21.

71 Chapman, "Your Telephone: A Timesaver or Troublemaker," 24.

72 Chapman, "Your Telephone: A Timesaver or Troublemaker," 24.

73 Jeffrey L. Brown, *The Complete Parents' Guide to Telephone Medicine: How, When, and Why to Call Your Child's Doctor* (New York: Perigee Books, 1982), ix.

74 Brown, *The Complete Parents' Guide to Telephone Medicine*, ix.

75 Brown, *The Complete Parents' Guide to Telephone Medicine*, 8.

76 Brown, *The Complete Parents' Guide to Telephone Medicine*, 127.

77 Brown, *The Complete Parents' Guide to Telephone Medicine*, 134.

78 Jo Simms and Reba McGear, *Telephone Triage and Management: A Nursing Process Approach* (Philadelphia: W. B. Saunders Co., 1988), v.

79 Margarete Sandelowski, *Devices and Desires: Gender, Technology, and American Nursing* (Chapel Hill: University of North Carolina Press, 2000).

80 On the role of nurses as researchers in these and other allied social sciences of health and medicine, see Dominique Tobbell, *Dr. Nurse: Science, Politics, and the Transformation of American Nursing* (Chicago: University of Chicago Press, forthcoming).

81 Simms and McGear, *Telephone Triage and Management*, 112–13.

82 Simms and McGear, *Telephone Triage and Management*, 62–63.

83 Marc Berg, *Rationalizing Medical Work: Decision Support Techniques and Medical Practices* (Cambridge, MA: MIT Press, 1997). On algorithms and the reproduction of existing power asymmetries and gender and racial disparities, see Safiya Umoja Noble, *Algorithms of Oppression: How Search Engines Reinforce Racism* (New York: NYU Press, 2018); Ruha Benjamin, *Race after Technology: Abolitionist Tools for the New Jim Code* (New York: Polity Press, 2019).

84 Anna B. Reisman and David L. Stevens, *Telephone Medicine: A Guide for the Practicing Physician* (Washington, DC: American College of Physicians, 2002).

85 Harvey Cushing, "One Hundred and Fifty Years," *New England Journal of Medicine* 2–4, no. 24 (1931): 1235–44.

86 Andrew Birt, "Latest News from Team CC," July 7, 2017, blog post, https://clinicloud.com; Andrew Lin, "Technologies and Trends Shaping the Future of Remote Patient Monitoring," March 16, 2017, https://www.linkedin.com.

Chapter Two

1 Rose M. F. L. da Silva, "Syncope: Epidemiology, Etiology, and Prognosis," *Frontiers in Physiology* 5 (2014): 471.

2 Natasha Schull, "Data for Life: Wearable Technology and the Design of Self-Care," *BioSocieties* 11, no. 3 (2016): 317–33; Lucie Gerber, "Faire une histoire sociale et cul-

turelle des thérapies comportementales," paper presented at "Cycle de séminaires de recherche et de conférences du FADO," University of Lausanne, 3 December 2019.

3 Norman J. Holter, "Radioelectrocardiography: A New Technique for Cardiovascular Studies," *Annals of the New York Academy of Sciences* 65, no. 6 (1957): 913–23, at 914–15, emphasis mine.

4 H. F. MacInnis, "The Clinical Application of Radioelectrocardiography," *Canadian Medical Association Journal* 70 (May 1954): 574–76, at 575.

5 Gerald J. Skibbins to Norman J. Holter, December 26, 1957, box 2, folder 5, Holter Research Foundation records, Montana Historical Society, Helena (hereafter HRF). Before becoming executive of the Opinion Research Council in Princeton, New Jersey, Skibbins had worked closely with Holter as chair of the Montana Chamber of Commerce. His writings on "Dynamic Conservatism" were read into the *Congressional Record* by Barry Goldwater in 1963 (*Congressional Record, 88th Congress, First Session—Senate*, October 1, 1963, p. 18453), and his later book, *Organizational Evolution: A Program for Managing Radical Change* (New York: AMACOM, 1974), emphasized the need for conservative thinkers to embrace new technologies in the evolution of free-market organizations, while ensuring that surveillance powers not be concentrated in any limb of the federal government.

6 Christopher Rowland, "With Fitness Trackers in the Workplace, Bosses Can Monitor Your Every Step—and Possibly More," *Washington Post*, February 19, 2019.

7 Phoebe V. Moore, *The Quantified Self in Precarity: Work, Technology, and What Counts* (London: Routledge, 2017); Junko Kitanaka, *Depression in Japan: Psychiatric Cures for a Society in Distress* (Princeton: Princeton University Press, 2011); Adrian Mackenzie, *Wirelessness: Radical Empiricism in Network Cultures* (Cambridge, MA: MIT Press, 2010).

8 "President's Reports to Board of Trustees, 1950, 1951, 1974," box 40, folder 10, HRF; quote from 1974 report.

9 Nicolas Rasmussen, "Of 'Small Men,' Big Science, and Bigger Business: The Second World War and Biomedical Research in the United States," *Minerva* 40, no. 2 (2002): 115–46; on the role of tinkerers in the development of radio technologies, see Kristin Haring, *Ham Radio's Technical Culture* (Cambridge, MA: MIT Press, 2006).

10 David D. Geddes to Ferdinand Schemm, March 8, 1954; N. J. Holter to David D. Geddes, March 15, 1954, box 2, folder 2, HRF. Although Holter does not appear in Angela Creager's *Life Atomic: A History of Radioisotopes in Science and Medicine* (Chicago: University of Chicago Press, 2013), her description of how physicians and scientists used radioisotopes as a new medium for tracing physiological processes applies equally to Holter's radioisotope and radio-broadcasting research.

11 Marco Piccolino, "Visual Images in Luigi Galvani's Path to Animal Electricity," *Journal of the History of the Neurosciences* 17, no. 3 (2008): 335–48.

12 J. A. Gengerelli and N. J. Holter, "Experiments on Stimulation of Nerves by Alternating Electrical Fields," *Proceedings of the Society for Experimental Biology and Medicine* 46 (1941): 532.

13 Norman J. Holter, "Genesis of Biotelemetry," unpublished MS (1976), 4, box 76, folder 6, HRF; J. A. Gengerelli and V. Kallejian, "Remote Stimulation of the Brain in the Intact Animal," *Journal of Psychology: Interdisciplinary and Applied* 29 (1950):

263–69; J. A. Gengerelli, "Patterns of Response to Remote Stimulation of the Brain of the Intact Animal," *Journal of Comparative and Physiological Psychology* 44 (1951): 535–42.

14 Holter, "Genesis of Biotelemetry," emphasis in original. See also Rebecca Lemov, *World as Laboratory: Experiments with Mice, Mazes, and Men* (New York: Hill & Wang, 2006).

15 Daniel Todes, *Pavlov's Physiology Factory* (Baltimore: Johns Hopkins University Press, 2002); Daniel Todes, *Ivan Pavlov: A Russian Life in Science* (Oxford: Oxford University Press, 2014).

16 Cornelius Borck, *Brainwaves: A Cultural History of Electroencephalography*, trans. Ann M. Hentschel (London: Routledge, 2018).

17 Norman J. Holter and J. A. Gengerelli, "Remote Recording of Physiological Data by Radio," *Rocky Mountain Medical Journal* 46, no. 9 (1949): 747–51.

18 Amanda Rees and Iwan Rhys Morus, eds., "Presenting Futures Past: Science Fiction and the History of Science," *Osiris* 34 (2019).

19 Grant Wythoff, *The Perversity of Things: Hugo Gernsback on Media, Tinkering, and Scientifiction* (Minneapolis: University of Minnesota Press, 2016).

20 Several physicists also speculated on the role of electricity in human thought, including J. J. Thomson (who discovered the electron) and William Crookes (who invented the cathode-ray tube). For more on their work see Andrew Gaedtke, *Modernism and the Machinery of Madness: Psychosis, Technology, and Narrative Worlds* (Cambridge: Cambridge University Press, 2017).

21 Hugo Gernsback, "The Thought Recorder," *Electrical Experimenter*, May 1, 1919, 2.

22 Gernsback, "The Thought Recorder," 2: "The writer, in suggesting the aurion as a thought-wave detector, does not do so because he thinks that it is suitable in all respects, or even feasible. His main idea is to set the stone rolling, and get other people to think about the problem, when sooner or later something surely will emerge."

23 Nikola Tesla, response to Hugo Gernsback, "The Thought Recorder," *Electrical Experimenter*, May 1, 1919, 2.

24 S. R. Winters, "Diagnosis by Wireless," *Scientific American* 124, no. 24 (1921): 465.

25 John Durham Peters, *Speaking into the Air: A History of the Idea of Communication* (Chicago: University of Chicago Press, 1999); Peter J. Bowler, *A History of the Future: Prophets of Progress from H. G. Wells to Isaac Asimov* (Cambridge: Cambridge University Press, 2017).

26 Borck, *Brainwaves.* Though the idea of the "brain wave" is older, Borck argues that Berger was the first to connect a galvanometer to the head of a human subject to document electrical activity. See also Katja Guenther and Volker Hess, "Soul Catchers: The Material Culture of the Mind Sciences," *Medical History* 60, no. 3 (2016): 301–7.

27 Borck, *Brainwaves.*

28 Allison Oswald, "At the Heart of the Invention: The Development of the Holter Monitor," blog post, O Say Can You See: Stories from the National Museum of American History (blog), November 16, 2011, http://americanhistory.si.edu.

29 Wilford R. Glasscock and Norman J. Holter, "Radioelectroencephalograph for Medical Research," *Electronics*, August 1952, 126–29, at 126.

30 DJH to N. J. Holter, n.d. (1950), box 1, folder 7, HRF.

31 OAP to J. A. Gengerelli, January 14, 1951; RHB to N. J. Holter, January 25, 1971, box 1, folder 7, HRF.

32 Edward Hunter, "Brain-Washing Tactics Force Chinese into Ranks of Communist Party," *Miami Daily News*, September 19, 1950; Rebecca Lemov, "Brainwashing's Avatar: The Curious Career of Dr. Ewen Cameron," *Grey Room* 45 (2011): 61–87.

33 HFC to N. J. Holter, March 25, 1952, box 1, folder 8, HRF.

34 N. J. Holter to HFC, April 8, 1952, box 1, folder 8, HRF.

35 LR to N. J. Holter, April 22, 1951, box 1, folder 7, HRF.

36 He specifically requested Holter to write an expert statement for his case for release on grounds of full mental competency. LR to N. J. Holter, July 8, 1952, box 1, folder 8, HRF.

37 Norman J. Holter to Drs. Klegfer and Wormington, November 24, 1952, box 1, folder 8, HRF. On broadcasting and delusions of reference, see John Durham Peters, "Broadcasting and Schizophrenia," *Media, Culture, and Society* 32, no. 1 (2010): 123–40.

38 P. D. White, *Heart Disease*, 4th ed. (New York: Macmillan, 1951), 182, 225; P. D. White to NJH, November 10, 1952, box 1, folder 8, HRF. Holter's network of collaborators grew to include locally the Great Falls Clinic in Montana, regionally the National Jewish Hospital in Denver, and then nationally the VA Hospital of Portland, Oregon, and the UCLA/Cedars of Lebanon Hospital of Los Angeles. See also J. Gordon Spendlove to N. J. Holter, December 2, 1952; Norman S. Blackman to N. J. Holter, October 15, 1952; N. J. Holter to Norman S. Blackman, November 20, 1952, all in box 1, folder 8, HRF.

39 Norman J. Holter to J. Franklin Yeager, October 29, 1954, box 2, folder 2, HRF. Holter also provided a detailed bibliography to a physician in the Navy Medical Corps who also claimed to have discovered RECG, documenting Holter's priority in the field. N. J. Holter to Norman L. Barr, November 4, 1955, box 2, folder 3, HRF.

40 Gerald J. Skibbins to Norman J. Holter, December 26, 1957, box 2, folder 5, HRF.

41 Herbert I. Winer to Norman J. Holter, November 19, 1958, box 2, folder 6, HRF.

42 Hugh D. Galusha to Norman J. Holter, November 30, 1956, box 2, folder 4, HRF.

43 N. J. Holter to J. Franklin Yeager, June 18, 1966, box 2, folder 4, HRF. Patent no. 631,288 was filed on Audio-Visual Superimposed ECG Presentation (AVSEP) December 28, 1956. Martin Kirkpatrick to Norman J. Holter, January 28, 1957, box 2, folder 5, HRF; Holter, "Radioelectrocardiography: A New Technique for Cardiovascular Studies."

44 Norman J. Holter, "New Method for Heart Studies," *Science* 134, no. 3486 (1961): 1214–20, at 1215.

45 Norman J. Holter, to J. A. Gengerelli, June 30, 1961, box 3, folder 2, HRF.

46 "Electronic Aids Help Doctors Diagnose Ills, Ease Nurses' Task," *Wall Street Journal*, June 26, 1961; Norman J. Holter to Carl Berkley, July 1, 1961, box3, folder 3, HRF.

47 F. Lowell Dunn and H. G. Beenken, "Short Distance Radio Telemetering of Physiological Information," *JAMA* 169, no. 14 (1959): 158–61, at 158.

48 "Long Distance Telemetering and Monitoring," *JAMA* 169, no. 13 (1959): 1486.

49 John F. Kennedy, "The President and the Press: Address before the American Newspaper Publishers Association, April 27, 1961," speech (sound recording, transcript), John F. Kennedy Presidential Library and Museum, https://www.jfklibrary.org.

50 Holter, "The Genesis of Biotelemetry," 6.
51 Albert Abramson, *Zworykin, Pioneer of Television* (Urbana: University of Illinois Press, 1995).
52 "Station PILL Broadcasts from Inside Human Body," *Chicago Tribune*, April 8, 1957, 1; "Radio Pill' Developed for Medical Research Work; Sends out FM Signals as It Passes through the Body," RCA press release, April 8, 1957, box 82, folder 46, DSRC. Follow-up studies on twenty-eight patients indicated that the pill had no untoward effect on digestion and was "not a source of patient discomfort." Vladimir K. Zworykin, "Adventures in Medical Electronics," talk given to Chicago Section, Institute of Radio Engineers, May 1959, 12, box 78, folder 2, DSRC.
53 Vladimir K. Zworykin, "An International Conference on Medical Electronics," n.d., 1, box 78, folder 2, DSRC.
54 R. Stuart Mackay and Bertil Jacobson, "Endoradiosonde," *Nature* 179 (1957): 1239–40.
55 Vladimir K. Zworykin to Bertil Jacobson, May 27, 1957, box 82, folder 46, DSRC; H. G. Nöller to Vladimir K. Zworykin, November 1, 1959, DSRC; Bertil Jacobson, "Tracking Radio Pills in the Human Body," *New Scientist* 14 (1962): 288–90; Akihito Uchiyama, Taketoshi Morimoto, and Hisato Yoshimura, "Continuous Recording of Blood pH and Its Radiotelemetry," *Japanese Journal of Physiology* 14 (1964): 630–37.
56 Vladimir Zworykin, "Prospects in Medical Electronics," in *Proceedings of the Third International Conference on Medical Electronics, London, 1960* (Springfield, IL: Charles C. Thomas, 1961), 6.
57 John T Farrar, Carl Berkley, and Vladimir K Zworykin, "Telemetering of Intraenteric Pressure in Man by an Externally Energized Wireless Capsule," *Science* 131, no. 3416 (1960): 9–12.
58 E. G. Ramberg to H. G. Dyke, "Interdepartmental Correspondence: RCA Laboratories," April 25, 1957, box 92, folder 46, DSRC.
59 "As I understand it," RCA's legal counsel argued, "it is not our policy to try to make money out of this medical invention; and even if we wanted to we might run into complications with our associate the Rockefeller Institute for Medical Research . . . The prestige has already ensued by the wide publication in *Life* and elsewhere." "Radio Pill," internal RCA memo, October 1, 1957, box 82, folder 46, DSRC. On the role of the pacemaker in the development of the medical device industry, see Kirk Jeffrey, *Machines in Our Hearts: The Cardiac Pacemaker, the Implantable Defibrillator, and American Health Care* (Baltimore: Johns Hopkins University Press, 2001).
60 M. H. Halpern to H. S. Dordick, "Medical Instrumentation Development and Possible Cooperative Marketing Agreement with Smith, Kline & French Labs., Phila. 5-29-61, 5-31-61," 3, box 81, folder 13, DSRC.
61 "At no additional cost to RCA but with the promise of additional income, RCA can benefit immediately in three ways: (1) by becoming an established name in biomedical electronics, (2) by entering the commercial market at an earlier date than anticipated, (3) by physician acceptance as a result of SKF endorsement." Halpern to Dordick, "Medical Instrumentation Development and Possible Cooperative Marketing Agreement with Smith, Kline & French Labs., Phila. 5-29-61, 5-31-61," 3.
62 "Electronic System for Monitoring Hospital Patients from Central Point Is Described by RCA Scientists," press release, Radio Corporation of America, July 20, 1961, 1–2, box 82, folder 40, DSRC.

63 Vladimir K. Zworykin and F. L. Hatke, "A Miniaturized Hospital Telemetering System," paper presented at the 4th International Conference and Exhibition on Medical Electronics, New York City, July 1961, 1, box 78, folder 41, DSRC.

64 The RCA radiotelemetry prototype boasted "a sensitive transmitter only slightly larger than a lump of sugar. Placed against the skin at the fingertips, on the arm, or elsewhere, the miniature unit detects heart action from the associated electrical charges in the body and transmits information to a receiving antenna elsewhere in the room for display as a visible cardiogram on a viewing tube or an inked pattern on recording paper—thus eliminating the connections from the patient to the recorder." Zworykin and Hatke, "A Miniaturized Hospital Telemetering System," 1.

65 Halpern to Dordick, "Medical Instrumentation Development and Possible Cooperative Marketing Agreement with Smith, Kline & French Labs., Phila. 5-29-61, 5-31-61," 2.

66 "Cesar Caceres, M.D., 1927–2020," *Washington Post*, February 21, 2020.

67 Thomas C. Gibson, William E. Thornton, William P. Algary, and Ernest Craige, "Telecardiography and the Use of Simple Computers," *New England Journal of Medicine* 267, no. 24 (1962): 1218–24; R. M. Farrier, "Problems of Electronic Patient Monitoring," *Hospitals*, April 1, 1963, 50–54.

68 Bruce Fye, *American Cardiology: The History of a Specialty and Its College* (Baltimore: Johns Hopkins University Press, 1996); David Jones, *Broken Hearts: The Tangled History of Cardiac Care* (Baltimore: Johns Hopkins University Press, 2013).

69 Peter Galison, "The Ontology of the Enemy: Norbert Wiener and the Cybernetic Vision," *Critical Inquiry* 21 (1994): 228–66; Ronald R. Kline, *The Cybernetics Moment: Or Why We Call Our Age the Information Age* (Baltimore: Johns Hopkins University Press, 2015).

70 Laika's vital functions were transmitted for seven days, after which the electrical power for the radio transponder and climate control units ceased as planned and she died of hypoxemia. Andrei G. Kousnetzov, "Some Results of Biological Experiments in Rockets and Sputnik II," *Journal of Aviation Medicine* 29, no. 11 (1958): 781–84; Eugene B. Konecci and A. James Shiner, "Uses of Telemetry in Space," in Cesar A. Caceres, ed., *Biomedical Telemetry* (New York: Academic Press, 1965), 321–50. On the history of telemetry in zoology and ecology, see Etienne Benson, *Wired Wilderness: Technologies of Tracking and the Making of Modern Wildlife* (Baltimore: Johns Hopkins University Press, 2010).

71 The telemetry system deployed on Alan Shepard's suborbital mission had initially been tested on Mercury chimpanzees. Rita Chow, "Patient Monitoring Is More Than Just a Dream," *American Journal of Nursing* 61, no. 11 (1961): 60–62; H. Strughold and O. Benson, "Space Medical Research," *New England Journal of Medicine* 10, no. 261 (1959): 494–502.

72 F. W. Fascenelli, "Electrocardiography by Do-It-Yourself Radiotelemetry," *New England Journal of Medicine* 273, no. 20 (1965): 1076–79, at 1076.

73 Morris Soled, "Electrocardiography by Radiotelemetry," *New England Journal of Medicine* 274, no. 9 (1966): 521.

74 Caceres, *Biomedical Telemetry*, xi.

75 R. Stuart Mackay, "Telemetering from within the Body: Endoradiosondes," in Caceres, ed., *Biomedical Telemetry*, 147–233, at 223.

76 Nelson T. Grisamore, James K. Cooper, and Cesar A. Caceres, "Evaluating Telemetry Systems," in Caceres, ed., *Biomedical Telemetry*, 351–76.
77 Holter, "New Method for Heart Studies."
78 Holter, "Radioelectrocardiography: A New Technique for Cardiovascular Studies," 921.
79 Cesar A. Caceres, Clarence A. Imboden, and Mary Alice Smith, "A Medical Monitoring System," in Caceres, ed. *Biomedical Telemetry*, 107–15, at 107.
80 Cesar A. Caceres, "Telemetry in Medicine and Biology," *Advances in Biomedical Engineering and Medical Physics* 1 (1968): 279–316.
81 David A. Davis, William Thornton, Doris C. Grosskreutz et al., "Radio Telemetry in Patient Monitoring," *Anesthesiology* 22, no. 6 (1961): 1010–13, at 1012.
82 Holter, "Radioelectrocardiography: A New Technique for Cardiovascular Studies."
83 Norman J. Holter to Harry L. Kirkpatrick, November 4, 1958, box 2, folder 6, HRF.
84 Norman J. Holter to J. A. Gengerelli, June 30, 1961, box 3, folder 2, HRF.
85 In recent years, however, some Holter monitors have become wireless devices once more. Joseph A. Walsh III, Eric J. Topol, and Steven R. Steinhubl, "Novel Wireless Devices for Cardiac Monitoring," *Circulation* 130, no. 7 (2014): 573–81.
86 Z. Danilevicius, "Telemetry—Best Detective in Tracing CHD," *JAMA* 229, no. 11 (1974): 1475–76.
87 Robert P. Grant, "Foreword," in Caceres, ed., *Biomedical Telemetry*, vii–ix, at vii.
88 Ceylan Yeginsu, "If Workers Slack Off, the Wristband Will Know," *New York Times*, February 1, 2018.
89 Thuy Ong, "The FDA Has Approved the First Digital Pill," *The Verge*, November 14, 2017, https://www.theverge.com.
90 R. A. Becker, "Dealing with the Health Data Deluge," *Nova*, May 20, 2015, https://www.pbs.org/wgbh/nova/article/health-data/.
91 Lily Hay Newman, "A New Pacemaker Hack Puts Malware Directly on Your Device," *Wired*, August 9, 2018, https://www.wired.com.
92 "Hospitals Are Encouraged to Do More to Avoid Medical Device Hacking," *NPR Morning Edition*, January 7, 2020.

Chapter Three

1 "A Day in the Life of Dr. Archer," *Motorola Newsgram* 23, no. 3 (1966): 10–15, at 15, Motorola Solutions Heritage Services and Archives, Schaumberg, IL (hereafter MSHSA). This story appeared in the pages of the *Motorola Newsgram*, an in-house publication circulated to researchers, executives, salesmen, and prominent distributors and customers of the Motorola corporation.
2 "A Day in the Life of Dr. Archer," 15.
3 On the importance of studying "everyday technologies" and "small technologies," see David Arnold, *Everyday Technology: Machines and the Making of India's Modernity* (Chicago: University of Chicago Press, 2013); Projit Bihari Mukharji, *Doctoring Traditions: Ayurveda, Small Technologies, and Braided Sciences* (Chicago: University of Chicago Press, 2016).
4 Gary Frost, *Early FM Radio: Incremental Technology in Twentieth-Century America* (Baltimore: Johns Hopkins University Press, 2010).
5 Samuel Shem (Stephen Bergman), *The House of God* (New York: Richard Marek

Publishers, 1978); Wayne King, "The People Beeper: It's Peace of Mind to Some, Electronic Leash to Others," *New York Times*, July 20, 1976.

6 For further analysis on the shift from pages to paging, see Benjamin Morton, "Broadcast for One: Paging and Network Communication," PhD diss., University of Iowa, 2018.

7 Shakespeare's *Romeo and Juliet*, originally cited in Morton, "Broadcast for One."

8 "When Computers Were People," in Martin Campbell-Kelly et al., *Computer: A History of the Information Machine*, 3rd ed. (Boulder: Westview Press, 2014), 3–19.

9 From an interview with Dr. Percy Fridenberg, dated February 19, 1938, recalling a time when the hospital was still at 67th Street, so pre-1904: "The arrival of the attendings had to be announced. There were no telephones or signal lights in those days and the question was what sound signal would be most appropriate. At first they used a steam whistle but this scared some of the patients to death, it was very soon abolished and, in its place, there was installed an enormous gong, the number of strokes on which indicated the arriving M.D. was a surgeon or a 'medical man.'" Archives and Records Management, Icahn School of Medicine (hereafter ISM), New York, courtesy Barbara Nissen.

10 "Use Phone Call System," *Modern Hospital* 17, no. 3 (1921): 456–58, at 456. Some hospitals gave each physician their own sequence of dots and dashes as a unique Morse code identifier, but "its success depends entirely upon the doctor's being able to distinguish his particular call." E. Newton-Wells, "Hospital Signal Systems," *Modern Hospital* 18, no. 6 (1922): 549–50; Harold J. Seymour, "Eliminating Noise from the Hospital," *Modern Hospital* 21, no. 2 (1923): 264.

11 "A Successful Call System for Doctors and Interns," *Modern Hospital* 32 (1929): 140–44.

12 "'Calling Doctor Smith'—By Telephone and Amplifier," *Modern Hospital* 47 (1936): 84.

13 "Calling Dr. Kildare! Calling Dr. Kildare! Calling Dr. Kildare! Where IS That Man?" *Modern Hospital* 71, no. 2 (1948): 89.

14 Nicholas Genes, "Why the *Beep* Do doctors Still Use Pagers?" *Telemedicine* 25 (June 2017), http://www.telemedmag.com.

15 "S. C. Amsden Dead; Led Message Plan," *New York Times*, November 11, 1958.

16 Bob Considine, "Aircall . . . Bring 'Em Back," *Fireman's Fund Record*, November 1952.

17 Aircall in New York was soon followed by Page-ette in Cincinnati and similar networks in Cleveland, St. Louis, Minneapolis, Columbus, Indianapolis, Portland, and Seattle. Michael J. Saada, "Walkie Hearies," *Wall Street Journal*, November 25, 1952.

18 Considine, "Aircall . . . Bring 'Em Back."

19 As cited at https://www.smecc.org/richard_florac_-_first_pager,_early_portable_fm_radios.htm. By 1952, the firm had a list of 400 clients that included salesmen, detectives, plumbers, and undertakers, as well as physicians.

20 "Buzzing Dr. Kildare," *Pfizer Spectrum*, 1952, "Radiopager, 1952–53," box 3, folder 8, Al Gross Papers, 1909–2000, Ms2001-011, Special Collections, Virginia Polytechnic Institute and State University, Blacksburg (hereafter AGP).

21 Al Gross, "Autobiography," 1, undated MS, box 5, folder 6, AGP. For more on the culture of tinkering among amateur radio enthusiasts, see Kristin Haring, *Ham Radio's Technical Culture* (Cambridge, MA: MIT Press, 2006).

22 E. K. Jett and Girard Chaput, "Phone Me by Air," *Saturday Evening Post*, July 28, 1945, 41–47.

23 "Each signal goes out on a complicated wave system, which only one receiver can unscramble and receive; all the other receivers remain mute." Charles F. Neergaard, "Radio Paging Goes into the Doctor's Pocket," *Modern Healthcare* 79, no. 2 (1952): 123–26.
24 Neergaard, "Radio Paging Goes into the Doctor's Pocket," 125–26.
25 Neergaard, "Radio Paging Goes into the Doctor's Pocket," 125–26.
26 Robert F. Brown, "Everybody Likes to Be Called—by Radio," *Modern Hospital* 86, no. 6 (1952): 55–56. Later in his career, Gross complained that "attempts to promote and sell this idea met with resistance from doctors and nurses . . . also there was little interest by potential industrial users." Gross, "Autobiography."
27 Remarks by Commissioner James H. Quello before PCIA's Personal Communications Showcase, San Francisco, September 18, 1996. Online archives of Quello Center, Michigan State University, http://quello.msu.edu.
28 "Death of the Pager?" *Forbes*, December 13, 2001. See also https://www.statista.com/statistics/214235/us-paging-service-revenue-since-1983/.
29 "World War II Product History," 9, c. 1948, Products/Communication Products/Sales/Product Literature (General), box 2, folder 9, MSHSA.
30 *Quarterback of Battle*, 19–23, promotional pamphlet, 1947, box 2 folder 8, MSHSA.
31 *Motorola, Inc., Annual Report, 1953*, MSHSA, courtesy Sue Topp.
32 "Rx Radio: Benefits Both Patient and Doctor in Routine As Well As Emergency Cases," *Motorola Newsgram*, January–February 1955, 13.
33 "'Powerful Medicine' for Rural Practitioners," *Motorola Newsgram*, January–February 1958, 16–17.
34 "Transistorized Handie-Talkie Radio Pocket Pager: New Transistorized Wireless Paging System Provides Truly Private Plant-Wide Calling," *Motorola Newsgram*, November–December 1955, 28–29.
35 As Motorola vice president David Noble claimed in a speech earlier that year, his company was "one of the world's most enthusiastic proponents of the transistor and of its promise." Daniel E. Noble, "Electronic Frontiers," *Motorola Newsgram*, May–June 1955, 4–5.
36 "Transistorized Handie-Talkie Radio Pocket Pager," 29.
37 "The vintage footage appeared in a 1956 Universal Newsreel, now housed at the National Archives. It errs in calling the pager a 'walkie-talkie.' But visually this is great footage of this pager in use. The 'Handie-Talkie paging receiver' received an alert tone from the control console, then the person who was alerted held down a 'push-to-listen' button to hear a voice message." Sue Topp, MSHSA, personal communication.
38 "Beeping Dr. Kildare," *Motorola Newsgram*, September–October 1957, 13. Another article featured the deployment of a similar system at St. Anthony of Padua hospital in Chicago: "Compact Radio Pager Delivers Quiet Voice Messages at St. Anthony's Hospital," *Motorola Newsgram*, January–February 1958, 6–7.
39 "Radio Paging at Mount Sinai Hospital," *Motorola Newsgram* 18, no. 2 (1961): 14–15, at 15.
40 "Radio Paging at Mount Sinai Hospital," 15.
41 Ira Eliasoph, personal communication, December 7, 2016.
42 Mort Weinberg, with Sally Ruth Bourrie, Oral History, November 29, 1988, 60, MSHSA.

43 Ed Bales, with Sally Ruth Bourrie, Oral History, September 12, 1988, 4–5, MSHSA.

44 *You're in Excellent Company with a Motorola "Handie-Talkie" Radio Paging System*, sales pamphlet, c. 1958, "Handie Talkie Pager Systems, 1955—" folder, Products/Paging & Data Products/Sales/Product Literature (A–Z), MSHSA.

45 *Motorola Inc., Annual Report, 1959*, 16, MSHSA.

46 "Patients Need Never Be Alone So . . . Recuperating's Almost a Pleasure," *Motorola Newsgram* 20, no. 1 (1963): 8–9, at 9.

47 *Total Hospital Communications*, c. 1967, Products/Communication Products/Sales/Product Literature (General), box 5 folder 1, MSHSA. Motorola had become known increasingly for its consumer electronics line of televisions, and after the Dahlberg acquisition the company boasted that it was the only firm that "makes a television line exclusively for hospitals." "A hospital television has to take much more than a home TV . . . higher degree of humidity, a wider range of temperatures and constant handling by non-owners . . . this is the 'patient-proof' TV set." *Only Motorola Makes a Television Line Exclusively for Hospitals*, promotional pamphlet, Promotional Literature Binder, including case studies, 1959–60, box 5, folder 1, MSHSA. For more on the social science and political valences of multimedia interfaces in the 1960s, see Fred Turner, *The Democratic Surround: Multimedia and American Liberalism from World War II to the Psychedelic Sixties* (Chicago: University of Chicago Press, 2013).

48 *Total Hospital Communications.*

49 These features included a "nurse reminder" described as a "memory device [which] constantly alerts the nurse or ward clerk at the control station and personnel in the corridor that a certain action is to be taken in the patient's room. Nurse reminder electronically protects the patient by making sure that the nurse does not forget!" *Motorola Manumatic: The Personal Nurse Call*, promotional pamphlet, Promotional Literature Binder, including case studies, 1959–60, box 5, folder 1, MSHSA.

50 As early as 1922, *Modern Hospital* recommended that "the modern hospital should be equipped with an electrically operated doctors' in and out register" connected to the telephone switchboard room, but usage of these devices was often inconsistent. Hartford Hospital had already instituted a device in its telephone switchboard such that a light would come on indicating any given physician's presence in the hospital, so that the switchboard operator knew they were "pageable" via public address and hospital telephone systems. L. A. Sexton, "A New In and Out Registering Device for the Attending Staff," *Modern Hospital* 16, no. 6 (1921): 552–53. Thanks to the new Radio-Register, however, the Motorola Radio Pager could serve simultaneously as wireless communications device and an electronic punch card.

51 "We always took AT&T and the Bell Labs on, as far as fighting them for spectrum. Because, you know, they tried to get into our business in 1956 and we bounced them out of the business. We kept them out of the business of supplying two-way radio to an average customer, rather than somebody who wanted telephones." Tom Kain, Oral History, 61–62, MSHSA. When Bell began testing its prototypes in southern Pennsylvania in 1957, the *Baltimore Sun* warned that "hundreds of similar radio signals may soon be hurtling through space, calling busy people to telephones wherever they may be at work, at play, or at leisure." E. Carroll, "Pocket Call System Is Being Tested," *Baltimore Sun*, October 24, 1957, 28.

52 Ed Bales, with Sally Ruth Bourrie, Oral History, September 12, 1988, 4, MSHSA. See

also *Hearings before the Antitrust Subcommittee (Subcommittee no. 5) of the Committee on the Judiciary, House of Representatives, Eighty-fifth Congress, Second Session, March-April-May 1958, Part II, Volume II, American Telephone & Telegraph Co.* (Washington, DC: GPO, 1958).

53 In the general consent decree of 1956, the Justice Department allowed AT&T to remain a monopoly, with the provision that "AT&T has to allow other companies to access any newly patented technology developed by AT&T's Bell Labs." See S. C. Strother, *Telecommunication Cost Management* (Norwood, MA: Artech House, 2002), 11.

54 After the consent decree "got [AT&T] out of the paging business," Motorola's Ed Bales recalled, "they needed somebody to build their products. So they designed the specifications for the Bell Boy. We did it and we won." Bales, Oral History, 18–19. See also Martin Watzinger, Thomas A. Fackler, Markus Nagler, and Monika Schnitzer, "How Antitrust Enforcement Can Spur Innovation: Bell Labs and the 1956 Consent Decree," January 9, 2017, https://economics.yale.edu/sites/default/files/how_antitrust_enforcement.pdf.

55 See *Motorola 1965 Annual Report*. As Tom Kain recalled, "Bell Labs was designing their own paging equipment. And I used to follow them in the Bell Lab Record, which is where, when Bell Labs does some work, they'll talk about it. And they were, I kept telling John Mitchell they're having problems with the way the battery gets charged . . . And then John would go back and charge everybody, you know, to make it smaller, lighter. And finally we went to AT&T, after they were in the business, you know, making their own. We got them to stop making their own and only use Motorola. That was in 1963 . . . the weekend we were supposed to go, that Monday we were supposed to go to show them our latest Bell Boy product, was the weekend that [President John F.] Kennedy got killed." Kain, Oral History, 61.

56 *Motorola 1965 Annual Report*, 17, MSHSA.

57 *Motorola 1965 Annual Report*, 21, MSHSA.

58 Mort Weinberg, with Sally Ruth Bourrie, Oral History, November 29, 1988, 92, MSHSA. Dick Carsello, who developed Motorola's early VHF pagers, saw the Pageboy as the firm's singular "breakthrough" product. "Because we put on the market a product that was substantially more reliable, by probably a factor of four or something like that. And much more cost effective for Motorola to build. So the selling price that the customer saw came down by maybe 20 percent reliability was enhanced substantially. And it really started the spurt of, at least tone and voice type paging in the wide area market." Dick Carsello, Oral History, December 1, 1988, 9, MSHSA.

59 Bales, Oral History, 38; "Bleep May Mean a Call from the Office—or a Shopping List from Home," *Chicago Daily Herald*, November 13, 1974, 21.

60 Bales, Oral History, 62. Motorola had every advantage as it spread the pager outside the hospital, and soon dominated the market for mobile VHF paging.

61 Erving Goffman, "On the Characteristics of Total Institutions," in *Asylums: Essays on the Social Situation of Mental Patients and Other Inmates* (New York: Anchor Books, 1961), 3–124; Howard Becker, "The Politics of Presentation: Goffman and Total Institutions," *Symbolic Interaction* 26, no. 4 (2003): 659–69.

62 Weinberg, Oral History, 166. Basch's year of service at the House of God Hospital was a thinly disguised version of the 1974 medical internship Bergman completed

at Boston's Beth Israel Hospital—an early adopter of the first Handie-Talkie systems in the late 1950s.

63 Shem, *The House of God.*

64 As Dr. C. W. Munger, director of the Grasslands Hospital of Valhalla, New York, noted, a successful paging system depends on the proper socialization of the person being paged toward the page. "A Successful Call System for Doctors and Interns," *Modern Hospital* 32 (1929): 140–44.

65 Eviatar Zerubavel, "Private Time and Public Time: The Temporal Structure of Social Accessibility and Professional Commitments," *Social Forces* 58, no. 1 (1979): 38–58, at 48.

66 Zerubavel, "Private Time and Public Time," 49.

67 Shem, *The House of God.*

68 Serial barking could also be performed from hospital conference rooms with multiple lines: a sophisticated resident parked in a conference room could use two phones with two lines each and rapidly send eight barking pages to an unwitting colleague. Joel Howell, personal communication, August 7, 2019.

69 Sue Shellenbarger, "Work at Home? Your Employer May Be Watching," *Wall Street Journal*, July 30, 2008.

70 Ron Eglash, "Broken Metaphor: The Master/Slave Analogy in Technical Literature," *Technology and Culture* 48, no. 2 (2007): 360–69; this point is also developed in Ruha Benjamin, *Race after Technology: Abolitionist Tools for the New Jim Code* (New York: Wiley, 2019).

71 Jim Wright, Oral History, November 28, 1988, 128–29, MSHSA.

72 "Hit Me Up: Hip-Hop's History with Pagers," Boombox.com, March 1, 2018, https://theboombox.com.

73 "Twilight of the Beeper," *American Medical News*, June 9, 2008, https://amednews.com.

74 Allison Bond, "Why Do Doctors Still Use Pagers? How Hospitals Got Stuck in the '90s," *Slate*, February 12, 2016, https://slate.com.

75 Michael B. Rothberg, Ashish Arora, Reva Kleppel et al., "Phantom Vibration Syndrome among Medical Staff: A Cross-Sectional Survey," *British Medical Journal* 341 (2010): c6914.

76 Agnes Arnold-Forster and Samuel Schotland, "COVID-19 Only Exacerbated a Longer Pattern of Health-Care Worker Stress," *Washington Post*, April 29, 2021.

Chapter Four

1 James Lieberman, "Introductory Remarks," *Annals of the New York Academy of Sciences* 142, no. 2 (1967): 341–42, at 342. On the historical narrative of antibiotic, anesthetic, or other biomedical revolutions, see Jeremy A. Greene, Elizabeth Siegel Watkins, and Flurin Condrau, *Therapeutic Revolutions: Pharmaceuticals and Social Change in the Twentieth Century* (Chicago: University of Chicago Press, 2016).

2 Kirsten Ostherr, *Medical Visions: Producing the Patient through Film, Television, and Imaging Technologies* (New York: Oxford University Press, 2013); Susan Murray, *Bright Signals: A History of Color Television* (Durham, NC: Duke University Press, 2018); Susan Murray, "The New Surgical Amphitheater: Color Television and Medical Education in Postwar America," *Technology & Culture* 61, no. 3 (2020): 772–97.

3 The Council on Medical Television seems to have exclusively concerned itself in

the 1950s and 1960s with the television as a medium for popular and professional health education. "About the Council on Medical Television," MS, n.d. (c. 1965); and Sam A. Agniello, "A History of the Council on Medical Television," MS, n.d. (c. 1965), box 5, folder 2, Reba Benschoter collection (unprocessed), National Library of Medicine, Bethesda, MD (hereafter ReBC).

4 Murray, "The New Surgical Amphitheater." On the broader political promises of interactive multimedia, see Fred Turner, *The Democratic Surround: Multimedia and American Liberalism from World War II to the Psychedelic Sixties* (Chicago: University of Chicago Press, 2013).

5 Television psychotherapy can be situated within a longer history of teletherapy; see Hannah Zeavin, *The Distance Cure: A History of Teletherapy* (Cambridge, MA: MIT Press, 2021).

6 Kenneth T. Bird, "The Amplified Doctor," conference paper (n.d.), Massachusetts General Hospital Telemedicine Center Records, 1968–96, AC 1, Massachusetts General Hospital Archives, Boston (hereafter MGHTC).

7 Kenneth T. Bird, "The Future: Man Augments His Senses: The Lowell Institute Lectures In Medicine," March 27, 1973, 4, box 3, folder 11, MGHTC.

8 Marshall McLuhan and Q. Fiore, *The Medium Is the Message* (New York: Bantam Books, 1966); "Telemedicine: A New Health Information System," n.d. (1970), box 2, folder 19, MGHTC.

9 On the role of psychiatry in the American World War II effort, and its relevance for the postwar peace, see Frank J. Sladen, *Psychiatry and the War* (Baltimore: Johns Hopkins University Press, 1943); William C. Menninger, *Psychiatry in a Troubled World: Yesterday's War and Today's Challenge* (New York: Macmillan, 1948).

10 Cecil Wittson and Ron Dutton, "Interstate Telecommunication," *Mental Hospitals* 2 (1957): 15–17; Zeavin, *The Distance Cure.*

11 Cecil Wittson and Ron Dutton, "A New Tool in Psychiatric Education: First Report from Nebraska Psychiatric Institute on Teaching Methods with Closed-Circuit Television," *Mental Hospitals* (November 1956): 11–14.

12 Outside of historically Black medical campuses like Howard and Meharry, only a very small number of African American physicians and scientists were able to find jobs at most academic medical centers in the United States, which through a combination of tacit and explicit exclusions restricted most of their training and hiring to white men. In the mid-twentieth century, much of the intellectual effort contributed by Black researchers in mainstream academic medical centers was hidden in the role of the "technician." Occasionally, as with Vivien Thomas at Johns Hopkins, the work of Black technicians in academic medicine would be acknowledged with minor publication credit or retrospective recognition, but countless others remain hidden figures in the history of science and medicine. Wittson acknowledged Johnson's work with coauthorship of one 1961 paper: C. L. Wittson, D. C. Affleck, and V. L. Johnson, "Two-Way Television in Group Therapy," *Mental Hospitals* 12, no. 11 (1961): 22–23. For more on Johnson, see "Obituary: Van Lear Johnson," *Lincoln Journal-Star*, October 24, 2009; on Thomas, see Stefan Timmermans, "A Black Technician and Blue Babies," *Social Studies of Science* 33, no. 2 (2002): 197–229.

13 Reba Benschoter, personal communication, November 3, 2016.

14 Reba Benschoter, "Modern Communications to Assist a State Hospital," grant application to NIMH, July 23, 1963, 1 R11 MH 1573-01, D2 H1, folder 2, Nebraska

Psychiatric Institute Files, Special Collections and Archives, McGoogan Health Sciences Library, University of Nebraska Medical Center, Omaha (hereafter NPI).

15 Benschoter's early research showed that "it was possible to observe reasonably detailed neurological examinations on TV, and EEGs can be read quite easily using inexpensive equipment." Reba Ann Benschoter, "Multi-Purpose Television," *Annals of the New York Academy of Sciences* 142, no. 2 (1967): 471–78. For more on the perceived tension between therapeutic and custodial roles of public mental hospitals in the mid-twentieth century, see Joel Braslow, *Mental Ills and Bodily Cures: Psychiatric Treatment in the First Half of the Twentieth Century* (Berkeley: University of California Press, 1997).

16 Communications Division, Nebraska Psychiatric Institute, "Utilization of Closed-Circuit Television and Videotape," 10, n.d., D2 H4, folder 2, NPI.

17 Benschoter, "Multi-Purpose Television," 475–76.

18 Despo Kritsotaki, Vicky Long, and Matthew Smith, eds., *Deinstitutionalisation and After: Post-War Psychiatry in the Western World* (London: Palgrave Macmillan, 2016); Joel T. Braslow and Luke Messac, "Medicalization and Demedicalization: A Gravely Disabled Homeless Man with Severe Psychiatric Illness," *New England Journal of Medicine* 379 (2018): 1885–88.

19 Cecil Wittson, interview with Ben Park, cited in Ben Park, *An Introduction to Telemedicine: Interactive Television for Delivery of Health Services* (New York University Alternate Media Center, 1974), 24.

20 Gerard Weidman, "Hundreds Rush to Rescue Survivors of Crash," *Boston Globe*, October 5, 1960, 21; E. G. McGrath, "Crash Site Nightmare of Lurid Light, Bodies," *Boston Globe*, October 5, 1960, 1; Seymour Linscott, "Blame Starlings for Crash: Now Believe Birds Sucked into Jets," *Boston Globe*, October 6, 1960, 1; Gloria Negri, "Airport's Medical Station a High-Flying Experiment," *Boston Globe*, July 21, 1963.

21 Loretta McLaughlin, "Big Brother, M.D., Is Watching over You," *Boston Globe*, June 30, 1968, D7.

22 James Hammond, "Doctors at Logan Airport—Both in Person and on TV," *Boston Globe*, April 27, 1969.

23 Michael Crichton, "The Doctor's Stethoscope Is Three Miles Long," *Boston Globe*, August 9, 1970, 3.

24 Michael Crichton, *The Andromeda Strain* (New York: Knopf, 1969), and *Five Patients: The Hospital Explained* (New York: Knopf, 1970). On the contextualization of Crichton's speculative fiction in relation to Cold War science, technology, and medicine, see Joanna Radin, "The Speculative Present: How Michael Crichton Colonized the Future of Science and Technology," *Osiris* 34 (2019): 297–315.

25 McLaughlin, "Big Brother, M.D., Is Watching over You." Elsewhere Bird referred to the Logan International Airport Medical Station of MGH as something "created as a direct physical extension of the Hospital into the community," which was "intended as a tool to explore new ways to deliver medical care to every segment of the population even though the current setting is within the aero-transportation community." "Telediagnosis: A New Community Resource," n.d. (1968), box 2, folder 15, MGHTC.

26 Park, *Introduction to Telemedicine*, 26.

27 McLuhan and Fiore, *The Medium Is the Message*; also cited in "Telemedicine: A New Health Information System," n.d. (1970), box 2, folder 19, MGHTC.

28 "Telemedicine: A New Health Information System," emphasis mine.

29 Stanley Krainin notes, box 2, folder 17, MGHTC.

30 For example, see J. L. Lesher, L. S. Davis, F. W. Gourdin et al., "Telemedicine Evaluation of Cutaneous Diseases: A Blinded Comparative Study," *Journal of the American Academy of Dermatology* 38, no. 10 (1998): 27–31; Sira P. Rao, Nikil S. Jayant, Max E. Stachura et al., "Delivering Diagnostic Quality Video over Mobile Wireless Networks for Telemedicine," *International Journal of Telemedicine and Applications* (2009): 406753. For a systematic review of this literature, see Rashid L. Bashshur, Gary W. Shannon, Brian R. Smith et al., "The Empirical Foundations of Telemedicine Interventions for Chronic Disease Management," *Telemedicine and E-Health* 20, no. 9 (2014): 769–800.

31 "I Don't Want No Tele-medicine," n.d., MGHTC.

32 McLaughlin, "Big Brother, M.D., Is Watching over You."

33 Reba Ann Benschoter, Progress Report: Exchange of Medical Information Agreement #EMI-68-001-G/P: Two-Way Closed Circuit Television, Nebraska VA Hospitals—Univ. of Nebraska Medical Center, June 1975, box 5, folder 12, ReBC.

34 Doctor X (Alan E. Nourse), *Intern* (New York: Harper & Row, 1965); quotation from "Books: Inside Story," *Time*, July 23, 1965.

35 John H. Knowles, ed., *Doing Better and Feeling Worse: Health in the United States* (New York: Norton, 1977); David Rothman, *Strangers at the Bedside: A History of How Law and Bioethics Transformed Medical Decision Making* (New York: DeGruyter, 2003); Wendy Kline, *Bodies of Knowledge: Sexuality, Reproduction, and Women's Health in the Second Wave* (Chicago: University of Chicago Press, 2010); Alondra Nelson, *Body and Soul: The Black Panther Party and the Fight against Medical Discrimination* (Minneapolis: University of Minnesota Press, 2011); Beatrix Hoffman, *Health Care for Some: Rights and Rationing in the United States since 1930* (Chicago: University of Chicago Press, 2012).

36 On the theorization of social space as a materialization of existing power relations, see Henri Lefebvre, *The Production of Space*, trans. Donald Nicholson-Smith (Malden, MA: Wiley-Blackwell, 1991 [1974]).

37 Barbara Korsch and Vida Francis Negrete, "Doctor-Patient Communication," *Scientific American* 227, no. 2 (1972): 66–75; Jay Katz, *The Silent World of Doctor and Patient* (New York: Free Press, 1985); Nancy Tomes, *Remaking the American Patient: How Madison Avenue and Modern Medicine Turned Patients into Consumers* (Chapel Hill: University of North Carolina Press, 2016).

38 Red Burns, "Using New Technologies to Enhance Human Communications," Second International Harvard Conference on Internet & Society, May 26—9, 1998, https://cyber.harvard.edu/cybercon98/wcm/burns.html.

39 Park, *Introduction to Telemedicine*; Joel J. Reich, *Telemedicine: The Assessment of an Evolving Health Care Technology*, Report R(T)-74/4-6 (St. Louis: Washington University Center for Development Technology and Program in Technology and Human Affairs, August 1974); Maxine L. Rockoff, "An Overview of Some Technological/Health-Care System Implications of Seven Exploratory Broad-Band Communication Experiments," *IEEE Transactions on Communications* 23, no. 1 (1975): 20–30; Rashid L. Bashshur and Patricia A. Armstrong, "Telemedicine: A New Mode for the Delivery of Health Care," *Inquiry* 13, no. 3 (1976): 233–44.

40 "Man's perception and use of space," they noted, "as a specialized aspect of his culture is barely understood." Kenneth T. Bird and Marie Kerrigan, "Telemedicine: A New Health Exchange System," paper presented at 1970 Medical Services Conference, American Medical Association, November 28, 1970, 5 (emphasis mine), box 2, folder 19, MGHTC.

41 Bird and Kerrigan, "Telemedicine: A New Health Exchange System," 5, emphasis mine.

42 Reich, *Telemedicine: The Assessment of an Evolving Health Care Technology*, 155. For more on the history of smell in medicine and public health, see Melanie Kiechle, *Smell Detectives: An Olfactory History of Nineteenth-Century Urban America* (Seattle: University of Washington Press, 2017).

43 Murray, *Bright Signals.*

44 Hughes Evans, "Losing Touch: The Controversy over the Introduction of Blood Pressure Instruments into Medicine," *Technology & Culture* 34, no. 4 (1993): 794–807. For analysis of the information loss in reducing the clinician's sensing hand into the numerical value of temperature, see Volker Hess, *Der wohltemperierte Mensch: Wissenschaft und Alltag des Fiebermessens, 1850–1900* (Frankfurt am Main: Campus Verlag, 2000); Christopher Hamlin, *More Than Hot: A Short History of Fever* (Baltimore: Johns Hopkins University Press, 2014).

45 Kenneth T. Bird, *Telemedicine: A New Health Information Exchange System*, 05 Annual Report (Boston: Massachusetts General Hospital, July 1973). As quoted in Reich, *Telemedicine: The Assessment of an Evolving Health Care Technology.*

46 Park, *Introduction to Telemedicine*, 37.

47 "Trip Report—Society for Industrial and Applied Mathematics 21st Annual Meeting and Meeting with Erving Goffman on Communication," memorandum, June 27, 1973, Maxine Rockoff Papers, private collection shared with author (hereafter MRP).

48 "Trip Report—Society for Industrial and Applied Mathematics 21st Annual Meeting and Meeting with Erving Goffman on Communication."

49 "Trip Report—Society for Industrial and Applied Mathematics 21st Annual Meeting and Meeting with Erving Goffman on Communication."

50 "This may result in the patient being more comfortable because the 'territory of the self' won't be violated," Rockoff noted after the conversation, but "if there is a third person touching the patient such as a nurse practitioner, and this third person is responding to the physician's instructions, this will be a complicated new kind of interaction—the touching person will have to incorporate feedback from the patient (e.g., grimace) as well as feedback from the physician." "Trip Report—Society for Industrial and Applied Mathematics 21st Annual Meeting and Meeting with Erving Goffman on Communication."

51 Ben Park, "Communications Aspects of Telemedicine," in R. Bashshur, P. A. Armstrong, and Z. I. Youssef, eds., *Telemedicine: Explorations in the Use of Telecommunications in Health Care* (Springfield, IL: Charles C. Thomas, 1975), 69, emphasis mine.

52 Elliott Friedsen, as quoted in Park, *Introduction to Telemedicine*, 54.

53 Park, *Introduction to Telemedicine.*

54 Edward Hall, *The Hidden Dimension* (Garden City: Doubleday, 1966). As cited in Park, *Introduction to Telemedicine.*

55 Park, *Introduction to Telemedicine*, 51.

56 Park, "Communications Aspects of Telemedicine," 81.

57 Park, *Introduction to Telemedicine*, 61.

58 "It is the absence of a television camera which prevents the recording of the distress on the patient's face. This then, is nothing new, and the use of television could very well spread this characteristic outside of hospital walls." Elliott Friedsen to Ben Park, as cited in Park, *Introduction to Telemedicine*, 61.

59 Park, *Introduction to Telemedicine*, 63.

60 Park, *Introduction to Telemedicine*, 63. See also Nathan Ensmenger, "Resistance Is Futile? Reluctant and Selective Users of the Internet," in William Aspray and Paul Ceruzzi, eds., *The Internet and American Business* (Cambridge, MA: MIT Press, 2010).

61 Michael T. Romano, "Health Science Education in the Space Age," *Annals of the New York Academy of Sciences* 142, no. 2 (1967): 348–56.

62 Park, "Communications Aspects of Telemedicine," 64–65. On subsequent design of medical technology to favor diagnostic thresholds for patients with lighter skin tones and miss diagnoses in patients with darker skin tones, see Amy Moran-Thomas, "How a Popular Medical Device Encodes Racial Bias," *Boston Review*, August 5, 2020; Roni Caryn Rabin, "Pulse Oximeter Devices Have Higher Error Rate in Black Patients," *New York Times*, December 22, 2020.

63 Park, "Communications Aspects of Telemedicine," 71.

Chapter Five

1 Roger O. Egeberg, "Health Care Faces Severest Demands," *New York Times*, January 12, 1970, 87.

2 C. Gerald Fraser, "Reuther Asks National Health System," *New York Times*, November 15, 1968, 28; Harry Schwartz, "What Health Crisis?" *New York Times*, October 18, 1971, 37; *Hearings before the Subcommittee on Health of the Subcommittee on Labor and Public Welfare, United States Senate, 92nd Congress, First Session, on "Elimination of the Health Care Crisis in America," May 17, 1971, San Francisco, California, May 18, 1971, Los Angeles California, Part II* (Washington, DC: GPO, 1971).

3 K. T. Bird, "Teleconsultation: A New Health Information Exchange System," 03 Annual Report, Veterans Administration, Washington, DC, Certificate of Award and Agreement No. EMI-69-OOI (Boston: Massachusetts General Hospital, 1971); quotation as cited in Joel J. Reich, *Telemedicine: The Assessment of an Evolving Health Care Technology*, Report R(T)-74/4-6 (St. Louis: Washington University Center for Development Technology and Program in Technology and Human Affairs, August 1974).

4 Rosemary Stevens, *In Sickness and in Wealth: American Hospitals in the Twentieth Century* (Baltimore: Johns Hopkins University Press, 1999); Andrea Park Chung, Martin Gaynor, and Seth Richards-Shubik, "Subsidies and Structure: The Lasting Impact of the Hill-Burton Program on the Hospital Industry," NBER Working Paper 22037, February 2016, https://www.nber.org.

5 John C. Norman, ed., *Medicine in the Ghetto* (New York: Appleton-Century-Crofts, 1969); Thomas McKeown, *The Role of Medicine: Dream, Mirage or Nemesis?* (London: Nuffield Provincial Hospitals Trust, 1976); John H. Knowles, ed., *Doing Better and Feeling Worse: Health in the United States* (New York: Norton, 1977); Alondra

Nelson, *Body and Soul: The Black Panther Party and the Fight against Medical Discrimination* (Minneapolis: University of Minnesota Press, 2011); Beatrix Hoffman, *Health Care for Some: Rights and Rationing in the United States since 1930* (Chicago: University of Chicago Press, 2012).

6 John Knowles, Introduction, in *Doing Better and Feeling Worse*, 2.

7 For historical accounts of "appropriate technology" in healthcare, see Heidi Morefield, "Developing to Scale: Appropriate Technology and the Making of Global Health," PhD diss., Johns Hopkins University, 2019; Peter Redfield, "On Band-Aids and Magic Bullets," in Stephen J. Collier, Jamie Cross, Peter Redfield, and Alice Street, eds., *Little Development Devices/Humanitarian Goods, Limn* 9 (2017), https://limn.it/on-band-aids-and-magic-bullets/.

8 Ralph Lee Smith, "The Wired Nation," *The Nation*, May 18, 1970, 582–606, at 582.

9 John McMurria, *Republic on the Wire: Cable Television, Pluralism, and the Politics of New Technologies, 1948–1984* (New Brunswick, NJ: Rutgers University Press, 2017).

10 Ralph Lee Smith, *The Wired Nation: Cable TV: The Electronic Communications Highway* (New York: Harper & Row, 1972). For further histories of televisual multimedia and democracy in the 1970s see Jennifer Light, *From Warfare to Welfare: Defense Intellectuals and Urban Problems in Cold War America* (Baltimore: Johns Hopkins University Press, 2003); Fred Turner, *The Democratic Surround: Multimedia and American Liberalism from World War II to the Psychedelic Sixties* (Chicago: University of Chicago Press, 2013).

11 Ben Park to John Knowles, March 24, 1972, box 317, folder 1953, "New York University—Telemedicine, 1972," Rockefeller Foundation, RG 1.3–RG 1.8 (A76-A82) (FA209), Subgroup 5: Projects (A78), Series 200: United States, Subseries A: United States—Medical Sciences, Rockefeller Archive Center, Tarrytown, NY (hereafter RAC).

12 Ben Park, "Investigation and Reporting of Current Experience in Telemedicine," February 22, 1973, box 381, folder 1954, RAC. In 1957 Park produced a well-regarded film about primary care medicine, *Doctor B*, with assistance from Smith, Kline & French Laboratories.

13 "We as a group," Knowles wrote to the health division at Rockefeller, "must be much more cognizant of the importance of communications media as it may affect all of our programs . . . perhaps we should keep a corner of our brains free for considering the communications aspects of any project we underwrite." John Knowles, "Ben Park's Video Papers," May 11, 1972, box 317, folder 1953, RAC. Knowles was convinced of the growing importance of networked television in the future of health and society.

14 Maxine L. Rockoff, "An Overview of Some Technological/Health-Care System Implications of Seven Exploratory Broad-Band Communication Experiments," *IEEE Transactions on Communications* 23, no. 1 (1975): 29.

15 Merlin Chowkwanyun, "The War on Poverty's Health Legacy: What It Was, and Why It Matters," *Health Affairs* 37, no. 1 (2018): 47–53; Reich, *Telemedicine: The Assessment of an Evolving Health Care Technology*, 14.

16 Konrad K. Kalba, *Communicable Medicine: Cable Television and Health Services* (New York: Alfred P. Sloan Foundation, 1971), 2.

17 Tim Wu, *The Master Switch: The Rise and Fall of Information Empires* (New York: Knopf, 2010); Light, *From Warfare to Welfare*.

18 Hyman H. Goldin, *Innovation and the Regulatory Agency: FCC's Reaction to CATV*, report prepared for Sloan Commission on Cable Communications, August 1970.

19 *On the Cable: The Television of Abundance*, report of Sloan Commission on Cable Communications (New York: Alfred P. Sloan Foundation, 1971).

20 *Report of the National Advisory Commission on Civil Disorders* (New York: Bantam Books, 1968). For more on the context of the Kerner Commission, see Thomas J. Sugrue, *The Origins of the Urban Crisis: Race and Inequality in Postwar Detroit* (Princeton: Princeton University Press, 1996); Fred Turner, *From Counterculture to Cyberculture: Stewart Brand, the Whole Earth Network, and the Rise of Digital Utopianism* (Chicago: University of Chicago Press, 2010).

21 Nicolas Johnson and Gary G. Gerlach, "The Coming Fight for Cable Access," *Yale Review of Law and Social Action* 2, no. 3 (1972): 217–25, at 218.

22 National Academy of Engineering, Committee on Telecommunications, *Communications Technology for Urban Improvement*, report to the Department of Housing and Urban Development under Contract H-1221, June 1971; Rashi Fein, *The Doctor Shortage: An Economic Diagnosis* (Washington DC: Brookings Institution, 1967).

23 Khalil Gibran Muhammad, *The Condemnation of Blackness: Race, Crime, and the Making of Modern Urban America* (Cambridge, MA: Harvard University Press, 2010); Antero Pietila, *Not in My Neighborhood: How Bigotry Shaped a Great American City* (Chicago: Ivan R. Dee, 2010); Mindy Thompson Fullilove, *Root Shock: How Tearing Up City Neighborhoods Hurts America, and What We Can Do about It* (New York: NYU Press, 2016).

24 For the same price as a single mile of highway, Massachusetts "could be the first to apply interactive television on a statewide scale large enough to make full use of our great human care resources." Kenneth T. Bird, "Telemedicine: Concept and Practice," in R. Bashshur, P. A. Armstrong, and Z. I. Youssef, eds., *Telemedicine: Explorations in the Use of Telecommunications in Health Care* (Springfield, IL: Charles C. Thomas, 1975), 107.

25 Dedicated to providing medical services to communities with limited access to specialty care, the INTERACT system, as it was called, attracted further speculation about the possibilities of a nationwide telemedical network. Dean J. Siebert, *A Decade of Experience Using Two-Way Closed Circuit Television for Medical Care and Education*, Final Report, Contract #1-LM-4-4704 (Bethesda, MD: Lister Hill National Center for Biomedical Communications, National Library of Medicine, November 1977). As cited in A. M. Bennett, W. H. Rappaport, and E. L. Skinner, "Telehealth Handbook," Publication no. 79-3210 (Washington, DC: US Dept. of Health, Education & Welfare, 1978), 128.

26 Light, *Warfare to Welfare*; Jill Lepore, *If Then: How the Simulmatics Corporation Invented the Future* (New York: Liveright, 2020).

27 Lockheed Missile & Space Company, "Integrated Medical & Behavioral Laboratory Measurement System (IMBLMS) Area Health Services Field Unit," February 1972, A-1, box 2, folder "Telemedicine—Lockheed Missiles and Space Co., IMBLMS, AHSFU," Rashid Bashshur Collection, NLM (hereafter RaBC). Norman Belasco and Sam L. Pool, "Space Technology and Remote Health Care," STARPAHC 3.3, Space Technology Applied to Papago Advanced Health Care (STARPAHC) Collections, 1970–79 and 1991, AHSL 2001-01, Arizona Telemedicine Archives, Arizona Health Science Library & Special Collections, University of Arizona, Tuc-

son (hereafter STARPAHC). "Site Selection Considerations for the NASA/HEW IMBLMS Area Health Services Field Test Project," STARPAHC 3.5, STARPAHC.

28 Bird, "Teleconsultation: A New Health Information Exchange System," 11–12.

29 For more on Waxman's work in the development of information technology in healthcare and biomedical research, see Joseph November, *Biomedical Computing: Digitizing Life in the United States* (Baltimore: Johns Hopkins University Press, 2012).

30 As health services researcher and historian of telemedicine Rashid Bashshur noted at a conference in the fall of 1973, the study of telemedicine required expertise across a variety of fields, including clinical medicine, economics, sociology, psychology, communications, systems analysis, geography, and engineering. Rashid L. Bashshur, "Telemedicine and Medical Care," in Bashshur, Armstrong, and Youssef, eds., *Telemedicine: Explorations in the Use of Telecommunications in Health Care*; Ben Park, *An Introduction to Telemedicine: Interactive Television for Delivery of Health Services* (New York: NYU Alternate Media Center, 1974), 35.

31 Daniel B. Schwartz, *Ghetto: The History of a Word* (Cambridge, MA: Harvard University Press, 2019). The term "inner city" originally designated the central portions of urban areas that served as hubs of commerce and transportation; now they were coded as racialized spaces of scarcity. *Documents of the Senate of the State of New York, 134th Session, Vol. IV, No. 10, Part 2* (Albany, 1911), 121–23.

32 Lynn E. Miller and Richard M. Weiss, "Revisiting Black Medical School Extinctions in the Flexner Era," *Journal of the History of Medicine and Allied Sciences* 67, no. 2 (2012): 217–43.

33 Lloyd A. Ferguson, "What Has Been Accomplished in Chicago?" in John C. Norman, ed., *Medicine in the Ghetto* (New York: Appleton-Century-Crofts, 1969), 87.

34 Reich, *Telemedicine: The Assessment of an Evolving Health Care Technology*, 30–31.

35 Reich, *Telemedicine: The Assessment of an Evolving Health Care Technology*, 32. See Alice O'Connor, *Poverty Knowledge: Social Science, Social Policy, and the Poor in Twentieth-Century U.S. History* (Princeton: Princeton University Press, 2002).

36 "White, Negro Doctors Assail Ghetto Medicine," *Los Angeles Times*, July 6, 1969, 8; Denton Cooley, "In Memoriam John C. Norman," *Texas Heart Institute Journal* 41, no. 6 (2014): 569–70. Norman was a pioneering cardiovascular surgeon credited for his work in developing the left-ventricular assist device (LVAD) among other innovations in artificial heart research. John C. Norman, "The Evolution of the Problem," in Norman, ed., *Medicine in the Ghetto*, 1.

37 Norman, "The Evolution of the Problem," 6.

38 Norman, "The Evolution of the Problem," 7.

39 Leon Eisenberg, "Preliminary Report of the Harvard Commission on Relations with the Black Community," April 11, 1969, Leon Eisenberg Papers, H MS c196, Center for the History of Medicine, Countway Medical Library, Boston (hereafter LEP).

40 Norman, "The Evolution of the Problem," 2–3.

41 Daniel Patrick Moynihan, *The Negro Family: The Case for National Action* (Washington, DC: Office of Policy Planning and Research, US Dept. of Labor, 1965).

42 Charles L. Sanders, "Does Separatism in Medical Care Have a Future?" in Norman, ed., *Medicine in the Ghetto*, 15.

43 Gabriel Mendes, *Under the Strain of Color: Harlem's Lafargue Clinic and the Promise*

of an Antiracist Psychiatry (Ithaca, NY: Cornell University Press, 2015); Dennis Doyle, *Psychiatry and Racial Liberalism in Harlem, 1936–1968* (Rochester, NY: University of Rochester Press, 2016).

44 M. Tolchin, "The Changing City: A Medical Challenge," *New York Times*, June 2, 1969, 1, as quoted in Kurt W. Deuschle, "What Is the Role of the Ghetto Hospital in Health Care Delivery?" in Norman, ed., *Medicine in the Ghetto*.

45 "Ghetto Medicine Fills Bill," *HEALTH-PAC Bulletin*, April 1970, 13–14.

46 "Empire Roundup: Caught in the Squeeze," *HEALTH-PAC Bulletin*, October 1970, 1–10; Nancy Hicks, "City Toughens Demands in Voluntary-Hospital Pacts," *New York Times*, June 18, 1972, 25; A. Schwarz et al., "A Ghetto Medicine Program," *Social Work* 18, no. 6 (1973): 90–96; Betty J. Bernstein, "What Happened to 'Ghetto Medicine' in New York State?" *American Journal of Public Health* (July 1971): 1287–93.

47 Walsh McDermott, "The Role of a Department of Community Medicine in a Medical School," lecture presented at Mount Sinai Medical Center, February 14, 1969. As cited in Deuschle, "What Is the Role of the Ghetto Hospital in Health Care Delivery?" 160.

48 Heidi Morefield, "Developing to Scale."

49 Morefield, "Developing to Scale." See also "Mt. Sinai to Work with E. Harlem Center," *New York Amsterdam News*, March 22, 1969, 18.

50 Carter L. Marshall and David Pearson, *Dynamics of Health and Disease* (New York: Appleton-Century-Crofts, 1972).

51 Marshall thought this system could also "have great impact in rural areas, where doctors are scarce." See "Child Clinic Gets Physicians via TV: Mt. Sinai Doctors Examine Patients in East Harlem," *New York Times*, June 6, 1973, 94.

52 Edward Wallerstein, Carter L. Marshall, and Alexander Rodney, "Pediatrics and Cable Television," paper presented at Annual Meetings of National Cable Television Association, 1973, 1–21.

53 "TelePrompTer Center in Joint CATV Development," *New York Amsterdam News*, May 8, 1971, 23. See also Tawala Kweli, "Cable TV and Blacks," *New York Amsterdam News*, March 3, 1973, D2; "Sickle Cell Disease Discussed on CATV," *New York Amsterdam News*, July 21, 1973, A11.

54 Chet Cummings. "Pulse of New York's Public: Pro-CATV," *New York Amsterdam News*, April 18, 1970, 16.

55 Park, *Introduction to Telemedicine*, 119.

56 Wallerstein, Marshall, and Rodney, "Pediatrics and Cable Television," 3.

57 As quoted in Park, *Introduction to Telemedicine*, 123.

58 "Tele-Communications Linked to Medicine," *New York Amsterdam News*, June 16, 1973, B10.

59 Wallerstein, Marshall, and Rodney, "Pediatrics and Cable Television," 3.

60 Jean Heller, "Syphilis Victims in U.S. Study Went Untreated for 40 Years," *New York Times*, July 26, 1972, 1; Nelson, *Body and Soul*; Hoffman, *Health Care for Some*; Samuel Kelton Roberts, "Why Must We Again Be the Guinea Pigs in This Genocidal Mentality? Understanding Black Popular Resistance to Harm Reduction," paper presented at University of Strasbourg, October 2, 2019.

61 "The provision of an emergency $13 million to the voluntary hospitals—and not to the impoverished municipal hospitals" under New York's new Ghetto Medicine

Program, the *New York Amsterdam News* warned, "underscores what is probably the most basic problem in the city's system of health care." "Health Task Force: Mobilizing to Meet Crisis," *New York Amsterdam News*, July 3, 1971, C10.

62 J. Zamgba Browne, "E. Harlem Council Hits Medicine Program," *New York Amsterdam News*, September 18, 1971, B16.

63 Cinnamon Dee, "Cable TV Is Accused of Ripping-Off Harlem," *New York Amsterdam News*, July 29, 1972, A5.

64 C. Muller, C. L. Marshall, M. Krasner et al., "Cost Factors in Urban Telemedicine," *Medical Care* 15, no. 3 (1977): 251–59, at 251.

65 Muller, Marshall, Krasner et al., "Cost Factors in Urban Telemedicine," 251.

66 Lisa Brown, "Clinic Is Losing Cable-TV Link to Doctors," *New York Times*, July 1, 1975.

67 John L. S. Holloman, Jr., "Future Role of the Ghetto Physician," in Norman, ed., *Medicine in the Ghetto*, 144.

68 J. T. English, "Is the OEO Concept—the Neighborhood Health Center—the Answer?" In Norman, ed., *Medicine in the Ghetto*, 263.

69 Benjamin Hedin, "From Selma to Black Power," *The Atlantic*, March 6, 2015.

70 J. C. Schwartz, "Rural Health Problems of Isolated Appalachian Communities," in R. L. Nolan and J. L. Schwartz, eds., *Rural and Appalachian Health* (Springfield, IL: Charles C. Thomas, 1973), 29–44.

71 Fein, *The Doctor Shortage*.

72 The first licensing examination for PAs had only been developed in 1973. AMA, *Physician Support Personnel, 1973–74*, HEW Publication No. NIH (74-318) (Washington, DC: US Dept. of Health, Education & Welfare, 1974). As cited in Reich, "Telemedicine: The Assessment of an Evolving Health Care Technology."

73 Julie Fairman, *Making Room in the Clinic: Nurse Practitioners and the Evolution of Modern Health Care* (New Brunswick, NJ: Rutgers University Press, 2009); Natalie Holt, "Confusion's Masterpiece: The Development of the Physician Assistant Profession," *Bulletin of the History of Medicine* 72, no. 2 (1998): 246–78.

74 Rockoff wrote Waxman that "the most appropriate effort in which to extend our present activities at Dartmouth would be to see whether *physician expansion* can take place via television." On a later site visit she added, "The physician expansion capacity of the telemedicine system is highly sensitive to the amount of autonomy achieved by the nurse practitioners, so that if they only handled 45 percent of the patients coming to the neighborhood health centers without physician intervention (as the Dartmouth MEDEXES using protocols do) instead of the 75 percent they now handle, the system would not be cost-effective." "Trip Report—Cambridge, Massachusetts—Meeting with Gordon Moore and AT&T Representatives (HSM 110-72-384)," memorandum, October 29, 1973, MRP.

75 "Visit to Matthew Thornton Health Plan and Dartmouth Medical School (HSM 110-72-387) on May 7, 8, 1973," memorandum, June 7, 1973, MRP. "Some of the tradeoffs that need to be studied include: cameraman vs. remote control vs. paramedic; nurse vs. MEDEX vs. no paramedic (or just a receptionist); color vs. black and white, with and without interaction; resolution with and without interaction (motion rendition)."

76 "Site Visit—Bird, June 28, 29, 1973," memorandum, July 5, 1973, MRP.

77 "Site Visit—Bird, June 28, 29, 1973."

78 Peter C. Goldmark, "Greening of the Pavement People," *New York Times*, August 2, 1974.

79 Rockoff added, "I believe this project could be truly significant as a test site for the development of a healthcare system . . . It appeals to me also because the project does not propose to provide health care for poor people exclusively, but rather addresses the issue of how to make existing health care resources more productive through the introduction of technology and related management and operational resources as required." "Site Visit for Neurath," "Development of a Differential White Cell Count System," and "Visit to CBS Laboratories, Stamford, Connecticut," memorandum, September 26, 1972, MRP.

80 "Site Visit for Neurath," "Development of a Differential White Cell Count System," and "Visit to CBS Laboratories, Stamford, Connecticut."

81 "Status Report on Logistics Program," memorandum, December 19, 1972, MRP.

82 "Trip Report—Lakeview Clinic, Waconia, Minnesota (HSM-110-72-386)," memorandum, March 20, 1973, MRP.

83 Park, *Introduction to Telemedicine*, 107.

84 Tess Lanzarotta and Jeremy Greene, "Communications Technologies as Community Technologies: Alaska Native Villages and the Satellite Health Trials of the 1970s," *Technology's Stories* 5, no. 2 (2017). On the ATS-1 and ATS-6 projects, see *Satellite House Call*, directed by Judy Irving, DVD (Palo Alto: Stanford University Department of Communication, 1975); Heather E. Hudson, *Connecting Alaskans: Telecommunications in Alaska from Telegraph to Broadband* (Fairbanks: University of Alaska Press, 2015). The film can be viewed in its entirety on the Community Health Aides Project Jukebox website. Dennis Foote, Edwin B Parker, and Heather E Hudson, *Telemedicine in Alaska: The ATS-6 Satellite Biomedical Demonstration* (Palo Alto: Institute for Communication Research, Stanford University, 1976).

85 Having seen an early draft of the AHSFU documents, Rockoff thought that NASA's approach to a rural telemedicine system was "extremely naïve and that the engineers who prepared it are totally unfamiliar with the sociological problems of health care delivery." "Trip Report: Travel to Boston, February 23 and 24, 1972," memorandum, March 3, 1972, MRP.

86 Jeremy Greene, Victor Braitberg, and Gabriella Maya Bernadett, "Innovation on the Reservation: Information Technology and Health Systems Research among the Papago Tribe of Arizona, 1965–1980," *Isis* 111, no. 3 (2020): 443–70; Andrew T. Simpson, Charles R. Doarn, and Stephen J. Garber, "Interagency Cooperation in the Twilight of the Great Society: Telemedicine, NASA, and the Papago Nation," *Journal of Policy History* 32, no. 1 (2020): 25–52.

87 "Space Technology Can Help Remote Indians in Arizona," *Sarasota Herald Tribune*, February 22, 1976, 14H.

88 Peter Ruiz, "STARPAHC: A Telemedicine Project: An Oral History Interview with Peter A. Ruiz," *Journal of the Southwest* (2021): 75–131.

89 "Papago" is used here as a historical term. The name, a colonial imposition first applied by the Spanish that means "bean-eater," was formally replaced in 1986 with "Tohono O'odham," which means "Desert People" in the O'odham language.

90 M. Fuchs, "Provider Attitudes toward STARPAHC: A Telemedicine Project on the Papago Reservation," *Medical Care* 17, no. 1 (1979): 59–68.

91 Brent Miller, "NAU Television Center Awaits Equipment," *The Lumberjack*, November 15, 1984, 10.

92 Francille A. Rusan, "What You See Is What You Get: Cable Television and Community Control," *Yale Review of Law and Social Action* 2, no. 3 (1972): 275–81, at 275. On the corporate consolidation of CATV, see Megan Mullen, *The Rise of Cable Programming in the United States: Revolution or Evolution?* (Austin: University of Texas Press, 2003).

93 As Rockoff noted, "Without such financial incentives, the prospects are dim for widespread adoption of either of these innovations, even assuming that careful research has demonstrated that certain health-care functions can be effectively and efficiently provided with these innovations." Rockoff, "An Overview of Some Technological/Health-Care System Implications of Seven Exploratory Broad-Band Communication Experiments," 29.

94 John L. S. Holloman, Jr., MD, "Future Role of the Ghetto Physician," in Norman, ed., *Medicine in the Ghetto*, 144.

95 Holloman, Jr., "Future Role of the Ghetto Physician."

96 Lisa Brown, "Clinic Is Losing Cable-TV Link to Doctors," *New York Times*, July 1, 1975.; C. Muller, C. L. Marshall, M. Krasner et al., "Cost Factors in Urban Telemedicine," *Medical Care* 15, no. 3 (1977): 251–59.

97 AMA, *Physician Support Personnel, 1973–74*.

98 Rockoff, "An Overview of Some Technological/Health-Care System Implications of Seven Exploratory Broad-Band Communication Experiments," 29.

Chapter Six

1 Vladimir Zworykin, "A Rapid Access Multi-Memory Unit for Medical Data Processing," 5, undated MS, 1960, box 78, folder 11, DSRC. Portions of this chapter appear in an abbreviated form in Jeremy A. Greene and Andrew Lea, "Digital Futures Past: The Long Arc of Big Data in Medicine," *New England Journal of Medicine* 381 (2019): 480–85.

2 Zworykin, "A Rapid Access Multi-Memory Unit for Medical Data Processing," 5. Zworykin cites the *Merck Manual*, 9th ed. (Rahway, NJ: Merck & Co., 1956); *Standard Nomenclature of Diseases and Operation* (New York: McGraw Hill, 1952).

3 Zworykin, "A Rapid Access Multi-Memory Unit for Medical Data Processing," 3–4.

4 Zworykin, "A Rapid Access Multi-Memory Unit for Medical Data Processing," 4.

5 Zworykin, "A Rapid Access Multi-Memory Unit for Medical Data Processing," 4.

6 Robert Steven Ledley, *Use of Computers in Biology and Medicine* (New York: McGraw-Hill, 1965), 42.

7 B. Kaplan, "The Medical Computing 'Lag': Perceptions of Barriers to the Application of Computers to Medicine," *International Journal of Technology Assessment in Health Care* 3, no. 1 (1987): 123–36; Nathan Ensmenger, "Resistance Is Futile? Reluctant and Selective Users of the Internet," in William Aspray and Paul Ceruzzi, eds., *The Internet and American Business* (Cambridge, MA: MIT Press, 2010).

8 G. Octo Barnett, "Automated Hospital Revolution Overdue," *Medical World News*, May 17, 1968, 54–55; David Theodore, "Towards a New Hospital: Architecture, Medicine, and Computation, 1960–75," PhD diss., Harvard University, 2014.

9 Carlos Vallbona, "Ten Years of Computers in Medicine—a Retrospective View," in *9th IBM Medical Symposium* (Poughkeepsie, NY: IBM, 1968), 189–94, at 189–90.

10 Bonnie Kaplan, "Computer Prescription: Medical Computing, Public Policy, and Views of History," *Science, Technology, and Human Values* 20, no. 1 (1995): 5–38. Historians of computing remind us that the computer was not a historically stable object but a highly dynamic entity in the middle decades of the twentieth century. The room-sized ENIAC brought different constraints and possibilities of computing into life in the 1940s than the refrigerator-sized IBM mainframes popularized in the 1950s, and the minicomputers and microcomputers that emerged in the 1960s and 1970s. Paul N. Edwards, *The Closed World: Computers and the Politics of Discourse in Cold War America* (Cambridge, MA: MIT Press, 1996); Christopher M. Kelty, *Two Bits: The Cultural Significance of Free Software* (Durham, NC: Duke University Press, 2008); Michael Sean Mahoney, *Histories of Computing* (Cambridge, MA: Harvard University Press, 2011); Nathan L. Ensmenger, *The Computer Boys Take Over: Computers, Programmers, and the Politics of Technical Expertise* (Cambridge, MA: MIT Press, 2012); John Harwood, *The Interface: IBM and the Transformation of Corporate Design, 1945–1976* (Minneapolis: University of Minnesota Press, 2016); Mar Hicks, *Programmed Inequality: How Britain Discarded Women Technologists and Lost Its Edge in Computing* (Cambridge, MA: MIT Press, 2017).

11 Joseph November, *Biomedical Computing: Digitizing Life in the United States* (Baltimore: Johns Hopkins University Press, 2012); Hallam Stephens, *Life Out of Sequence: A Data-Driven History of Bioinformatics* (Chicago: University of Chicago Press, 2013).

12 Robert S. Ledley, "Digital Electronic Computers in Biomedical Sciences," *IRE* 130, no. 3384 (1959): 1225–34, at 1230; J. A. Shannon and C. V. Kidd, "Medical Research in Perspective," *Science* 124 (1956): 1185–90. See also J. A. Greene and S. H. Podolsky, "Keeping Modern in Medicine: Pharmaceutical Promotion and Physician Education in Postwar America," *Bulletin of the History of Medicine* 83 (2009): 331–77.

13 Cheryl Rae Dee, "The Development of the Medical Literature Analysis and Retrieval System (MEDLARS)," *Journal of the Medical Library Association: JMLA* 95, no. 4 (2007): 416–25, doi:10.3163/1536-5050.95.4.416.

14 F. B. Rogers, "Stresses in Current Medical Bibliography," *New England Journal of Medicine* 267, no. 14 (1962):704–8.

15 Lisa Gitelman, *Paper Knowledge: Toward a Media History of Documents* (Durham, NC: Duke University Press, 2014); Markus Krajewski, *Paper Machines: About Cards and Catalogs, 1548–1929*, trans. Peter Krapp (Cambridge, MA: MIT Press, 2011); Lars Heide, *Punched-Card Systems and the Early Information Explosion, 1880–1945* (Baltimore: Johns Hopkins University Press).

16 Martin M. Cummings, "A Library-Based Computer System for Indexing, Storing, and Retrieving Literature in the Cardiovascular Field," paper presented at the Conference on Computation for Cardiovascular Research, New York Academy of Sciences, Waldorf Astoria Hotel, New York, October 30, 1964, box 24, folder 1, National Library of Medicine (US), Office of the Director, Martin M. Cummings Papers, MS C 554, National Library of Medicine, Bethesda, MD (hereafter MCP).

17 President's Commission on Heart Disease, Cancer, and Stroke, "A National Program to Conquer Heart Disease, Cancer, and Stroke," n.d. (c. 1964), box 24, folder 1,

"MEDLARS Miscellaneous, 1964," MCP; Dee, "The Development of the Medical Literature Analysis and Retrieval System (MEDLARS)."

18 "Push-Button Brains Produce Faster Facts for Doctors," *Courier-Journal Magazine*, November 22, 1964, 25–31, at 25 (found in box 24, folder 2, MCP).

19 Los Angeles, Boston, New York, Durham, New Orleans, Chicago, Houston, and Seattle soon mirrored the MEDLARS collection on their own computers. Quotation from National Library of Medicine Board of Regents Meeting, November 6, 1964, "Materials Relating to Discussion of MEDLARS Decentralization Policy," box 25, folder 7; also "MEDLARS Decentralization," memorandum, April 1, 1964, box 25, folder 8; "MEDLARS Search Centers: Consolidated Monthly Report: April 1967," box 24, folder 5, all in MCP.

20 "The Library and Specialized Information Services Components of the Biomedical Communications System, Its Impact on the MEDLARS System," box 24, folder 5, MCP.

21 William H. Mills, "International Exchange of Medical Literature with the United Kingdom," memorandum, October 13, 1965, box 24 folder 3, MCP.

22 Joseph Leiter to R. H. C. Wells, March 24, 1967, box 24, folder 5, MCP.

23 Alice E. Luethy to Scott Adams, December 14, 1964, box 25, folder 3; James A. Olson to Martin M. Cummings, February 23, 1967; Martin M. Cummings to James A. Olson, March 10, 1967; Sunil K. Pandya to Martin M. Cummings, June 12, 1967, box 24, folder 5, all in MCP.

24 F. W. Lancaster, "MEDLARS: Report on the Evaluation of Its Operating Efficiency," *American Documentation* 20, no. 2 (1969): 119–42, at 121.

25 M. M. Cummings, US House, Departments of Labor and Health, Education, and Welfare and Related Agencies Appropriations for 1967, Part 3: National Institutes of Health, Subcommittee of the Committee on Appropriations, House, Hearing 1966 Mar 6, 784–806, as cited in Dee, "The Development of the Medical Literature Analysis and Retrieval System (MEDLARS)."

26 Lancaster, "MEDLARS: Report on the Evaluation of Its Operating Efficiency."

27 Joseph Leiter, "Cleverdon Comments on MEDLARS Evaluation Document," memorandum, January 16, 1968, box 24, folder 8, MCP.

28 Charles J. Austin to Martin M. Cummings, "After MEDLARS, What?" memorandum, December 26, 1963, box 24, folder 2, MCP.

29 Joseph Leiter to Martin M. Cummings, December 12, 1966, box 24, folder 8, MCP.

30 Dee, "The Development of the Medical Literature Analysis and Retrieval System (MEDLARS)"; Irwin Pizer, "The Application of Computers in the State University of New York Biomedical Communication Network," in *8th IBM Medical Symposium* (Poughkeepsie, NY: IBM, 1967), 57–65.

31 Martin M. Cummings, "BOOKS and COMPUTERS (From BILLINGS to MEDLARS)," 14–15, speech delivered at dedication ceremony of Vanderbilt University's new medical library, November 19, 1964, box 24, folder 1, MCP.

32 F. W. Lancaster and L. Estabrook, "Reflections: An Interview with F. W. Lancaster," *Library Trends* 56, no. 4 (2008): 968–74, at 973.

33 Vladimir K. Zworykin, "New Frontiers in Medical Electronics: Electronic Aids for Medical Diagnosis," 3, 6, paper presented at 39th Anniversary Congress, Pan American Medical Association, February 19, 1964, box 78, folder 59, DSRC.

34 Athelstan Spilhaus, "Our New Age" (comic strip), January 17, 1960.

35 Andrew Lea, "Computerizing Diagnosis: Keeve Brodman and the Medical Data Screen," *Isis* 110, no. 2 (2019): 228–49.

36 Adrianus J. van Woerkom, "Program for a Diagnostic Model," *IRE Transactions on Medical Electronics* ME-7, no. 2 (1960): 220.

37 Keeve Brodman, Adrianus J. van Woerkom, Albert J. Erdmann, and Leo S. Goldstein, "Interpretation of Symptoms with a Data-Processing Machine," *Archives of Internal Medicine* 103, no. 5 (1958): 776–82.

38 "With the use of diagnostic criteria it developed itself," Brodman explained. "The machine evaluated correctly patients' symptoms in as many instances as did a physician, who used the same information and referred to criteria established by the accumulated experience of the medical profession." Brodman, van Woerkom, Erdmann, and Goldstein, "Interpretation of Symptoms with a Data-Processing Machine," 82.

39 John C. Burnham, "American Medicine's Golden Age: What Happened to It?" *Science* 215, no. 4539 (1982): 1474–79; John Harley Warner, "The Aesthetic Grounding of Modern Medicine," *Bulletin of the History of Medicine* 88, no. 1 (2014): 1–47.

40 Greene and Podolsky, "Keeping Modern in Medicine."

41 F. A. Nash, "Diagnostic Reasoning and the Logoscope," *The Lancet* 276, no. 7166 (1960): 1442–46.

42 A. McGehee Harvey and James Bordley, *Differential Diagnosis: The Interpretation of Clinical Evidence* (Philadelphia: Saunders, 1955).

43 Existing paper media like books and journals were, in Nash's estimation, insufficient to meet the information needs of the day. "The inadequacy of these tools," Nash reflected, "is inherent in the structure of the *page* as the unit holder of data." F. A. Nash, "Differential Diagnosis: An Apparatus to Assist the Logical Faculties," *The Lancet* 263, no. 6817 (1954): 874–75, at 874.

44 R. W. Pain, "Limitations of the Nash Logoscope or Diagnostic Slide Rule," *Medical Journal of Australia* 2, no. 18 (1975): 714–15.

45 A version was distributed with support from Schering, but the pharmaceutical manufacturer backed out soon after the project was criticized by the deans of several prominent medical schools. Letter from Schering Corporation, quoted by Mr. Lane in discussion of F. A. Nash, "The Mechanical Conservation of Experience, Especially in Medicine," *IRE Transactions on Medical Electronics* ME-7, no. 4 (1960): 240–43, at 243.

46 Nash, "The Mechanical Conservation of Experience, Especially in Medicine," 240. The Montpelier ophthalmologist François Paycha had devised a mechanical card-sorting device instead of a slide rule to automate diagnostic practice. Like Nash, Paycha noted, "Physicians are so used to making a diagnosis that they do not know how they do it." François Paycha, "Diagnosis, Therapeutics, Prognosis, and Computers," *IRE Transactions on Medical Electronics* ME-7, no. 4 (1960): 288–90.

47 Ruth Schwartz Cowan, *More Work for Mother: The Ironies of Household Technology from the Open Hearth to the Microwave* (New York: Basic Books, 1983).

48 Matthew L. Jones, "How We Became Instrumentalists (Again): Data Positivism since World War II," *Historical Studies in the Natural Sciences* 48, no. 5 (November 2018): 673–84; Stephanie Dick, "Artificial Intelligence," *Harvard Data Science Review*, June 22, 2019.

49 Martin Lipkin, and James D. Hardy, "Mechanical Correlation of Data in Differential Diagnosis of Hematological Diseases," *JAMA* 166, no. 2 (1958): 113–25, at 118.

50 Ralph L. Engle, "Medical Diagnosis," in John A. Jacquez, ed., *The Diagnostic Process: The Proceedings of a Conference Sponsored by the Biomedical Data Processing Training Program of the University of Michigan.* (Ann Arbor, 1964), 7–18.

51 R. Ebald and R. Lane, "Digital Computers and Medical Logic," *IRE Transactions on Medical Electronics* ME-7, no. 4 (1960): 283–88, at 283.

52 M. Lipkin, "Correlation of Data with a Digital Computer in the Differential Diagnosis of Hematological Diseases," *IRE Transactions on Medical Electronics* ME-7, no. 4 (1960): 243–46, at 246.

53 Taylor and Baruch in discussion of Robert S. Ledley, "Using Electronic Computers in Medical Diagnosis," *IRE Transactions on Medical Electronics* ME-7, no. 4 (1960): 274–80, at 279–90.

54 Taylor and Baruch in discussion of Ledley, "Using Electronic Computers in Medical Diagnosis," 279–90.

55 T. Tanimoto and R. G. Loomis, "The Application of Computers to Clinical Medical Data (including Machine Demonstration)," in *1st IBM Medical Symposium* (Poughkeepsie, NY: IBM, 1959), 93.

56 The relationship between Davis and Tanimoto reflected a robust interplay between disease classification and botanical classification that had been at work since the sixteenth and seventeenth centuries, when Thomas Sydenham and François Boisson de Sauvages made their appeals to organize the world of diseases according to the logic of plant classification. David J. Rogers and Taffee T. Tanimoto, "A Computer Program for Classifying Plants," *Science* 132, no. 3434 (1960): 1115–18; A. J. Cain, "Thomas Sydenham, John Ray, and Some Contemporaries on Species," *Archives of Natural History* 26, no. 1 (1999): 55–83. See also Anne Kveim Lie and Jeremy Greene, "From Ariadne's Thread to the Labyrinth Itself: Nosology and the Infrastructure of Modern Medicine," *New England Journal of Medicine* 382, no. 13 (2020): 1273–77.

57 Davis, "The Application of Computers to Clinical Medical Data (including Machine Demonstration)," 185.

58 William S. Middleton, "Medicine before Automation," *Archives of Internal Medicine* 109, no. 3 (1961): 251–55, at 255.

59 Ralph L. Engle and B. J. Davis, "Medical Diagnosis: Present, Past, and Future: I. Present Concepts of the Meaning and Limitations of Medical Diagnosis," *Archives of Internal Medicine* 112 (1963): 512–19, at 518; Lee B. Lusted and Walter R. Stahl, "Conceptual Models of Diagnosis," in Jacquez, ed., *The Diagnostic Process*, 157–74, at 157; Martin Lipkin, "The Role of Data Processing in the Diagnostic Process," in Jacquez, ed., *The Diagnostic Process*, 255–73.

60 Richard Taylor, "Major Problems in the Use of Computing Machines," *IRE Transactions on Medical Electronics* ME-7, no. 4 (1960): 253–54, at 254.

61 Vladimir K. Zworykin, "Electronics and Health Care," *Proceedings of the Institute of Radio Engineers*, May 1962, proof copy, box 78, folder 39, DSRC.

62 Zworykin, "Electronics and Health Care."

63 A standardized 195-question patient history could be coded in roughly 40 seconds. Joseph E. Schenthal, James W. Sweeney, and Wilson J. Nettleton, "Clinical Application of Large-Scale Electronic Data Processing Apparatus: I. New Concepts in

Clinical Use of the Electronic Digital Computer," *JAMA* 173, no. 1 (1960): 6–11, at 10.

64 Joseph E. Schenthal, "Clinical Concepts in the Application of Large Scale Electronic Data Processing," in *2nd IBM Medical Symposium* (Endicott, NY: IBM, 1960), 391–99, at 398–99. See also Joseph E. Schenthal, James W. Sweeney, and Wilson J. Nettleton, "Clinical Application of Electronic Data Processing: II. New Methodology in Clinical Record Storage," *JAMA* 178, no. 3 (1961): 267–70, at 270.

65 Joseph E. Schenthal, James W. Sweeney, Wilson J. Nettleton, and Richard D. Yoder, "Clinical Applications of Electronic Data Processing Apparatus III: System for Processing Medical Records," *JAMA* 186, no. 2 (1963): 101–5, at 101.

66 "Electronic Data Processing Apparatus," *JAMA* 173, no. 1 (1960): 58–59, at 59.

67 Jordan J. Baruch, "Doctor-Machine Symbiosis," *IRE Transactions on Medical Electronics* ME-7, no. 4 (1960): 290–93; Theodore, "Towards a New Hospital: Architecture, Medicine, and Computation, 1960–75"; G. Octo Barnett, "History of the Development of Medical Information Systems at the Laboratory of Computer Science at Massachusetts General Hospital," n.d. MGHTC.

68 Barnett, "History of the Development of Medical Information Systems at the Laboratory of Computer Science at Massachusetts General Hospital."

69 "Medicine Faces the Computer Revolution," *Medical World News*, July 14, 1967, 46–50, at 46.

70 "Switch-Over to Electronic Care," *Medical World News*, May 17, 1968, 42–47.

71 Rachel Plotnick, "Computers, Systems Theory, and the Making of a Wired Hospital," *Journal of the American Society for Information Science and Technology* 61, no. 6 (2010): 1281–94.

72 Plotnick, "Computers, Systems Theory, and the Making of a Wired Hospital," 1287; W. Hammond, "Patient Management Systems: The Early Years," in B. I. Blum, ed., *Proceedings of the ACM Conference on the History of Medical Informatics* (New York: ACM Press, 1987), 152–64.

73 G. Octo Barnett, "From 'Farm Boy' to Director of the Laboratory of Computer Science: 2004 Interview of G. Octo Barnett," in Rebecca M. Goodwin, Joan S. Ash, and Dean F. Sittig, eds., *Conversations with Medical Informatics Pioneers: An Oral History Collection* (Bethesda: National Library of Medicine, 2015).

74 Lawrence. L. Weed, "Medical Records That Guide and Teach," *New England Journal of Medicine* 278 (1968): 593–600, 652–57, at 593.

75 In an earlier 1964 publication, "Medical Records, Patient Care, and Medical Education," Weed had suggested that the medical record merited more regular and systematic auditing, with more regular and standardized collecting of data, and "in addition, records on a large scale would become available for computer analysis." Lawrence L. Weed, "Medical Records, Patient Care, and Medical Education," *Irish Journal of Medical Science* 39, no. 6 (1964): 271–82.

76 Weed, "Medical Records That Guide and Teach," 595.

77 Weed, "Medical Records That Guide and Teach," 657.

78 Lawrence L. Weed, *Medical Records, Medical Education, and Patient Care: The Problem-Oriented Record as a Basic Tool* (Chicago: Year Book Medical Publishers, 1969).

79 Weed, *Medical Records, Medical Education, and Patient Care*, 288–89.

80 Weed, *Medical Records, Medical Education, and Patient Care*, 120–21.

81 George E. Nelson, "Experiences with the Problem-Oriented Record at the University of Vermont," in J. Willis Hurst and H. Kenneth Walker, eds., *The Problem-Oriented System* (New York: Medcom, 1972), 72–76.
82 J Willis Hurst, "Ten Reasons Why Lawrence Weed Is Right," *New England Journal of Medicine* 284 (1978): 51–52.
83 Lawrence L. Weed, "Questions Often Asked about the Problem-Oriented Record—Does It Guarantee Quality?" in Hurst and Walker, eds., *The Problem-Oriented System*, 51–56, at 51.
84 Lawrence L. Weed, "Background Paper for Concept of National Library Displays," in Hurst and Walker, eds., *The Problem-Oriented System*, 258–68, at 265.
85 Bonnie Kaplan, "Development and Acceptance of Medical Information Systems: A Historical Overview," *Journal of Health and Human Resources Administration* 11, no. 1 (1988): 9–29.
86 As with many other scientific computing paradigms, the cognitive impact of PROMIS far exceeded its limited uptake in practice. For a comparable example in organic chemistry, see Evan Hepler-Smith, "A Way of Thinking Backwards: Computing and Method in Synthetic Organic Chemistry," *Historical Studies in the Natural Sciences* 48, no. 3 (2018): 300–337.
87 Jordan J. Baruch, "The Generalized Medical Information Facility," *Inquiry* 5, no. 3 (1968): 17–23.
88 Baruch, "The Generalized Medical Information Facility."
89 Warner V. Slack, "Ongoing Trial and Pilot Programs," in *10th IBM Medical Symposium* (Poughkeepsie, NY: IBM, 1972), 107.

Chapter Seven

1 https://obamawhitehouse.archives.gov/precision-medicine.
2 American Medical Informatics Association, https://www.amia.org/fact-sheets/what-informatics.
3 Natasha Schull, "Data for Life: Wearable Technology and the Design of Self-Care," *BioSocieties* 11, no. 3 (2016): 317–33.
4 Vladimir K. Zworykin, "Electronics and Health Care," *Proceedings of the Institute of Radio Engineers*, May 1962, proof copy, box 78, folder 39, DSRC.
5 "By the time the patient turns in this last questionnaire," clinic founder Morris Collen declared, "the 'on-line' computer processing has been completed and supplemental tests and appointments are 'advised' by the programmed rules of the computer, and these are arranged for the patient." Morris Collen, *Detection and Prevention of Chronic Disease Utilizing Multiphasic Health Screening Techniques: Hearing before the Subcommittee on Health of the Elderly of the Special Committee on Aging, United States Senate, Eighty-ninth Congress, Second Session, September 20, 21, and 22, 1966* (Washington, DC: GPO, 1966), 215–16.
6 Morris F. Collen, "Computer Medicine: Its Application Today and Tomorrow," *Minnesota* 11 (1966): 1705–7, at 1705.
7 Leonard A. Stevens, statement before Senate Special Committee on Aging, in *Detection and Prevention of Chronic Disease Utilizing Multiphasic Health Screening Techniques*, 223; Leonard A. Stevens, "Now—The Automated Physical Checkup," *Reader's Digest*, July 4, 1966, 95–98.

8 Tom Debley and Jon Stewart, *The Story of Dr. Sidney R. Garfield: The Visionary Who Turned Sick Care into Health Care* (Oakland, CA: Permanente Press, 2009).

9 Here Breslow echoed several other prominent public health officials across the country to make the problem of "chronic disease" increasingly visible to American municipal and state public health programs starting in the 1930s and 1940s. See George Weisz, *Chronic Disease in the Twentieth Century: A History* (Baltimore: Johns Hopkins University Press, 2014).

10 Lester Breslow and Malcolm H. Merrill, "Chronic Disease: The Chronic Disease Study of the California Department of Public Health," *American Journal of Public Health* 39 (1946): 593–97. See also Weisz, *Chronic Disease in the Twentieth Century*; Jeremy Greene, *Prescribing by Numbers: Drugs and the Definition of Disease* (Baltimore: Johns Hopkins University Press, 2007).

11 Lester Breslow, "Multiphasic Screening," *American Journal of Public Health* 40, no. 3 (1950): 324–25.

12 Lester Breslow, "Multiphasic Screening Examinations an Extension of the Mass Screening Technique," *American Journal of Public Health* 40 (1950): 274–78. See also Statement of Dr. Nemat O. Borhani, in *Detection and Treatment of Chronic Disease*, 176.

13 Breslow, "Multiphasic Screening Examinations an Extension of the Mass Screening Technique," 278. Not long afterward, the AMA showed interest in conducting multiphasic screenings of physicians at its annual conference. See James Don Collom, "The Premise and Promise of Computerized Multiphasic Screening/Testing: An Advocate's Perspective," PhD diss., University of Iowa, 1971.

14 Morris Collen, "Health Care Information Systems: A Personal Historical Review," in Association for Computing Machinery (ACM), *ACM Conference on the History of Medical Informatics: Conference Proceedings* (Baltimore: ACM, 1987), 123–36. Oral History with Morris Collen conducted by Martin Meeker, 2005, "Morris Collen, M.D., Kaiser Permanente Oral History Project II," Regional Oral History Office, Bancroft Library, University of California, Berkeley (hereafter ROHO), https://digitalassets.lib.berkeley.edu.

15 E. Richard Weinerman, Lester Breslow, Nedra B. Belloc et al., "Multiphasic Screening of Longshoremen with Organized Medical Follow-Up," *American Journal of Public Health* 42 (1952): 1552–67.

16 Statement of Dr. Nemat O. Borhani, in *Detection and Treatment of Chronic Disease*, 179–80.

17 Collen, "Health Care Information Systems: A Personal Historical Review," 124. IBM proposed a novel configuration of an IBM 1440/1311/1050 system, a computer with 12K operating memory. "An Online, Teleprocessing, Random Access, Medical Physical Check-Up System," memorandum, n.d., box 5.1, folder 3, Morris F. Collen Collection, Kaiser Permanente Archives, Oakland, CA (hereafter MFCC).

18 "An Online, Teleprocessing, Random Access, Medical Physical Check-Up System."

19 "Medicine: Push-Button Hospital," *Time*, June 29, 1953.

20 With the exception of Papanicolau smears, sigmoidoscopies, and interpretations of chest X-rays and electrocardiograms, which required direct physician role.

21 Morris F. Collen, "Machine Diagnosis from a Multiphasic Screening Program," in *5th IBM Medical Symposium, October 7–11, 1963* (Endicott, NY: IBM, 1963), 131–53, at 131–32, emphasis mine. M. F. Collen, L. Rubin, J. Neyman et al., "Automated Multi-

phasic Screening and Diagnosis," *American Journal of Public Health* 54, no. 5 (1964): 741–50, at 743.

22 "The only data necessary for this method," Collen continued, "are the determinations (for each disease to be screened) of the proportions of individuals with the disease who have certain selected combinations of symptoms, and the proportions of individuals without this disease who have identical combinations of symptoms." Collen. "Machine Diagnosis from a Multiphasic Screening Program," 131–32.

23 R. E. Bonner, C. J. Evangelisti, H. D. Steinbeck, and L. Cohen, "DAP—A Diagnostic Assistance Program," in *6th IBM Medical Symposium, October 5–9, 1964* (Poughkeepsie, NY: IBM, 1964), 81–108.

24 "Periodic inventory will be taken of internist's diagnoses stored in the computer," Collen explained in the *American Journal of Public Health* the following year, "so as to continually enlarge and improve the diagnostic ability of the program." Collen, Rubin, Neyman et al., "Automated Multiphasic Screening and Diagnosis," 743.

25 *Detection and Prevention of Chronic Disease Utilizing Multiphasic Health Screening Techniques*, 1.

26 *Detection and Prevention of Chronic Disease Utilizing Multiphasic Health Screening Techniques*, 2–3.

27 *Detection and Prevention of Chronic Disease Utilizing Multiphasic Health Screening Techniques*, 54–55.

28 Morris F. Collen, "Computer Medicine: Its Application Today and Tomorrow," *Minnesota* 11 (1966): 1705–7; see also Morris F. Collen, "The Multitest Laboratory in Health Care of the Future," *Hospitals, Journal of the American Hospital Association* 41 (1967): 119–25.

29 "Training Program in Automated Multiphasic Screening: Phase 1—For Project Directors," November 1966, unpublished transcript, box 4.3, folder 5, MFCC.

30 Morris F. Collen to James A. Reynolds, May 5, 1967, box 4.1, folder 7, MFCC.

31 *Summary of Legislative Activities of Interest to the National Institutes of Health, 89th Congress, Second Session* (Bethesda, MD: National Institutes of Health, 1967).

32 *Detection and Prevention of Chronic Disease Utilizing Multiphasic Health Screening Techniques*, 137; H. R. Oldfield, Jr., "Automated Multiphasic Health Testing: A Diagnosis in Three Parts," *Journal of Clinical Engineering* 3, no. 2 (1978): 1–5.

33 James L. Craig, "Adaptation of Occupational Medicine Techniques to Community Health Care," *Journal of Occupational Medicine* 14, no. 1 (1972): 50–54 at 54. See also James L. Craig, "The Role of AMHT in Health Care Systems," in *9th IBM Medical Symposium* (Poughkeepsie, NY: IBM, 1968); James L. Craig, "Automated Multiphasic Health Testing: The TVA Experience," *Archives of Environmental Health* 27, no. 4 (1973): 264–68.

34 Specifically, "to speed the attainment of equity in access for the American public without adding unreasonably to the costs of care." Eleanor F. Smith, "The Utilization of Multiphasic Screening in Public Health Centers," *Journal of Occupational Medicine* 11, no. 7 (1969): 364–68; AMHTS Advisory Committee to the National Center for Health Services Research and Development, *Automated Multiphasic Health Testing and Services: Volume 1* (Washington, DC: US Dept. of Health, Education, and Welfare, 1970), ix.

35 *Detection and Prevention of Chronic Disease Utilizing Multiphasic Health Screening Techniques*, 270–71. The automation of clinical intake would play a crucial role as

"a means of entry into a health-care system for persons who have had little or no medical attention." Leo Gitman, "Automated Multiphasic Health Screening: A Conceptual Model and Description of a Program," *Journal of Occupational Medicine* 11, no. 12 (1969): 669–73.

36 *Detection and Prevention of Chronic Disease Utilizing Multiphasic Health Screening Techniques*, 272–73.

37 "The Computers That Will Keep You Healthy," 10, unpublished MS, *Look Magazine*, box 4.1, folder 7, MFCC.

38 Morris Collen, "Possibilities for Adapting Multiphasic Testing Techniques at the State or Local Level," in US Public Health Service, *Programming Chronic Disease Laboratory Services: Summary of Multiregional Seminars* (Washington, DC: US Public Health Service, 1967), 139–47, at 147.

39 Norman Roberts et al., "Conference on Automated Multiphasic Screening: Panel Discussion, Morning Session," *Bulletin of the New York Academy of Medicine* 45, no. 12 (1969): 1326–37.

40 Joseph H. Chadwick, J. J. de Lang, and Murray Greyson, *Cost and Operations Analytics of Automated Multiphasic Health Testing*, PB-193881 (Menlo Park, CA: Stanford Research Institute, December 1969).

41 "An Online, Teleprocessing, Random Access, Medical Physical Check-Up System," 10.

42 H. R. Oldfield, Jr., "Automated Multiphasic Health Testing: A Diagnosis in Three Parts," *Journal of Clinical Engineering* 3, no. 2 (1978): 1–5; "Multiphasic Testing, 1971," *Socio-Economic Report* (Bureau of Research and Planning, California Medical Association) 9, no. 1 (January 1970).

43 Morris F. Collen, "Value of Multiphasic Health Checkups," *New England Journal of Medicine* 280, no. 19 (1969): 1072–73, at 1073.

44 Emerson Day to Morris F. Collen, May 10, 1971, box 3.1, folder 1, MFCC.

45 Emerson Day, "Automated Health Services—Reprogramming the Doctor," in *9th IBM Medical Symposium* (Poughkeepsie, NY: IBM, 1968), 105–12; Peter Brown, "Computer Gives You a Physical: Patient Provided a Complete Medical Profile in 2 Hours," *San Diego Union*, March 5, 1971.

46 James B. McCormick and Joseph B. Kopp, "A Review of AMHT Centers," undated MS, box 3.3, folder 4, MFCC.

47 Health Screening Services, Limited, pamphlet, n.d., box 5.2, folder 3, MFCC.

48 Takuji Miwa, "Prehistory and Institution of Ningen Dokku in Japan," in Toshio Yasaka, ed., *Progress in Health Monitoring (AMHTS): Proceedings of the International Conference on Automated Multiphasic Health Testing and Services, Tokyo, October 4–6, 1980* (Amsterdam: Excerpta Medica, 1981); Norio Sasamori, "The Present Outlook of the Human Dry Dock in Japan and Its Outlook for the Future," *Japan Hospitals*, July 1, 1982, 49–55.

49 "To make such a system practicable," a group of Toshiba physicians and engineers postulated in the early 1970s, ". . . it should make use of systems-engineering techniques, automated devices, and the exploitation of computer technology." Tohru Kobayashi, Yutaka Moriyama, and Yoshisuke Iwai, "Health Test Systems in Japan," *Progress in Technology* 1 (1972): 26–34, at 26.

50 Yasaka, ed., *Progress in Health Monitoring (AMHTS)*, 146.

51 These changes reflected the greater concern for fish-borne parasites and GI can-

cers in the epidemiological profile of Japan: as its architects summed up in a 1972 publication, "the system is specifically adapted to the Japanese nationality and to the needs of the Toshiba Company." Wendell Leo Chappell, "Improving Physician Acceptance of Automated Multiphasic Health Testing," thesis, Naval Postgraduate School, Monterey, CA, 1975, 16–17; Hioyuki Tohma, "Remarks by the Chairman," in Yasaka, ed., *Progress in Health Monitoring (AMHTS)*, 144; Kobayashi, Moriyama, and Iwai, "Health Test Systems in Japan," 26.

52 Morris F. Collen, "Advanced Medical Systems: The International Picture," 21–22, paper presented at SAMS Annual Scientific Meeting, Baltimore, November 11, 1974, box 2.3, folder 1, MFCC. The mobile automated "dry dock" could be mounted on five, 8-ton trucks using a fifteen-member crew including five drivers and ten doctors, technicians, and nurses.

53 Collen, "Advanced Medical Systems: The International Picture," 21. See also Sasamori, "The Present Outlook of the Human Dry Dock in Japan and Its Outlook for the Future," 50.

54 Kobayashi, Moriyama, and Iwai, "Health Test Systems in Japan," 29–30.

55 *Detection and Prevention of Chronic Disease Utilizing Multiphasic Health Screening Techniques*, 54–55.

56 Sidney R. Garfield, "Multiphasic Health Testing and Medical Care as a Right," *New England Journal of Medicine* 280, no. 20 (1970): 1087–89.

57 Collom, "The Premise and Promise of Computerized Multiphasic Screening/Testing: An Advocate's Perspective."

58 Garfield, "Multiphasic Health Testing and Medical Care as a Right."

59 Sidney R. Garfield, "The Delivery of Medical Care," *Scientific American* 222 (1970): 15–23.

60 Malcolm S. M. Watts, "AMHTS and the Health Care System," in AMHTS Advisory Committee to the National Center for Health Services Research and Development, *Automated Multiphasic Health Testing and Services: Volume 3: Proceedings of the Invitational Conference on AMHTS* (Washington, DC: US Dept. of Health, Education, and Welfare, 1970), 3–9, at 3.

61 Jerrold S. Maxmen, *The Post-Physician Era: Medicine in the Twenty-First Century* (New York: John Wiley & Sons, 1976).

62 "Better Prepare for the Machinicare Era!" 2, unpublished MS, *Medical Economics*, 1967, box 4.12, folder 7, MFCC. Later version published as Richard C. Bates, "Get Ready for Mass Health Screening!" *Medical Economics*, October 16, 1967, 75–83; rebuttal "Mass Health Screening: The Experts' Views," *Medical Economics*, October 16, 1967, 84–85.

63 Arnold Relman, "The New Medical-Industrial Complex," *New England Journal of Medicine* 303 (1980): 963–70.

64 "Multiphasic Health Screening: '. . . A Fad, a Gimmick, or the Significant Innovation,'" *Group Practice* (March 1970): 7–11.

65 "The Computers That Will Keep You Healthy," 5.

66 "The Computers That Will Keep You Healthy," 11.

67 "The Computers That Will Keep You Healthy," 4–5.

68 Morris F. Collen, "Preventive Medicine and Automated Multiphasic Screening," in *9th IBM Medical Symposium* (Poughkeepsie, NY: IBM, 1968), 81–97.

69 Collen, "Health Care Information Systems: A Personal Historical Review," 129.

"Multiphasic health testing provided not only more information on more patients, but more information on each patient, thereby producing greater individualization of patient care."

70 J. L. Cutler, S. Ramcharan, R. Feldman et al., "Multiphasic Checkup Evaluation Study: 1. Methods and Population," *Preventive Medicine* 2 (1973): 197–206; L. G. Dales, S. Ramcharan, R. Feldman et al., "Multiphasic Checkup Evaluation Study: 3. Outpatient Clinic Utilization, Hospitalization, and Mortality Experience after Seven Years," *Preventive Medicine* 2 (1973): 221–35; E. G. Knox, "Multiphasic Screening," *The Lancet* 2 (1974): 1434–36; S. R. Garfield, M. F. Collen, R. Feldman et al., "Evaluation of an Ambulatory Medical-Care Delivery System," *New England Journal of Medicine* 294 (1976): 426–31; Loring G. Dales, Gary D. Friedman, and Morris F. Collen, "Evaluating Periodic Multiphasic Health Checkups: A Controlled Trial," *Journal of Chronic Disease* 32 (1979): 385–404; G. D. Friedman, M. F. Collen, and B. H. Fireman, "Multiphasic Health Checkup Evaluation: A 16-Year Follow-Up," *Journal of Chronic Disease* 39 (1986): 453–63.

71 For a review of this literature, see Paul K. J. Han, "Historical Perspectives in the Objectives of the Periodic Health Examination," *Annals of Internal Medicine* 126, no. 10 (1997): 910–17; South-East London Screening Study Group, "A Controlled Trial of Multiphasic Screening in Middle-Age: Results of the South-East London Screening Study," *International Journal of Epidemiology* 6 (1977): 357–63.

72 A. L. Cochrane and P. C. Elwood, "Medical Scientists Look at Screening," in C. L. E. H. Sharp and H. Keen, *Presymptomatic Detection and Early Diagnosis: A Critical Appraisal* (London: Pitman Medical Publishing Co., 1968), 359–66; Greene, *Prescribing by Numbers.*

73 J. M. G. Wilson and G. Jungner, *Principles and Practice of Screening for Disease* (Geneva: WHO, 1968).

74 E.g., J. E. Devitt, "Unnecessary Morbidity from Breast Surgery," *Canadian Medical Association Journal* 106, no. 1 (1972): 37–39; L. R Tancredi and J. A. Barondess, "The Problem of Defensive Medicine," *Science* 200, no. 4344 (1978): 87—82; Ilana Löwy, *Preventive Strikes: Women, Precancer, and Prophylactic Surgery* (Baltimore: Johns Hopkins University Press, 2009); Robert Aronowitz, *Unnatural History: Breast Cancer and American Society* (Cambridge: Cambridge University Press, 2013).

75 Ivan Illich, *Medical Nemesis: The Expropriation of Health* (New York: Pantheon, 1976).

76 H. R. Oldfield, Jr., "Automated Multiphasic Health Testing: A Diagnosis in Three Parts," *Journal of Clinical Engineering* 3, no. 2 (1978): 1–5, at 4. See also box 1, folder 3, MFCC.

77 Oldfield, Jr., "Automated Multiphasic Health Testing: A Diagnosis in Three Parts."

78 "Whatever Happened to Automated Multiphasic Health Testing?" 3, unpublished MS, box 5, folder 1, MFCC.

79 Wilson and Junger, *Principles and Practice of Screening for Disease.*

80 Morris F. Collen to Leonard Rubin, C. C. Cutting, R. Feldman, and D. Crawford, August 29, 1969, 1–2, memorandum, "Complaints about Multiphasic Program," box 4.2, folder 2.

81 "Only in Japan has there been a steady growth of AMHT centers as an accepted part of the country's health care delivery system, supported primarily by industry."

"Whatever Happened to Automated Multiphasic Health Testing?" 3, unpublished MS, box 5, folder 1, MFCC.

82 Oldfield also accused the AMA lobby of opposing federal funding for AMHT. "Many physicians in the United States have rejected Automated Multiphasic Health Testing" he complained, ". . . because the AMA lobby was successful in eliminating the U.S. Public Health Service program of the late '60s and early '70s to the point where the four large research AMHT centers—New Orleans, Brooklyn, Providence, and Milwaukee—were abandoned just at the point when their research was about to begin." Homer R. Oldfield, Jr., "A New Health Management Information System for Industry," in Yasaka, ed., *Progress in Health Monitoring (AMHTS): Proceedings of the International Conference on Automated Multiphasic Health Testing and Services, Tokyo, October 4–6, 1980*, 301. Advocates of AMHT services in the United States, including Collen and Oldfield, "have had hard going in the U.S. and have had to proceed entirely without Government funding for either system development or epidemiological research," and because "most insurance carriers refuse to reimburse the patient for such procedures though they will cheerfully pay thousands of dollars for the subsequent surgical or therapeutic procedures." "Accordingly," Oldfield continued, "it only survived in the United States in some health maintenance organizations (HMOs) and in some industries." Morris F. Collen, "AMHTS—Past, Present, & Future," 2, paper presented at IHEA Tokyo Conference and 22nd JMHTS Conference, Tokyo, May 1994, box 2.3, folder 3, MFCC. Takuji Miwa, "Prehistory and Institution of Ningen Dokku in Japan," in Miwa, ed., *Progress in Health Monitoring*, 203–13.

83 Yukito Shinohara, "Message from the President," July 2016, Japan Society of Ningen Dock, https://www.ningen-dock.jp/en/society/message.

84 Norio Sasamori, "The Significance of Human Dry Dock Activities Relative to the Prolonged Average Life Span of the Japanese," *Japan Hospitals* 6, no. 7 (1987): 59–64.

85 Collen, "AMHTS—Past, Present, & Future," 6.

86 Collen, "AMHTS—Past, Present, & Future," 9.

87 Collen, "AMHTS—Past, Present, & Future," 8.

88 Collen, "Health Care Information Systems: A Personal Historical Review."

89 See, for example, H. Gilbert Welch, Lisa M. Schwartz, and Steven Woloshin, *Overdiagnosed: Making People Sick in the Pursuit of Health* (Boston: Beacon Press, 2011).

Conclusion

1 Erika Frye and Fred Schulte, "Death by a Thousand Clicks: Where Electronic Medical Records Went Wrong," *Fortune*, March 18, 2019.

2 US Census statistic on broadband internet access: https://www.census.gov/newsroom/press-releases/2021/computer-internet-use.html. Smartphone use: https://www.pewresearch.org/internet/fact-sheet/mobile/.

3 Shift from paper to computer records: https://www.cdc.gov/nchs/data/nhsr/nhsr115.pdf; https://www.cdc.gov/nchs/products/databriefs/db187.htm.

4 https://www.epic.com/about.

5 On the masculinization of computer programming in the United States and United Kingdom in the late twentieth century, see Nathan Ensmenger, *The Computer Boys Take Over: Computers, Programming, and the Politics of Technical Expertise* (Cambridge, MA: MIT Press, 2012); Mar Hicks, *Programmed Inequality: How Britain Discarded Women Technologists and Lost Its Edge in Computing* (Cambridge, MA: MIT Press, 2017).

6 Mark Eisen, "Epic Systems: Epic Tale," *Isthmus*, June 20, 2008, https://isthmus.com; "Q&A: Epic CEO Faulkner Tells Why She Wants to Keep Her Company Private," *Modern Healthcare*, March 14, 2015, https://www.modernhealthcare.com.

7 "Q&A: Epic CEO Faulkner Tells Why She Wants to Keep Her Company Private"; Dave Zweifel, "Plain Talk: Doctor Left Enduring Mark on Madison," *Capitol Times*, July 20, 2018, https://madison.com.

8 Katelyn Ferral and Erik Lorenzsonn, "Her Way: Epic Systems CEO Talks about Trusting Her Vision," *Capitol Times*, April 12, 2017, https://madison.com; Tracy Kidder, *The Soul of a New Machine* (New York: Little, Brown, 1981).

9 *Mother Jones* reported that Faulkner earned a seat on the President's Council on Healthcare Information Technology, from which she helped shape the contours of new legislation. Patrick Caldwell, "We've Spent Billions to Fix Our Medical Records and They're Still a Mess: Here's Why," *Mother Jones*, November–December 2015, https://www.motherjones.com.

10 Hospital adoption of electronic health records: http://www.healthit.gov/sites/default/wp-content/uploads/data-brief/2014HospitalAdoptionDataBrief.pdf, referenced in Caldwell, "We've Spent Billions to Fix Our Medical Records and They're Still a Mess: Here's Why." For a broader analysis of the conception and implementation of HITECH, see Robert Wachter, *The Digital Doctor: Hype, Harm, and Hope at the Dawn of Medicine's Information Computer Age* (New York: Simon & Schuster, 2017).

11 Caldwell, "We've Spent Billions to Fix Our Medical Records and They're Still a Mess: Here's Why;" Joseph November, "Ask Your Doctor . . . about Computers," *IEEE Spectrum* (n.d., MS shared with author).

12 Robert Morrow, *Sesame Street and the Reform of Children's Television* (Baltimore: Johns Hopkins University Press, 2006); Michael Davis, *Street Gang: The Complete History of Sesame Street* (New York: Viking, 2008); Jill Lepore, "How We Got to Sesame Street," *New Yorker*, May 4, 2020; David Kamp, *Sunny Days: The Children's Television Revolution That Changed America* (New York: Simon & Schuster, 2020).

13 Note that after 2015, these new episodes have been first aired on the Home Box Office cable channel and only later made available through PBS.

14 Kamp, *Sunny Days*.

15 Lepore, "How We Got to Sesame Street."

16 Sanjeev Arora, Karla Thornton et al., "Outcomes of Treatment for Hepatitis C Virus Infection by Primary Care Providers," *New England Journal of Medicine* 364 (2011): 2199–207.

17 David Barash, "Project ECHO: Force Multiplier for Community Health Centers," *Health Affairs*, July 20, 2015.

18 J. Agley, J. Delong, A. Janota et al., "Reflections on Project ECHO: Qualitative Findings from Five Different ECHO Programs," *Medical Education Online* 26, no. 1 (2021).

19 Rama A. Salhi, Mashid Abir, and Bisan A. Salhi, "No Patient Left Behind: Considering Equitable Distribution of Telehealth," *Health Affairs*, April 20, 2021.

20 Marshall McLuhan, *Counterblast* (Toronto: McClelland & Stewart, 1969), 5.

21 John L. S. Holloman, Jr., MD, "Future Role of the Ghetto Physician," in Norman, ed., *Medicine in the Ghetto* (New York: Appleton-Century-Crofts, 1969), 144.

INDEX

Page numbers in italics refer to figures.

ABC, 106, 147

access: automation and, 214–15, 220, 222–28, 238–39; barriers to, 3–4, 131, 153–54, 165, 171; community access television (CATV), 5, 106, 146–48, 159–60, 173; computers and, 180, 184–88, 200–201, 205; equity in, 3–4, 145, 148, 176, 222, 246, 249–52, 255; ethnicity and, 15, 138–39, 153, 238, 248, 252; healthcare disparities and, 3–4, 131, 141–50, 153–55, 158–59, 165–77, 238, 242, 248–52, 255; insurance, 4; internet, 242; medium of care and, 242, 246, 248–56; pagers and, 83–84, 93–99, 103–4; race and, 3–4, 30, 32, 139, 144, 148, 154–55, 158, 165, 171, 173, 242, 249; rural areas and, 5 (*see also* rural areas); telephones and, 5, 11, 14–16, 22, 26, 29–33, 43, 46; television and, 108, 119, 123, 125, 131, 139; urgent care, 2; wireless technology and, 52, 71, 74, 76–77

accountability, 28, 100–101, 104, 246, 251

Adrian, Edgar Douglas, 58

Adult Health Protection Act, 221–22

advertising, 246; Archer character, 79–81, 95–97, 104; pagers and, 84, *91*, 100–102; pharmaceutical, 190; telephones and, 26, 30, *31*; television and, 105; wireless technology and, 56

Advisory Task Force on CATV and Telecommunications, 147

African Americans: academic medicine and, 110, 282n12; access and, 4, 155; Children's Television Workshop (CTW) and, 250; Du Bois, 154; Ferguson, 155; healthcare disparities, 154–60; Johnson, 110–13, 116, 118, 282n12; medium of care and, 250; Tuskegee Syphilis Study and, 125, 134, 162

"Afro Health Talk" (Jackson), 32

Agency for Healthcare Research and Quality (AHRQ), 252

Aircall, 84, 86, 96

Alaska Native Health Service, 90

algorithms: automation and, 214, 235, 237–39; Bayesian, 199; CliniCloud, 15; electrocardiograms and, 7; healthcare disparities and, 166; medium of care and, 244; multiphasic testing and, 214, 235–37; new technologies and, 8; telephones and, 15, 40–46; television and, 121; triage, 40–43, 46; wireless technology and, 74

Alinsky, Saul, 157

Allen, Scott I., 36

Alternative Media Center, 125–26

Amazing Stories magazine, 56

Amazon, 52, 76, 247

ambulances, 79–80, 85, 102, 117, 152
American Association for the Advancement of Science (AAAS), 62
American Association of Medical Colleges, 231
American College of Physicians (ACP), 45–46, 207, 220
American Hospital Association (AHA), 86, 181, 201
American Journal of Nursing, 71
American Journal of Public Health, 32
American Medical Association (AMA), 305n82; automation and, 215–16; code of medical ethics, 99; healthcare disparities and, 155, 157, 165, 176; *JAMA*, 14, 26, 64, 70, 75, 201; telephones and, 35, 37
American Medical Informatics Association, 238
American Medical News, 103
American Recovery and Reinvestment Act, 2
American Sociological Association, 99
AMHT suites, 225, 235–37, 305n82
AM radio, 81, 88, 97
Amsden, Sherman, 83–84
Andromeda Strain, The (Crichton), 119
answering machines, 187
answering services, 34, 47, 81, 83–84
Apple Health, 247
"Appreciation of the Telephone" (Firebaugh), 27–28
Archer, Eliot, 79–81, 95–97, 104
Archives of Internal Medicine, 196
Area Health Services Field Unit (AHSFU), 149–52, 170
Arizona Northern University, 171
artificial intelligence (AI), 11, 193, 199
asymptomatic disease, 49–50, 216–19, 227–234
AT&T, 96, 152, 279n51, 280nn54–55
Audio-Visual Superimposed ECG Presentation (AVSEP), 73–74
Automated Multiphasic Health Testing Center: AMHT suite sales, 225, 235–37, 305n82; Kaiser Permanente and, 212–17, 225–26, 235–37, 305n82; Oldfield and, 235–36; reimbursement for, 235–36
Automatic Nurse Call, 93
automation: access and, 214–15, 220, 222–28, 238–39; algorithms and, 214, 235, 237–39; American Medical Association (AMA) and, 215–16; assessment of, 236–37; blood pressure, 216; Collen and, 216–30, 233, 236–39, 299n5, 301n22; computers and, 212–30, 233, 235, 237–39; concern over, 220–24; cost of, 230–34; decline of, 255; diagnostics and, 214–21, 226–27, 229, 234; efficiency and, 217, 220–21, 227–30, 238; electrocardiograms (ECGs), 212, 216; equity and, 222; ethnicity and, 223, 238; examinations, 211–13, 219–22, 226, 235; failures of, 239; fee-for-service, 227, 231; Garfield and, 215–17, 227–28, 233; health records, 237; IBM and, 212, 217, 219, 224, 227–28; information technology (IT) and, 227; insurance and, 224, 228, 230; Kaiser Permanente and, 212–37; mammograms, 212, 228, 234, 237; markets and, 211, 215, 224–27, 231, 234–36, 239; multiphasic testing, 212–37, 255; networks and, 237; nurses and, 225, 229; physical exams, 211–12, 219, 221, 230; prevention and, 212–28, 231, 233–39, 247–48; promise of, 219–24; "Push-Button Hospital" and, 217; remaining elements of, 235–39; storage and, 212; symptoms and, 216, 227, 234; telemedicine and, 222, 236; telemetry and, 221; X-rays, 212, 216, 226; Zworykin and, 211, 217
autonomy: computers and, 188; healthcare disparities and, 163, 165–69, 176; nurses and, 42, 45, 47, 108, 119, 163, 165, 169, 176, 247; pagers and, 80–82, 96, 98, 102; telephones and, 42, 45, 47; television and, 108, 119

Baldwin, James, 23, 25, 28, 154, 157
Bales, Ed, 97, 280n54
Baltimore Afro-American magazine, 30, 32
Barnett, Guy Octo, 182, 201–4, 206, 244
Baruch, Jordan, 195, 201–2, 208–9
Bashshur, Rashid, 126, 289n30
Bates, Richard, 230

Bayesian algorithms, 199
beepers, 81, 98–104
"Beepers" (Sir Mix-A-Lot), 103
Bell, Alexander Graham, 5, 13, 19, 21, 36
Bellboy, 83, 96, 100–102
Bell Labs, 152, 279n51, 280n55
Bell telephone, 14, 25, 37, 96
Ben Casey (TV Show), 125
Benschoter, Benny, 109
Benschoter, Reba: background of, 109; Nebraska Psychiatric Institute (NPI) and, 109–10, 123, 149; New York Academy of Sciences and, 105, 112, 116, 136; psychology and, 109–10, 112, 149; television and, 105–19, 123, 126, 136–37, 141, 149, 248, 283n15; University of Nebraska Medical Center (UNMC) and, 106, 110–12; Wittson and, 109–13, 116, 118–19, 149
Berger, Hans, 55, 58
Bergman, Stephen, 98
Bernard, Claude, 17, 55
Best Buy, 13, 15, 48
Beth Israel Hospital, 89–90, 98
Big Brother, 123
Big Science, 53
Billings, John Shaw, 183–84
biomedical communications, 106–110, 149
Biomedical Telemetry (Caceres), 76
Bird, Kenneth: Crichton and, 119; healthcare disparities and, 141–44, 149, 152, 176; Kerrigan and, 123, 126; Knowles and, 117, 144; Krainin and, 121–22, 127; Logan Medical Station and, 117–23; McLuhan and, 107, 120, 176; NASA and, 152; television and, 107–8, 117–23, 126–28, 131, 137, 141–42, 288n24
Bizmatic computers, 180, 194
Black Center for Strategy and Community Development, 160
Black Panthers, 125, 162
Black people: access to healthcare, 4, 30, 32, 139, 144, 148, 154–55, 165, 242, 249; "Afro Health Talk" and, 32; civil rights and, 125, 155–56, 164–65, 255; ghettos and, 147–48, 154–60, 163–65, 168, 173–75, 256; Harlem and, 3, 144, 154, 157–64; Harvard study and, 154–56; healthcare disparities, 144, 148, 154–65, 173; medium of care and, 242; Moynihan Report and, 157; New York State Ghetto Medicine Program and, 158, 160, 163; *Sesame Street* and, 250; telephones and, 30, 32. *See also* African Americans
Blake, Clarence John, 19–21, 36
blood pressure: at-home cuffs for, 1; automation and, 216; CliniCloud and, 47; kymograph and, 17; telephones and, 39; wireless technology and, 55, 69
blood tests, 33, 194, 216
Bolt, Beranek & Newman, 201–2
Boston Globe, 89, 118, *124*, 156–57
Boston Medical and Surgical Journal, 17, 26
Bowditch, Henry Pickering, 17–18
Boys' Life magazine, 29
brainwashing, 60
brain waves: electroencephalograms (EEGs), 7, 55, 58–59, 61; Holter and, 59–61; mind control and, 60; telephones and, 17, 35; Thought Recorder and, 57, *58*; wireless technology and, 52, 55, 57–61, 78
Bramwell, Byrom, 20–21
Breslow, Lester, 216, 225, 300n9
British Broadcasting Corporation (BBC), 58
British Medical Association, 32
British Medical Journal, 14, 21, 23
British Society for Psychical Research, 58
British Telecom, 26
broadcasting physiological data, 49–58, 61–64, 67–75, 78
Brobeck, A. L., 23, 25, 28
Brodman, Keeve, 189–90, 193–94, 296n38
Brookdale Hospital Center, 222–23
Brown, Jeffrey L., 40–41
Brown, William, 20
burnout, 8, 44, 98, 104, 252–54
Businessweek, 64

Cable News Network (CNN), 173
cable television, 5, 106, 126, 143–52, 160–69, 174–76, 252
Caceres, Cesar: background of, 70; *Biomedical Telemetry*, 76; Public Health Service and, 70, 73; wireless technology and, 69–77; Zworykin and, 70

Calandra, Robert, 2
Caldwell, Patrick, 245
California Department of Public Health, 218
California Medical Association, 228
California State Journal of Medicine, 27–28, 30
cancer: breast, 228, 234; cervical, 155, 234; cloud-enabled pills and, 77; colorectal, 237; drugs for, 9; false positives and, 226–27; gastric, 226; precancerous lesions, 216; prostate, 234; War on Cancer and, 53
cardiology, 35, 62, 70, 73, 176
Carnegie Foundation, 155, 248
Case Western Reserve University, 205, 207
CBS, 106–7, 118, 121, 147, 167
Center for Medical Electronics, 7, 65, 179–80, 188–89, 194
Chapman, Stu, 38
Children's Television Workshop (CTW), 248–51
cholesterol, 216, 234
Chow, Rita, 71
chronic disease, 46, 49, 62, 70, 73, 214–218, 233, 252
cigarettes, 30, *31*, 87–88
Cincinnati Lancet and Clinic magazine, 13–14
Cisco Systems, 103
citizens band (CB) radio, 85
citizenship, 4
Citizens' Radio Corporation, 85
civil rights, 125, 155–56, 164–65, 255
Claremont General Hospital, 149
class: middle, 5, 15–16, 30, 103, 133, 138–39, 250; pagers and, 103; race and, 5, 15, 134, 139, 248; Skibbins and, 50, 63; telephones and, 15–16, 30; wireless technology and, 50, 63
CliniCloud: Best Buy and, 13, 15, 48; blood pressure and, 47; demise of, 48; social media and, 48; stethoscopes and, 13; telephones and, 13–15, 47–48; thermometers and, 13
cloud technology: automation and, 211, 235; CliniCloud, 13–15, 47–48; wearable technology and, 49; wireless technology and, 49, 77
Cold War, 50, 52, 60–61, 64, 70, 76–77, 185
Collen, Morris: American Medical Informatics Association and, 238; automation and, 216–30, 233, 236–39, 299n5, 301n22; Bates and, 230; Garfield and, 216–17, 228, 233; Japan and, 225–26; multiphasic testing and, 216–30, 233, 236–37; Oldfield and, 236; Training Program in Automated Multiphasic Screening and, 222; Zworykin and, 217
colon, 213, 237
Columbia space shuttle, 171
Committee on Telecommunications, 147
Commonwealth Fund, 252
communications technologies, 2, 5–9, 15–17, 22–25, 29–33, 70, 106, 109, 118, 133, 136, 146–7, 153, 164, 167–8, 217, 253
Communications Technology for Urban Improvement, 167
Communist Party, 60
community access television (CATV), 5, 106, 146–48, 159–60, 173
community health medics (CHMs), 165–67
Complete Parents' Guide to Telephone Medicine, The (Brown), 40–42
computers: access and, 180, 184–88, 200–201, 205; algorithms, 7–8, 15, 40–46, 74, 121, 166, 193–96, 199, 214, 235, 237–39, 244; automation, 212–30, 233, 235, 237–39; autonomy and, 188; Baruch and, 195, 201–2, 208–9; BB&N systems, 201–2; Bizmatic, 180, 194; cloud technology, 211, 235; community access television (CATV), 5, 106, 146–48, 159–60, 173; diagnostics and, 8, 180–82, 188–99, 205, 208, 210, 218, 221; Eclipse, 244; efficiency and, 196–97, 200, 205, 208–9; electrocardiograms (ECGs) and, 7, 200; emergencies and, 182; ENIAC, 294n10; examinations and, 8, 179, 194, 200, 206; failures, 181, 186, 201–2, 209–10; Honeywell 800, 184; Hospital Computer Project, 201–2; IBM, 182, 187, 189, 195–96, 199, 201, 209, 294n10; impact of, 11, 243, 246; information technology (IT),

183; INTERNIST-I, 199; mainframe, 5, 35, 181–89, 194, 199, 201, 208–9, 219, 235, 241–46, 254; Medical Literature Analysis and Retrieval System (MEDLARS), 184–87, 209–10; Medical PROBE, 201; medium of care and, 241–44, 254–55; MYCIN, 199; networks of, 5, 16, 36, 181, 185, 201, 241; nurses and, 201–3; pagers and, 82; paper and, 179–89, 193, 202–6, 208–10, 242; physical exams and, 191, 194, 200; prevention and, 213; psychiatry and, 189, 198; punch cards, 36, 182, 194, 212, 217–18, 222, 224; records and, 180–82, 200–210, 243–44; resistance to, 188–99; Role of Computers in Biology and Medicine commission, 181; screen fatigue, 179; screening and, 193, 214–37; Slack and, 243–44; software, 138, 208, 244; storage, 180, 182–88, 200; symptoms and, 181, 188–96, 206; telemedicine and, 189; telemetry and, 189; telephones and, 16, 35–36, 181, 187, 193, 204; Turing test for, 190; Zworykin and, 179–81, 187–88, 194, 199–200, 205–8
Cook, Thomas D., 248–49
Cooney, Joan Ganz, 248
Cornell University, 40, 158, 189, 194, 197–99
COVID-19, 1, 4, 241–42
Cowan, Ruth Schwartz, 28
Crichton, Michael, 119
Cunningham, Nicholas, 163–64
Current List (NLM), 183
cybernetics, 72

Dahlberg Company, 90, 93, 279n47
d'Arsonval, Jacques-Arsène, 18
Dartmouth-Hitchcock Medical Center, 149
data: crude or raw, 190, 196; deluge of 73–78, 182–87, 193, 219–21, 238–39; health as, 7, 40, 50, 56, 62, 69, 71–73, 180, 200–206, 211–19, 233, 238–39, 245–46; privacy of, 51–52; storage of, 16, 33–35, 73–78, 180
Data General, 244
Dataphone, 35–36, 46
Davis, B. J., 195–96, 297n56
dead spots, 89
Dee, Cinnamon, 163
dehumanization, 7, 16, 38–39, 98
De Long, A. J., 28–30, 47, 83, 268n43
demonstration projects, 3, 78, 144–45, 153, 169–77, 222, 236, 242, 247–49
Department of Housing and Urban Development (HUD), 167
dermatology, 121, 127, 176, 191, 198
Deuschle, Kurt W., 158–59, 164
diagnosis: automation and, 214–21, 226–27, 229, 234; Baruch and, 195, 201–2, 208–9; computers and, 8, 180–82, 188–99, 205, 208, 210, 218, 221; Davis and, 195–96; Engle and, 197–99; false positives, 226–27; healthcare disparities and, 141–42, 159, 167, 176; letters and, 10; multiphasic testing, 212–37, 255; Nash and, 190–93, 198–99; new technologies and, 8–11; pagers and, 104; physical exams, 7–8, 41, 43, 107, 117–18, 121, 191, 194, 200, 211–12, 219, 221, 230; radio, 11, 61–62, 69, 107, 121, 142, 176, 256; screening, 1, 69, 193, 212–37, 255; telemetry, 69, 74, 189, 256; telephones and, 14, 16, 18–19, 22, 33, 37, 40, 43, 46; television and, 107, 111, 113, 117–24, 127; Training Program in Automated Multiphasic Screening, 222; true positives, 227; wireless technology and, 61–62, 69, 74; Zworykin on, 8
Dictaphones, 33–34, 36
dictation, 33–34, 57
Differential Diagnosis (Harvey), 191
Digital Health Innovation Plan, 47–48
disability, medical technologies and, 29–30, 125, 233, 268
disparities, health. *See* health disparities
disruptive technology, 8, 10, 13, 132, 146
DIY approaches, 64, 69, 71–72, 77
Doctor on Demand, 13
doctors: Archer character and, 79–81, 95–97, 104; epistolary medicine and, 10; evasion by, 82; healthcare disparities and, 141–42, 153, 166, 169; human connection to patients, 7, 11, 24, 43, 46; opposition to Zworykin, 7–8; power relations, 129–36; public image of, 123, 125; telepresence, 121–22, 127–37. *See also specific technologies*

Doing Better and Feeling Worse (Knowles), 142–43
Doppler flow detectors, 72
Dr. Kildare (TV show), 125
drug dealers, 81, 100, 103
dry dock, 225, 303n52
Dwyer, Jeanne, 39
Dwyer, Walter, 39
dystopian concerns, 47, *124*, 188, 200, 253–54

Ebony magazine, 156–57
Eclipse minicomputer, 244
Edison, Thomas, 21
Educational Testing Service, 249
efficacy, 116, 145, 164, 234
efficiency: automation and, 217, 220–21, 227–30, 238; computers and, 196–97, 200, 205, 208–9; healthcare disparities and, 164, 168; medium of care and, 253–54, 256; pagers and, 98; telephones and, 21–22, 24, 40–41, 136; wireless technology and, 50, 62–63
Einthoven, Willem, 55
El Camino Hospital, 202–3
Electrical Experimenter magazine, 56–57
electrocardiograms (ECGs): algorithms and, 7; automation, 212, 216; computers and, 7, 200; multiphasic testing, 212, 216; radioelectrocardiograms (RECGs), 50, *51*, 59, 61–64, 67, 71–72; telephones and, 34–35; television and, 118; wireless technology and, 50, *51*, 55, 71, 74–76; Zworykin and, 7
"Electrocardiography by Do-It-Yourself Radiotelemetry" (Fascenelli), 71
electroencephalograms (EEGs), 7, 55, 58–59, 61
electronic health record (EHR), 2–3, 180–81, 200–205, 238–53
electronic leash, 98–101, 103–4
Elektrenkephalogramm technique, 58
Eliasoph, Ira, 89
emergencies, 3; air, 117; ambulances, 79–80, 85, 102, 117, 152; blocked vehicles and, 117; computers and, 182; constraints of telemedicine and, 2; healthcare disparities and, 148; HITECH and, 245; pagers and, 80–88, 97, 100; syncope and, 49; telephones and, 14, 26, 44; television and, 106, 118; wireless technology and, 49
Emory University, 207
employers, 51–52, 76, 225–26, 253
empowerment: failed promises of, 255; Faulkner on, 244; healthcare disparities and, 143, 146, 166; new technologies and, 8; telephones and, 13–15, 40, 43, 45–47; television and, 108–9, 125–26, 137–38
Engle, Ralph, 197–99
English, Joseph T., 165
ENIAC computer, 294n10
entrepreneurship, 29–30, 47, 65, 224
Epic Systems, 204, 243–47
epistolary medicine, 10
equity: access and, 3–4, 145, 148, 176, 222, 246, 249–52, 255; automation and, 222; healthcare disparities and, 145, 148, 175–76; increasing, 5, 148, 222; markets and, 2–3, 246; medium of care and, 244–52, 255; Project ECHO and, 251–52; proxemics and, 134; *Sesame Street* and, 248–50; Telemedicine for Health Equity Toolkit, 252
ER scheduling software, 244
ethics, 9, 22, 99, 125
ethnicity: access and, 15, 138–39, 153, 238, 248, 252; automation and, 223, 238; healthcare disparities and, 153–54; immigrants, 134, 158; medium of care and, 248, 252; socioeconomic divides and, 139
examinations: automation and, 211–13, 219–22, 226, 235; computers and, 8, 179, 194, 200, 206; Harvey on, 191; malpractice and, 39; power issues and, 108; telephones and, 39; television and, 113, 117–18, 121; theory of nursing and, 43; Zworykin and, 7
Executive Physical programs, 225
experimental media, 5–6, 17–22, 111, 235

expertise, 14–15, 35, 108, 125, 131, 159, 188–190, 289n30
expert system, 190–94, 199, 219

Facebook, 47
"Face of God" problem, 129, 133, 169
false positives, 226–27
Fanon, Frantz, 157
Farrar, John T., 65–67
Fascenelli, Frederick W., 71–72
Faulkner, Judith, 243–45
fax machines, 34, 36
FBI, 60
fear of technological change, 6, 10–17, 20–24, 41, 45–46, 52, 58, 61, 76–78, 118, 124, 134, 137, 188, 197, 210, 214, 220, 230, 247, 251–53
Federal Communications Commission (FCC), 61, 96, 144, 147–48, 173
fee-for-service market, 227, 231
Ferguson, Lloyd, 155
Firebauch, Ellen "Mary," 26–28, 45
Fitbits, 52
Five Patients (Crichton), 119
Flint water supply, 245
FM radio, 9, 49–50, 59, 69, 81, 88, 97
Fogarty, John, 220
Forbes magazine, 243
forgetting, 6, 9, 16, 46–47, 254
Fortune magazine, 65
France, 17–18, 225
French's Differential Diagnosis, 180
Fridenberg, Percy, 277n9
Friedsen, Elliott, 131–35, 138

Galvani, Luigi, 54–55
Galvin Manufacturing Corporation, 86
garbage in, garbage out (GIGO), 221
Garfield, Sidney, 215–17, 227–28, 233
gastrointestinal tract, 67, 226, 302n51
GE Healthcare, 204
Geiger counters, 72
gender, 28, 93, 95, 134, 241
General Medical Services Board, 32
Gengerelli, Joseph, 54, 56, 64, 75
George Washington University, 70
Germany, 17–18, 55, 67, 225
Gernsback, Hugo: Holter and, 60, 77, 241; publications, 56–57; radio and, 56–60, 65, 72, 77, 241; Tesla and, 57; Thought Recorder, 57, *58*, 272n22; wireless technology and, 56–60, 65, 72, 77, 241
ghetto medicine: Deuschle and, 158–59, 164; Harlem and, 3, 144, 154, 157–64; medical care, 154–57, 164–65, 256; New York State Ghetto Medicine Program, 158, 160, 163, 290n61; poverty and, 147–48, 154–60, 163–65, 168, 173–75, 256; rural, 147, 154–56, 164–65, 168, 173, 175; Wagner Homes projects, 159–66, 174, 176. *See also* medical ghettos
Gilbert, Frederick I., 231
Gitelman, Lisa, 15–16
Gitman, Leo, 223
GlaxoSmithKline, 68
glucose, 77, 212
Goffman, Erving: power relations and, 129–35; *Presentation of Self in Everyday Life*, 129–30; Rockoff and, 129–30; television and, 129–35; total institution of hospital and, 98
Goldmark, Peter, 167
Google, 104, 247
Greist, John, 244
Grim Beeper, 98–100, 104
Gross, Al, 85–86
Group Health Cooperative of Puget Sound, 42, 228
Grundrisse (Marx), 24
gynecology, 35, 207, 213, 216, 234, 237

Hall, Edward, 133–35
H&H Automat, 212
Handie-Talkies, 86–89, *91*, 278n37
Hanna-Attisha, Mona, 245
Harlem: New York State Ghetto Medicine Program, 158, 160, 163; poverty and, 3, 144, 154, 157–64
Harvard Medical School, 119, 154, 156–57, 244
Harvey, A. McGehee, 191–93
headaches, 39, 65, 212

Head Start program, 249
healthcare crisis, 141, 144
Health Care Technology Division (HCTD), 126, 153–54
health disparities: access and, 3–4, 141–50, 153–55, 158–59, 165–66, 168–77, 238, 242, 248–52, 255; AHSFU and, 149–52; algorithms and, 166; American Medical Association (AMA) and, 155, 157, 165, 176; autonomy and, 163, 165–69, 176; Bird and, 141–44, 149, 152, 176; close/far phase and, 133; diagnostics and, 141–42, 159, 167, 176; doctors and, 141–42, 153, 166, 169; efficiency and, 164, 168; emergencies and, 148; equity, 145, 148, 175–76; ethnicity and, 153–54; "Face of God" problem and, 129, 133, 169; ghettos and, 147–48, 154–60, 163–65, 168, 173–75, 256; Hall on, 133–35; Hill-Burton Act and, 142; increasing, 4–5; information technology (IT) and, 143, 146, 148, 153, 155; intensive care units, 142, 150; Kerner Commission and, 143–44, 148, 154, 156, 159; Knowles and, 142–44; markets and, 146–47, 163, 173–74; mortality rates, 141, 233, 237; neglect, 143; New York State Ghetto Medicine Program, 158, 160, 163; nurses and, 160, 162–66, 169, 176, 255; operating rooms, 142, 150; patient empowerment and, 143, 146, 166; power relations and, 129–36; prevention, 141; psychiatry and, 149, 176; race and, 144–48, 153–65, 168–75, 238, 255; radio and, 142, 150, 152, 169, 176; Rockoff and, 152–53, 166–70, 174, 177, 292n79, 293n93; rural areas and, 3, 142, 144–47, 149–50, 153–56, 164–76, 215, 222, 242; telehealth and, 3, 142, 145, 154, 162, 167, 169; telemedicine and, 141–45, 149, 152–53, 162, 164, 166, 168–71, 174–77; television and, 121–22, 127–37, 141–50, 153–54, 159–63, 166–69, 173–76; Wagner Homes projects, 159–66, 174, 176; wearable technology and, 175
Health Information Technology for Economic and Clinical Health (HITECH), 2, 244–47
Health Information Technology Policy, 245
Health Insurance Plan of Greater New York, 228
Health Maintenance Act, 228
health maintenance organizations (HMOs), 42, 228
heartbeat: Holter monitor, 49–52, 63, 73–78, 241; telephones and, 16, 21; television and, 117
heart disease, 62, 73, 75, 216
Heart Disease (White), 62
hematology, 127, 194–99
Henry, John, 186, 187
Hidden Dimension (Hall), 133
Hildreth, John L., 24–25, 29
Hill-Burton Act, 142
Holloman, John, 164, 174–75, 255–56
Holter, Norman "Jeff": background of, 52–54; brain waves and, 59–61; broadcasting data and, 50–51; funding of, 53; Gengerelli and, 54, 56, 64, 75; Gernsback and, 60, 77, 241; International Conference and Exhibition on Medical Electronics and, 65, 67; monitor of, 49–52, 63, 73–78, 241; psychiatry and, 60–61; psychology and, 54; radioelectrocardiograms (RECGs) and, 62, 64, 67, 72; radioelectroencephalograms (REEGs) and, 59; White and, 54, 62; Zworykin and, 67–69, 75–77, 241
Holter Research Foundation, 62, 73
Holter Satchel Clubs, 62–63
Home Box Office (HBO), 173
Honeywell 800 computer, 184
Hoover, Edgar J., 60
Hospital (Wiseman), 138
Hospital Computer Project, 201–2, 244
Hospital for Sick Children, 34
House of God, The (Bergman), 98, 100
hubris, 108, 131, 138, 182
human connection, 7, 11, 24, 43, 46
Human Services Computing, 244
hypertension, 234

IBM: automation, 212, 217, 219, 224, 227–28; computers, 182, 187, 189, 195–96, 199, 201, 209, 294n10; Davis and, 195–96;

dominant market position of, 247; 1440 Data Processing System, 217, 219, 300n17; Kaiser Permanente and, 212, 217; Medical Symposia of, 182, 209; Model 650, 201; Model 704, 189, 196; Tanimoto and, 195–96; Taylor and, 195, 199
Illich, Ivan, 157, 234–35
imaging, 193, 206
immigrants, 134, 158
Index Medicus (NLM), 181, 183
Indiana University Medical Center, 33
Indian Health Service, 169, 171–73, 203
inflation, 40, 141
information technology (IT): automation and, 227; computers and, 183; health-care disparities and, 143, 146, 148, 153, 155; HITECH, 2, 244–47; promises of, 3–4, 254; resistance to new, 8, 17; telephones and, 17, 36
information theory, 72
infrastructure, 17, 33–35, 74–77, 90, 97, 106, 171, 246, 251
insulin, 77–78
insurance: access, 4; annual physical exam and, 230; automation and, 224, 228, 230; employee information, 51; fee-for-service, 227, 231; Medicaid, 141, 158, 215, 222–23, 227, 230; Medicare, 141, 215, 220, 222–23, 227, 230, 236; national, 228; Permanente Health Plan, 215–16; prepaid plans, 231; wireless technology and, 51
Integrated Medical and Behavioral Life Monitoring System (IMBLMS), 169, 171–72
intensive care units (ICUs), 70, 74–75, 142, 150
INTERACT program, 166
Interex 1024, 224
interface, 4, 74, 94, 126, 134, 185, 200, 207–9, 214, 218, 237–38, 244–47, 294
International Conference and Exhibition on Medical Electronics, 64–67
International Longshoremen & Warehousemen's Union, 215–16
internet, 11, 146, 241–42, 244
Internet of Things, 11
INTERNIST-I computer, 199
Introduction to Telemedicine, An (Park), 126

Japan, 67, 225–28, 236–37, 303n52, 304n81
Jetsons, The (animated sitcom), 35
Jewish Hospital of Philadelphia, 34
Johns Hopkins University, 191
Johnson, Lyndon B., 143–44, 147
Johnson, Nicolas, 148
Johnson, Van, 110–13, 116, 118, 282n12
Journal of the American Medical Association (JAMA), 14, 26, 64, 70, 75, 201
Journal of the Indiana State Medical Association, 28–29
Justice, James, 171

Kain, Tom, 280
Kaiser Permanente: automation and, 212–37; computers and, 212–30, 233, 235, 237; facilities design of, 217–18; fee-for-service, 227, 231; Garfield and, 215–17, 227–28, 233; Health Plan, 215–16, 222; IBM and, 212, 217; market of, 224–27; Medicaid and, 215, 222–23, 230, 237; Medicare and, 215, 220, 222–23, 227; model of, 214–15, 220–23, 225, 230; multiphasic testing, 212–37, 255; Preventicare and, 215, 220–21, 227, 230, 236; Senate and, 215, 219–21, 227; Total Health and, 214, 227–28; Training Program in Automated Multiphasic Screening and, 222; X-rays, 212, 216, 226
Kansas State Tuberculosis Sanatorium, 34
Karolinska Institute, 67
Kennedy, John F., 64
Kerner Commission, 143–44, 148, 154, 156, 159
Kerrigan, Marie, 123, 126
Kidder, Tracy, 244
kidneys, 16
King, Martin Luther, Jr., 148
Knowles, John, 117, 142–44, 287n13
Komaroff, Anthony, 166
Krainin, Stanley, 121–22, 127

Laika, 71, 275n70
Lancaster, F. W. "Wilf," 186

Lancet, The (journal), 20, 32, 47
Land Policy and Planning Assistance Act, 168
Latino communities, access to healthcare, 4, 139, 144, 154, 162, 250
lead poisoning, 245
Ledley, Robert, 181–82
Legal Aspects of Medical Practice (Chapman), 38
legal issues: pagers and, 99; telephones and, 32, 37–38, *39*; wireless technology and, 61
letters, 10, 59–61, 123, 224
liberalism, 50, 145, 154–55, 174
Life magazine, 54
Lin, Andrew, 48
Lindberg, Donald, 204
Lindsay, John, 147, 157–58, 160
Lipkin, Martin, 194
Lister Hill Center for Biomedical Communications, 149
Lockheed Missile and Space Company, 149–50, 169, 171, 202
Lodge, Oliver, 58
Logan Medical Station, 117–23
Logoscope, 191–93
Long Island Jewish Hospital, 85–86
Look magazine, 223
Lopez, Rosemary, 170
Low End Theory (Tribe Called Quest), 103
Ludwig, Carl, 17, 55

machine learning, 193–97
Mackay, R. Stuart, 67, 72
magnetic disks, 33
magnetic tape, 36, 75, 185, 194, 201
malpractice, 16, 32–33, 37, 39, 134
mammography, 212, 228, 234, 237
Managed Care Magazine, 2
Manhattan Project, 53
Marey, Étienne-Jules, 17
markets: advertising, 26, 30, *31*, 56, 84, *91*, 100–101, 190, 246; AMHT suites, 225, 235–37, 305n82; automation and, 211, 215, 224–27, 231, 234–36, 239; cigarette, 30, *31*, 87–88; equity, 2–3, 246; fee-for-service, 227, 231; healthcare disparities and, 146–47, 163, 173–74; Holter devices and, 64; human connection and, 24; increase of telehealth, 2–3; medium of care and, 242, 245–50, 253; monopolies, 23, 25, 77, 96, 146, 148, 163, 173; multiphasic testing, 224–27; pagers and, 83–86, 89–90, 93, 96–98; pharmaceutical, 68–69, 106, 156, 190, 215, 224, 230; stakeholder differences, 6; telephones and, 24, 30; television and, 106; tobacco, 30, *31*, 87–88; wireless technology and, 64–65, 68–69, 73, 77. *See also specific companies*
Marshall, Carter L., 159–60
Marvin, Carolyn, 15
Marx, Karl, 24–25
Massachusetts General Hospital (MGH), 3; Barnett and, 182, 202–4, 244; computers and, 181–82, 201–4, 209; Knowles and, 117, 142–44; television and, 117–26
Maxmen, Jerrold S., 229–30
Mayo, 235
McDermott, Walsh, 158, 171
McGear, Reba, 42–45
McKendrick, John, 19
McLuhan, Marshall, 107, 120, 176, 252–53
Medequip Corporation, 224
media history, 6, 15, 243, 264nn7–8
Medicaid, 141, 158, 215, 222–23, 227, 230
Medical Data Screen, 189
Medical Economics magazine, 230–31, 233
medical ghettos, 154–57, 164–65, 256
Medical Information System, 202
Medical Literature Analysis and Retrieval System (MEDLARS): Automated Multiphasic Health Testing Center, 212–17, 226; computers and, 184–87, 209–10
Medical Nemesis (Illich), 234–35
Medical PROBE, 201
Medical Records, Medical Education, and Patient Care (Weed), 207
medical technology and device industry, 2–18, 52, 64–69, 224–25
Medical World News, 202
Medicare, 141, 215, 220, 222–23, 227, 230, 236
medication: digital pills, 77; mistimed, 49; radio pills, 52, 65–68, 72, 75, 77, 180

Medicine in the Ghetto (Norman), 154
Meditech, 204, 206, 244
medium of care: access and, 242, 246, 248–56; computers and, 241–44, 254–55; efficiency and, 253–54, 256; equity and, 244–52, 255; ethnicity and, 248, 252; health records and, 242–47; lack of neutrality, 242–43; markets and, 242, 245–50, 253; nurses and, 246–47, 255; prevention and, 247, 254; race and, 242, 248, 250, 252; symptoms and, 246; telehealth, 242, 250–52; telemedicine, 241–42, 247–52; Zworykin and, 241
MEDLINE, 210
mental health, 252; Benschoter and, 248 (*see also* psychiatry); early remote medicine and, 3; prisons and, 98, 116, 145; Wittson and, 109, 116. *See also* psychology
Merck Manual of Diagnosis and Therapy, 180
Mercury program, 71
Miami-Dade County Correctional System, 145
Miami Daily News, 60
microscopes, 18, 53, 168, 221
microwaves, 3, 28, 112, 118, 123, 126, 150
Middleton, William S., 196–97
Miller, Joseph, 117
Mills, Mara, 21
Milwaukee Journal, 61
mind control, 60
"Miniaturized Hospital Telemetering System," 68–69
MIT, 203
Mitchell, Donald H., 86
Mitchell, John, 280n55
MITRE Corporation/Urban Institute Symposium on Urban Cable Television, 147
Mobile Health Unit, 171
mobility, technology and, 86, 89, 93, 98, 102–3
Modern Electronics magazine, 56
Modern Hospital magazine, 279n50
monopolies, 23, 25, 77, 96, 146, 148, 163, 173
More Work for Mother (Cowan), 28
Morrisett, Lloyd, 248
Morse telegraph, 20, 277n10
mortality rates, 141, 233, 237
Mother Jones magazine, 245, 306n9
Motorola: Archer character, 79–81, 95–97, 104; Bales and, 280n54; Bellboy, 83, 96, 100–102; brand name of, 86; Dahlberg Company and, 90, 93, 279n47; dominant market position of, 247; Galvin Manufacturing Corporation and, 86; Handie-Talkies, 86–89, *91*; Hospital Communications Division, 90, 92–94; Pageboy, 79, 83, 96–98, 100–104; Radio Pagers, 87–94; Radio-Register, 81, 94; RADIO W-E-L-L, 90, 92; television and, 80, 92; Total Hospital Communications system, 80, 92–95; two-way radio and, 86–92, 97; Weinberg and, 89–90, 97, 99; wireless technology, 96–99, *101*, 103–4, 247; Wright and, 103
Motorola Annual Report, 97
Motorola Newsgram, 79, *80*, 87, *88*
Mount Sinai Hospital, 37, 82, 89–90, 158–63, 169, 176
Moynihan Report, 157
multiphasic testing: algorithms and, 214, 235–37; AMA and, 215–16; assessment of, 227–37; Breslow and, 216, 225; Collen and, 216–30, 233, 236–37; computers and, 212–30, 233, 235, 237; cost of, 227–34; description of procedure, 212–13; early adopters of, 222–24; electrocardiograms (ECGs), 212, 216; expansion of, 222–27; facility design for, 217–18; fee-for-service, 227, 231; Garfield and, 215–17, 227–28, 233; Interex 1024 and, 224; mammograms, 212, 228, 234, 237; market of, 224–27; Medicaid and, 215, 222–23, 230, 237; Medicare and, 215, 220, 222–23, 227; pap smears, 213, 216, 234, 237; physical exams, 211–12, 219, 221, 230; Preventicare and, 215, 220–21, 227, 230, 236; screening, 212–37, 255; Senate and, 215, 219–21, 227; Training Program in Automated Multiphasic Screening, 222; X-rays, 212, 216, 226
MUMPS (Massachusetts General Hospital Utility Multi-Programming System), 204, 206, 244
Murray, R. Milne, 20–21

MyChart, 244
MYCIN computer, 199
Myers, Jack, 199

Nash, Firmin, 190–94, 198–99, 296n43
National Academy of Engineering, 147, 167, 224
National Academy of Sciences (NAS), 181–82
National Advisory Commission on Civil Disorders, 143–44, 148, 154, 156, 159
National Aeronautics and Space Administration (NASA): health disparities and, 149–53, 169–71, 174; Mercury program, 71; Rockoff and, 153, 170, 174, 292n85; STARPAHC and, 170–73; Weed and, 207
National Cable Television Association, 162
National Center for Health Services Research, 164
National Health Services Corps, 168
National Heart Institute (NHI), 62–63
National Institutes of Health (NIH), 36, 106, 182–84, 203
National Library of Medicine (NLM), 183–87
National Science Foundation (NSF), 154
Native Americans: access to healthcare, 3, 139, 158, 171, 173; Alaskan, 90, 169; healthcare disparities, 145, 158, 169–73; Indian Health Service, 169, 171, 203; Navajo, 158; Papago, 169–73; STARPAHC and, 170–73
Navajo people, 158
NBC, 106, 147
Nebraska Psychiatric Institute (NPI), 109–16, 123, 149
Neergaard, Charles E., 85–86
neglect: healthcare disparities and, 143; mental hospitals and, 111; telephones and, 16, 32–33, 40; television and, 1, 119, 130
networks: automation and, 237; Big Science, 53; computer, 5, 16, 36, 181, 185, 201, 241; decentralized communications, 146; growth of, 149, 169; medium of care and, 245, 252; national, 145, 152; Nebraska Psychiatric Institute and, 110; point-to-point, 123; regional, 142; rural, 116, 165; social media, 241; telephone, 16–17, 19, 25, 34–37, 81; television, 106, 143, 146–47, 167, 173; use of physicians' time, 163
Neuberger, Maxine, 226–27
New England Journal of Medicine, 32, 70, 230
new media, 15, 47, 105, 215, 248, 254–55
New Rural Society Project, 167–68
Newsweek magazine, 156
New York Academy of Sciences, 105, 112, 116, 136
New York Amsterdam News, 162–63
New York Botanical Gardens, 195–96
New York Doctors' Telephone Service, 83
New York Hospital, 194
New York State Ghetto Medicine Program, 158, 160, 163, 290n61
New York Times, 7, 159, 164
New York University, 126
Nixon, Richard M., 141, 149, 158, 228
Nobel Prize, 55, 58
Norfolk State Mental Hospital, 112–18, 123, 149
normal values, 50, 56, 216–19, 222–26, 233
Norman, John C., 154–56
nuclear weapons, 60
nurse practitioners (NPs), 165–67, 176
nurses: Automatic Nurse Call, 93; automation and, 225, 229; autonomy, 42, 45, 47, 108, 119, 163, 165, 169, 176, 247; burnout, 43–45; computers and, 201–3; healthcare disparities and, 160, 162–66, 169, 176, 255; medium of care, 246–47, 255; pagers and, 79–80, 89–94, 99, 102; telephones and, 37, 39, 41–47; television and, 107–8, 112, 117, 119, 123, 127–28, 137; wireless technology and, 71, 74

Obama, Barack, 211, 245
obsolescence, 7, 107, 157, 181, 187, 229
obstetrics, 35
Office of Research and Development, 169
Ohio State Medical Journal, 23
Oldfield, H. R., 235–36

operating rooms, 142, 150
oscilloscopes, 57, 73
Otten, John, 87–88
Otten, Michael, 36
"Our New Age" (Spilhaus), 188

pacemakers, 9, 68, 78
Packard Motor Car Company, 56
Pageboy: Archer character and, 79–81, 95–97, 104; wireless technology and, 79, 83, 96–98, 100–104
Page One, 97
pagers, 11; access and, 83–84, 93–99, 103–4; adoption of, 82–83; advertising of, 84, *91*, 100–102; Aircall, 84, 86, 96; Amsden and, 83–84; Archer character and, 79–81, 95–97, 104; AT&T and, 280n54; autonomy and, 80–82, 96, 98, 102; barking and, 100; as beeper, 81, 98–104; Bellboy, 83, 96, 100–102; class and, 103; computers and, 82; diagnostics and, 104; as electronic leashes, 98–104; emergencies and, 80–88, 97, 100; Google and, 104; Gross and, 85–86; Handie-Talkies, 86–89, *91*; impact of, 252; legal issues, 99; markets and, 83–86, 89–90, 93, 96–98; Motorola and, 79–81, 86–104; Neergaard and, 85–86; nurses and, 79–80, 89–94, 99, 102; Pageboy, 79–83, 95–104; power relations and, 129, 131, 133–36; privacy and, 253–54; radio and, 7, 79–100, 204, 252, 254; Radio Pager, 87–94; Radio-Register, 81, 94; Royalcall, 84–86, 88, 96; rural areas and, 87; surveillance and, 81, 96, 102; telephones and, 82–84, 98, 100, 104; Zworykin and, 7
Palese, Robert, 163
Papago Tribe, 169–73. *See also* Tohono O'odham Nation
Papanicolau smears, 213, 216, 234, 237
paper, 264n7; cards, 187, 224; charts, 179, 202–3, 205–6, 210; computers and, 179–89, 193, 202–6, 208–10, 242; drums, 16, 21, 55; gowns, 125, 212; MEDLARS and, 209–10; nostalgia for, 179; PROMIS and, 207–8; punch cards, 36, 182, 194, 212, 217–18, 222, 224; records, 181, 205–10; SOAP note, 206; tape, 182, 184–85; technologies, 193, 209; written data and, 16
Pappalardo, Neal, 204, 244
Park, Ben: Alternative Media Center, 125–26; *An Introduction to Telemedicine*, 126; lens choice and, 133; Rockoff and, 129; television and, 125–26, 129–39; typical patient assumption and, 138
Patterns of Time in Hospital Life (Zerubavel), 100
Pavlov, Ivan, 17, 55
Paycha, François, 296n46
pharmaceutical firms, 68–69, 106, 156, 190, 215, 224, 230
pharmacies, 38, 203, 207, 210, 244
Philadelphia Ledger, 24
physical exams: automation, 211–12, 219, 221, 230; computers and, 191, 194, 200; multiphasic testing, 211–12, 219, 221, 230; screening, 7–8, 41, 43, 107, 117–18, 121, 191, 194, 200, 211–12, 219, 221, 230; telephones and, 41, 43; television and, 107, 117–18, 121
physician assistants (PAs), 165–67, 176
physiology: Galvani and, 54–55; Skibbins and, 50, 62–63, 73; surveillance and, 49–52, 63, 73–78, 241, 253; telephones and, 16–21, 35; wireless technology and, 49–58, 61–64, 67–75, 78
picture archiving and communication systems (PACS), 34
Popular Science magazine, 84
Potential Uses of Television in Preschool Education, The (Cooney), 248
poverty, 156, 165, 174, 222, 235, 242, 249, 251; access and, 4 (*see also* access); ghettos and, 147–48, 154–60, 163–65, 168, 173–75, 256; Harlem and, 3, 144, 154, 157–64; Head Start program, 249; Rockoff on, 292n79; *Sesame Street* and, 248–49
power relations, 6, 42, 47, 105–8, 126–38, 145
precision medicine, 211–14, 233, 237
prepaid plans, 231
presence, 11, 24, 33, 41–43, 79, 106–8, 114, 121–38, 200, 209, 264, 279

Presentation of Self in Everyday Life (Goffman), 129–30
Preventicare, 215, 220–21, 227, 230, 236
prevention: automation, 212–28, 231–39, 247–48; community action, 143; computers and, 213; cost-effectiveness, 231, 233; healthcare disparities and, 141; medium of care and, 247, 254; pop-up fairs, 218–19; US Preventive Medicine Task Force, 234; wireless technology and, 49, 59, 70, 72
prison, 98, 116, 145
privacy, 58, 78, 88, 118, 121, 175, 253–54
Problem-Oriented Medical Information System (PROMIS), 207–8, 299n86
Project ECHO (Extension for Community Healthcare Outcomes), 251–52
proxemics, 133–38
psychiatry: Berger, 58; Bergman, 98; Brodman, 189; computers and, 189, 198; healthcare disparities and, 149, 176; Holter, 60–61; Johnson, 110–13, 116, 118; Maxmen, 229; Nebraska Psychiatry Institute (NPS), 109–16, 123, 149; Norfolk State Mental Hospital, 112–18, 123, 149; television and, 109–19, 123; wireless technology and, 58, 60–61; Wittson and, 109–13, 116, 118–19, 149
psychology: Benschoter, 109–10, 112, 149; Cook, 248–49; "Face of God" problem and, 129, 133, 169; Gengerelli, 54; Holter and, 54; power relations and, 129–36; telepresence and, 121–22, 127–37; television and, 109–10, 112, 138, 248
public address (PA) systems, 82
PubMed, 210
pulse oximeters, 1, 47
punch cards, 36, 182, 194, 212, 217–18, 222, 224

race and racism: access to healthcare and, 3–4, 30, 32, 139, 155, 158, 171, 173; African Americans, 4, 110, 134, 154–60, 250 (*see also* Black people); civil rights and, 125, 155–56, 164–65, 255; class and, 5, 15, 134, 139, 248; ethnicity and, 134 (*see also* ethnicity); ghettos and, 147–48, 154–60, 163–65, 168, 173–75, 256; healthcare disparities and, 144–48, 153–65, 168–75, 238, 255; Latinos, 4, 139, 144, 154, 162, 250; medium of care and, 242, 248, 250, 252; Moynihan Report and, 157; Native Americans, 3, 90, 139, 145, 158, 169–73, 203; New York State Ghetto Medicine Program, 158, 160, 163; police and, 136; *Sesame Street* and, 248–49; socioeconomic divide and, 139; telephones and, 15; Tuskegee Syphilis Study and, 125, 134, 162; typical consumer and, 138
radio: AM, 81, 88, 97; citizens band (CB), 85; De Long and, 268n43; diagnostics and, 11, 61–62, 69, 107, 121, 142, 176, 256; FM, 9, 49–50, 59, 69, 81, 88, 97; Gernsback and, 56–60, 65, 72, 77, 241; healthcare disparities and, 142, 150, 152, 169, 176; impact of, 11, 243, 246, 252–53; pagers and, 7, 79–100, 204, 252, 254; remote control and, 54, 60; servos and, 241; shortwave, 85; television and, 106–7, 109, 117, 120–21, 127; Telimco kits, 56; two-way, 5, 81, 86–92, 97, 106–8, 115; very high frequency (VHF), 96, 280n58; wireless technology and, 49, 52, 54–78
radioelectrocardiograms (RECGs): Holter and, 62, 64, 67; wireless technology and, 50, *51*, 59, 61–64, 67, 71–72
radioelectroencephalograms (REEGs), 59, 61
Radio News magazine, 56
radio pills, 52, 65–68, 72, 75, 77, 180, 274n59
Radio Pocket Pager, 87–89
Radio-Register, 81, 94
RadioShack, 71
radiotelemetry, 7; DIY kits and, 71; Laika and, 71; market possibilities of, 64; NASA and, 71; Rockefeller Institute and, 69; Uchiyama and, 67; wireless technology and, 64, 67, 69–76, 253
RADIO W-E-L-L, 90, 92
Rappaport, Arthur E., 220–21
RCA Corporation: Bizmatic computers, 180, 194; Cornell and, 194; dominant market position of, 247; magnetic drives of, 181; radio pills and, 66, 68, 72,

180, 274n59; Zworykin and, 7, 65–66, 68, 180, 194, 211, 275n64
Reader's Digest, 214
records: automation of, 237; computers and, 180–88, 200–210, 243–44; data entry volume of, 180–81; electronic, 2, 8–9, 34, 75, 180–88, 200–210, 237, 241–47; HITECH and, 2; *Index Medicus* and, 181, 183; limits of technology and, 241–42; Medical Literature Analysis and Retrieval System (MEDLARS) and, 184–87, 209–10; Medical PROBE and, 201; medium of care and, 242–47; MEDLINE and, 210; MUMPS and, 204, 206, 244; National Library of Medicine (NLM) and, 183–87; paper, 181, 205–10; physician burnout and, 8; PROMIS and, 207–8; PubMed and, 210; sharing, 34, 208, 242; storage of, 212 (*see also* storage); Technicon and, 202–3; telephones and, 34; Weed and, 205–8; wireless technology and, 49, 75
Reich, Joel, 126–28, 155
Relman, Arnold, 230
remote control, 54, 60
remote sensors, 1, 60
research and development (R&D), 7, 21–22, 59, 65, 77, 90, 157, 169, 180
revolutionary technologies, 3–14, 47, 144, 182, 202, 209, 243
Robert Wood Johnson Foundation, 252
Robinson, Matt, 250
Rockefeller Foundation, 3, 142, 144, 155, 185
Rockefeller Institute for Medical Research, 7, 65, 68–69, 189
Rockey, A. E., 26
Rockoff, Maxine, 126, 285n50; background of, 153; Goffman and, 129–30; Goldmark and, 167–68; healthcare disparities and, 152–53, 166–70, 174, 177, 292n79, 293n93; HEW and, 126, 144, 168; Komaroff and, 166; Logistics Unit of, 153–54; NASA and, 153, 170, 174, 292n85; Park and, 129; poor people and, 292n79; STARPAHC and, 170–73; television and, 126, 129–30, 144, 174, 177; Waxman and, 152–53, 291n74
Rocky Mountain Medical Journal, 62
Rogers, Frank, 184
Role of Computers in Biology and Medicine (NAS), 181
Romeo and Juliet (Shakespeare), 82
Royal, Harry, 85
Royalcall, 84–86, 88, 96
rural areas: ghettos and, 147, 154–56, 164–65, 168, 173, 175; Goldmark and, 167–68; healthcare disparities and, 3, 142, 144–47, 149–50, 153–56, 164–76, 215, 222, 242; Land Policy and Planning Assistance Act and, 168; New Rural Society Project and, 167–68; pagers and, 87; Project ECHO and, 251–52; telephones and, 23, 25, 34–35; television and, 109, 116, 118, 139
Rusan, Francille, 173

Sandelowski, Margarete, 42
Sanders, Charles L., 157
satellites, 70, 105, 144, 150, 169
Schaefer, Theodor, 24, 29
schizophrenia, 61
science fiction, 52, 56–58, 119–20, 211
Scientific American journal, 20, 57
screening: computers and, 193, 214–37; COVID-19, 1; diagnostics and, 1, 69, 193, 212–37, 255; false positives, 226–27; multiphasic testing, 212–37, 255; physical exams, 7–8, 41, 43, 107, 117–18, 121, 191, 194, 200, 211–12, 219, 221, 230; Training Program in Automated Multiphasic Screening, 222; true positives, 227; wireless technology and, 69
Searle Metidata, 225, 231, 237
Searle Pharmaceuticals, 224–25
secrecy, 52, 64–65, 77
Senate Special Committee on Aging, 219–20
servos, 241
Sesame Street (TV show), 248–50
sexuality, 134
sex workers, 81, 102
Shannon, Claude, 72
Shepard, Alan B., 71
Shortliffe, Edward, 199
Short Message Service (SMS), 103–4

shortwave radio, 85
sigmoidoscopy, 213
Simms, Jo, 42–45
Sir Mix-A-Lot, 103
Skibbins, Gerald, 50, 62–63, 73
Slack, Warner, 209, 243–44
sleep, 14, 26, 32, 51, 88, 97, 246
Sloan Commission on Cable Communications, 145, 147, 159
Smith, Kline & French, 68–69, 106, 247
Smith, Ralph Lee, 143
SOAP note, 206
social media, 47–48, 241
Société de Biologie of Paris, 18
sociology, 98–99, 126–39, 154–55, 289n30
software, 138, 208, 244
Soul of a New Machine, The (Kidder), 244
Soviets, 60, 70, 76, 225
Space Race, 70
Space Technology Applied to Rural Papago Advanced Healthcare (STARPAHC), 169–73
Spilhaus, Athelstan, 188
Sprigle, Herbert, 248–49
square raindrops, 52
Squire, A. J. Balmanno, 21–22
Standard Nomenclature of Diseases and Operations, 180
Stanford University, 199
stenographers, 33–34
stethoscopes: CliniCloud and, 13; concern over, 11, 221; Crichton on, 119; electronic, 13, 168; introduction of, 5, 10–11, 221; as media, 10; mediate ausculation and, 10; symbolism of, 81, 104; telephones and, 13–14, 19–20, 47; wireless technology and, 76
storage: automation and, 212; computers and, 180–88, 200–210; data entry volume and, 180–81; *Index Medicus* and, 181, 183; magnetic, 16, 33, 36, 75, 185, 194, 201; Medical Literature Analysis and Retrieval System (MEDLARS) and, 184–87, 209–10; Medical PROBE and, 201; National Library of Medicine (NLM) and, 183–87; Technicon and, 202–3; wireless technology and, 73–75, 78; Zworykin and, 75, 180–81
Story of a Doctor's Telephone, The (Firebaugh), 26–28
strokes, 39, 216
surveillance: Amazon and, 76; Big Brother, 123; Cold War, 50, 52, 61, 76–77; employers and, 52, 76, 253; Holter monitor, 49–52, 63, 73–78; pagers and, 81, 96, 102; physiology and, 49–52, 63, 73–78, 241, 253; television and, 123–24; wireless technology and, 49–53, 61, 69–78
symptoms: automation and, 216, 227, 234; computers and, 181, 188–96, 206; COVID-19, 1; letters and, 10; medium of care and, 246; telephones and, 14, 39, 41, 117; wireless technology and, 49
syncope, 49

Tanimoto, Taffee, 195–96, 297n56
Task Force on Communications Policy, 144, 147
Taylor, Richard, 195, 199
TCP/IP protocols, 244
"Teaching Methods and Patient Care with Emphasis on the Weed System" (Emory), 207–8
Technicon Medical Information Systems, 202–3
technological fix, as solution to social problems, 1, 108, 145, 175, 249; historical approaches to, 6, 102, 263n5, 264n6; hubris, 108, 131, 138, 182; naïveté, 132, 137–8, 145, 148, 175, 182, 256, 292n85
Telanserphone, 83
Tele-ECG, 46
telegraphs, 15, 20, 23–24, 241, 277n10
telehealth: advantages of more data, 211–12; healthcare disparities and, 3, 142, 145, 154, 162, 167, 169; increased sophistication of, 9; investor returns on, 3; limits of, 2; medium of care and, 242, 250–52
telemedicine: access and, 4 (*see also* access); automation and, 222, 236; benefits of, 3–4; coining of term, 3; computers and, 189; constraints of, 2; healthcare dispar-

ities and, 4–5, 141–45, 149, 152–53, 162, 164, 166, 168–71, 174–77; investing in, 2–3; medium of care and, 241–42, 247–52; radiotelemetry and, 7, 64, 67–76; television and, 106–9, 116, 119–39
Telemedicine for Health Equity Toolkit, 252
telemetry: automation and, 221; computers and, 189; radiotelemetry, 7, 64, 67, 69–76, 253; volume of data produced by, 73–78; wireless technology and, 62, 64, 67–78
Telephone Medicine (Brown), 40
telephones: access and, 5, 11, 14–16, 22, 26, 29–33, 43, 46; addiction to, 38; adoption rates of, 25; advertising of, 26, 30, *31*; algorithms and, 40–46; American Medical Association (AMA) and, 35, 37; answering, 25–26, 28, 30, 32–34, 47, 81, 83, 187; AT&T, 25, 96, 152, 279n51, 280nn54–55; autonomy and, 42, 45, 47; Baldwin and, 23, 25, 28; Bell and, 5, 13–14, 19, 21, 25, 36–37, 96–97, 152–53; Bell Labs, 152, 279n51, 280n55; Bell telephone, 14, 25, 37, 96; Black people and, 30, 32; Blake and, 19–21, 36; blood pressure and, 39; Bowditch and, 17–18; brain waves and, 17, 35; Brobeck and, 23, 25, 28; class and, 15–16, 30; CliniCloud and, 13–15, 47–48; computers and, 16, 35–36, 181, 187, 193, 204; concern over, 16–17, 23–29, 37–41, 46, 253; convenience of, 14, 23, 25–26; Dataphone, 36, 46; decline of, 255; dehumanization and, 16, 38–39; De Long on, 28–30, 47, 83; diagnostics and, 14, 16, 18–19, 22, 33, 37, 40, 43, 46; Dictaphone, 33–34, 36; Doctor on Demand, 13; early medical application of, 17–22; early promise of, 11–12; education and, 248–51; efficiency and, 21–22, 24, 40–41, 136; electrocardiograms (ECGs) and, 34–35; emergencies and, 14, 26, 44; examinations and, 39; exchanges and, 22–25, 29–30; guidebooks for, 40–46; health records and, 34; heartbeat and, 21; impact of, 11, 241, 243, 246, 252–53; legal issues and, 32, 37–38, *39*; malpractice and, 16, 32–33, 37, 39; markets and, 24, 30; Marx on, 24–25; McLuhan and, 107, 120, 176, 252–53; neglect and, 16, 32–33, 40; networks and, 16–17, 19, 25, 34–37, 81; New York Doctors' Telephone Service, 83; nurses and, 37, 39, 41–47; pagers and, 82–84, 98, 100, 104; patient empowerment and, 13–15, 40, 43, 45–47; physical exams and, 41, 43; physiology and, 16–21, 35; as platform for other technologies, 33–36; privacy and, 253–54; proper medical communication over, 40–41; race and, 15; radio and, 83–84; rural areas and, 23, 25, 34–35; signal quality and, 20; smartphones, 7, 13–15, 48–49, 51, 102, 104, 137, 242; as sound amplifier, 18–22; Squire and, 21–22; stethoscopes and, 13–14, 19–20, 47; as symbol, 35–36; symptoms and, 14, 39, 41, 117; as taken for granted, 15–16; thermometers and, 13–14, 47; triage and, 40–46; twentieth-century use of, 22–45; Wittson and, 109
Telephone Triage and Management (Simms and McGear), 42–45
TelePrompTer Corporation, 160, 163, 173–74
television: ABC, 106, 147; access and, 108, 119, 123, 125, 131, 139; advertising, 105; algorithms and, 121; Alternative Media Center, 125–26; augmented signals and, 126; automation and, 224; autonomy and, 108, 119; Benschoter and, 105–19, 123, 126, 136–37, 141, 149, 248, 283n15; Bird and, 107–8, 117–23, 126–28, 131, 137, 141–42, 288n24; cable, 5, 106, 126, 143–51, 159–63, 167–69, 173–76; CBS, 106–7, 118, 121, 147, 167; Children's Television Workshop (CTW), 248–51; closed-circuit, 3, 7, 92, 105–6, 109–12, 118, 120, 175, 256; community access television (CATV), 5, 106, 146–48, 159–60, 173; Cooney report on, 248; diagnostics and, 107, 111, 113, 117–24, 127; education and, 105–6, 109, 112, 118, 121, 126, 248–51; electrocardiograms (ECGs) and, 118; emergencies and, 106, 118; examinations and,

television (*continued*)
113, 117–18, 121; "Face of God" problem and, 129, 133, 169; Friedsen and, 131–35, 138; Goffman and, 129–35; healthcare disparities and, 141–50, 153–54, 159–63, 166–69, 173–76; impact of, 11, 36, 241, 243, 246, 248, 252, 254, 256; Johnson and, 110–13, 116, 118; Logan Medical Station and, 117–23; makeup and, 43; markets and, 106; microwaves and, 112, 118, 123, 126, 150; Motorola and, 80, 92; NBC, 106, 147; neglect and, 1, 119, 130; networks and, 106, 143, 146–47, 167, 173; nurses and, 107–8, 112, 117, 119, 123, 127–28, 137; Park and, 125–26, 129, 131, 133–39; patient empowerment and, 108–9, 125–26, 137–38; physical exams and, 107, 117–18, 121; possibilities of, 107–8; power relations over, 129–36; psychiatry and, 109–19, 123; psychology and, 109–10, 112, 138, 248; radio and, 106–7, 109, 117, 120–21, 127; reception issues, 146; Rockoff and, 126, 129–30, 144, 174, 177; rural areas and, 109, 116, 118, 139; satellite, 105, 144, 150, 169; *Sesame Street*, 248–50; Sloan Commission, 145, 147; surveillance and, 123–24; Task Force on Communications Policy, 144, 147; telemedicine and, 106–9, 116, 119–39; telepresence and, 121–22, 127–37; two-way, 106–8, *115*, 123, 128, 153, 175–76, 256; Wittson and, 109–13, 116, 118–19, 149
Telimco, 56
Tennessee Valley Authority (TVA), 222
Tesla, Nikola, 57
texting, 103–4
thermometers: CliniCloud and, 13; electronic, 13; nurses and, 42; telephones and, 13–14, 47; wireless technology and, 71, 76
Thompson, Henry, 18
Thought Recorder, 57, *58*, 272n22
tilt-table tests, 212
tinkering, 29, 53, 56, 59, 64, 69–70, 72, 77, 271n9, 277n21
Time magazine, 217
Titicut Follies (Wiseman), 138
tobacco, 30, *31*, 87–88
Tohono O'odham Nation, 169–73
Toshiba Corporation, 225–26, 302n49
Training Program in Automated Multiphasic Screening, 222
transparency, 16, 49, 69, 72, 74, 76, 247, 253
triage, 40–46
Tribe Called Quest, 103
true positives, 227
Truman, Harry S., 60
Tulane University, 201
Turing test, 190
Turner, Ted, 173
Tuskegee Syphilis Study, 125, 134, 162
Twain, Mark, 56

Uchiyama, Akihito, 67
ultrasound, 10–11, 18, 22
Understanding Media (McLuhan), 107, 176
University College Hospital of London, 18
University of California, Berkeley, 67, 72
University of California, Los Angeles (UCLA), 54
University of California-Irvine, 203
University of California-San Francisco, 252
University of Chicago, 155, 203
University of Edinburgh, 21
University of Glasgow, 19
University of Michigan, 126
University of Nebraska Medical Center (UNMC), 106, 109–12, 123
University of New Mexico, 251
University of Pennsylvania, 82, 129, 153
University of Pittsburgh, 199
University of Washington, 155
University of Wisconsin, 243–44
Urban League, 157
urinalysis, 216
US Agency for International Development (USAID), 173
US Air Force, 86
US Army, 86
US Department of Health, Education, and Welfare (HEW), 3; automation and, 228–29; healthcare disparities and, 144, 149, 152, 165, 168–69, 171; Rockoff and, 126, 144, 168

US Food and Drug Administration (FDA), 47, 77
US House of Representative, 215, 221
US Navy, 86
US Preventive Medicine Task Force, 234
US Public Health Service, 8, 86; automation and, 215, 217, 222–23, 236; Caceres and, 70, 73; computers and, 180, 202; healthcare disparities and, 158, 162; television and, 118, 125; wireless technology and, 70, 73
US Senate, 215, 219–21, 227
US State Department, 86, 185
utopian hopes, 35, 134, 188, 200, 253–54

vaccines, 1, 159, 237
vacuum tubes, 64, 87, 241
Ventnor Clinic, 34
Verne, Jules, 56
very high frequency (VHF) radio, 96, 280n58
Veterans Affairs (VA), 35, 86, 123, 126, 149, 196, 203–4, 207
Vivian, C. T., 160
Voss, Betsy, 163–64

Wagner Homes, 159–66, 174, 176
waiting rooms, 2, 13, 108
Waldorf Astoria, 64, 68
Wall Street Journal, 64
War on Cancer, 53
Washington University, 126
Waxman, Bruce, 152–53, 291n74
wealth, 15, 246
wearable technology, 7; Amazon Halos, 52; cloud technology and, 49; Fitbits, 52; healthcare disparities and, 175; Holter monitor, 49–52, 63, 73–78, 241
Weed, Lawrence, 205–8, 298n75
Weinberg, Mort, 89–90, 97, 99
Weiner, Norbert, 72
Wells, H. G., 56
Wempner, Jon, 168
Western Electric, 96
Western Union, 20
WGBH, 118
White, Paul Dudley, 54, 62
Williams, Harrison A., 219–20
wireless technology: access and, 52, 71, 74, 76–77; advertising of, 56; algorithms and, 74; Archer character and, 79–81, 95–97, 104; blood pressure and, 55, 69; brain waves and, 52, 55, 57–61, 78; Caceres and, 69–77; class and, 50, 63; cloud technology, 49, 77; diagnostics and, 61–62, 69, 74; efficiency and, 50, 62–63; electrocardiograms (ECGs) and, 50, *51*, 55, 71–76; emergencies and, 49; Gernsback and, 56–60, 65, 72, 77, 241; Handie-Talkies, 86–89, *91*, 278n37; health records and, 49, 75; Holter and, 49–56, 59–78, 241; impact of, 241; insurance and, 51; International Conference and Exhibition on Medical Electronics, 65; legal issues, 61; markets, 64–65, 68–69, 73, 77; Motorola and, 96–99, *101*, 103–4, 247; nurses and, 71, 74; pagers, 98 (*see also* pagers); physiology and, 49–58, 61–64, 67–75, 78; prevention and, 49, 59, 70, 72; psychiatry and, 58, 60–61; radio and, 49, 52, 54–78; radio-electrocardiograms (RECGs) and, 50, *51*, 59, 61–64, 67, 71–72; radiotelemetry and, 7, 64, 67–76, 253; screening and, 69; Space Race and, 70; stethoscopes and, 76; storage and, 73–75, 78; surveillance and, 49–53, 61, 69–78; symptoms and, 49; telemetry and, 62, 64, 67–78; Telimco kits, 56; thermometers and, 71, 76; wearable technology, 49–52, 63, 73–78; Zworykin and, 7–8, 65–70, 75, 77
Wiseman, Frederick, 138
Wittson, Cecil, 109–13, 116, 118–19, 149
World Congress of Cardiology, 62
World Health Organization (WHO), 185, 234
World War II, 85, 105, 109, 215
worried well, 227–28, 233
Wright, Jim, 103

X-rays, 6–7, 107, 168, 212, 216, 226

Yale School of Forestry, 63
Young Lords, 125, 162

Zerubavel, Eviatar, 99–100
Zigler, Edward, 249
Zworykin, Vladimir: automation and, 211, 217; background of, 65; Caceres and, 70; Center for Medical Electronics, 7, 65, 179–80, 188–89, 194; Collen and, 217; computers and, 179–81, 187–88, 194, 199–200, 205–8; criticism of, 7–8; on diagnostics, 8; Holter and, 67–69, 75–77, 241; immigration of, 7; pagers and, 7; PROMIS and, 207–8; RCA Corporation and, 7, 65–66, 68, 180, 194, 211, 275n64; storage and, 75, 180–81; wireless technology and, 7–8, 65–70, 75, 77; X-rays and, 6–7